INTERMOLECULAR AND SURFACE
FORCES

INTERMOLECULAR AND SURFACE FORCES

SECOND EDITION

JACOB N. ISRAELACHVILI

*Department of Chemical & Nuclear Engineering
and Materials Department
University of California, Santa Barbara
California, USA*

ACADEMIC PRESS
Harcourt Brace & Company, Publishers
London   San Diego   New York   Boston
Sydney   Tokyo   Toronto

ACADEMIC PRESS LIMITED
24–28 Oval Road
London NW1 7DX

United States Edition published by
ACADEMIC PRESS INC.
San Diego, CA 92101

First Edition published in 1985
Second Edition 1991
Sixth Printing 1997
Seventh Printing 1998

**A catalogue record for this book is available
from the British Library**

ISBN 0-12-375181-0

98 99 00 01 02 SE 9 8 7

Typeset by P&R Typesetters Ltd, Salisbury, Wiltshire
Printed by St Edmundsbury Press Limited, Bury St Edmunds, Suffolk

Contents

PART TWO
The Forces Between Particles and Surfaces

PART THREE

Fluid-Like Structures and Self-Assembling Systems: Micelles, Bilayers and Biological Membranes

LIST OF TABLES

PREFACE TO SECOND EDITION

Since 1985 when the first edition of this book appeared there has been much experimental and theoretical progress in this multidisciplinary subject. The nature of some 'old' forces have been clarified while 'new' forces have been discovered. The subject has matured into a rigorous discipline and a unifying area of chemistry, physics and biology, and many university courses now routinely contain material on molecular and surface interactions. On the more practical side, many industrial and chemical engineering processes are now beginning to be understood and controlled at the fundamental level. It is with these developments in mind, together with the feedback I received from numerous colleagues, that the second edition was prepared.

The second edition is basically an updated version of the first, but it contains more than 100 problems. Most appear at the end of each chapter, but some appear as worked examples in the text. These problems should enhance the suitability of the book as an advanced undergraduate or graduate textbook. The problems have been devised to stimulate the mind; many are based on genuine research problems, others are tricky, some are extensions of the text into more advanced areas, and a few are open-ended to invite further reading, discussion and even speculation.

The text itself has been expanded to include recent developments in the areas of surface-force measurements, solvation and structural forces, hydration and hydrophobic forces, ion-correlation forces, thermal fluctuation forces, and particle and surface interactions in polymer melts and polymer solutions.

I am grateful to many colleagues who commented on the first edition, and have used their suggestions in writing the second. In particular, my thanks go to Hans Lyklema, Håkan Wennerström, Jacob Klein, to Helen Vydra and Josefin Israelachvili for typing the manuscript, to Dottie McLaren for drawing many of the figures, to my wife Karina who supported me throughout, and finally to my students who have sat through my lectures and by their questions have unwittingly contributed the most.

PREFACE TO FIRST EDITION

Intermolecular forces embrace all forms of matter, and yet one finds very few university courses devoted to all aspects of this important subject. I wrote this book with the aim of presenting a comprehensive and unified introduction to intermolecular and surface forces, describing their role in determining the properties of simple systems such as gases, liquids and solids but especially of more complex, and more interesting, systems. These are neither simple liquids nor solids, but rather a myriad of dissolved solute molecules, small molecular aggregates, or macroscopic particles interacting in liquid or vapour. It is the forces in such systems that ultimately determine the behaviour and properties of everyday things: soils, milk and cheese, paints and ink, adhesives and lubricants, many technological processes, detergents, micelles, biological molecules and membranes, and we ourselves—for each of us is one big biocolloidal system composed of about 75% water, as are most living organisms.

This subject therefore touches on a very broad area of phenomena in physics, chemistry, chemical engineering and biology in which there have been tremendous advances in the past 15 years. These advances can be viewed in isolation within each discipline or within a broader multidisciplinary framework. The latter approach is adopted in this book, where I have tried to present a general view of intermolecular and surface forces with examples of the various and often seemingly disparate phenomena in which they play a role.

Because of the wide range of topics covered and the different disciplines to which the book is addressed, I have presumed only a basic knowledge of the 'molecular sciences': physics (elementary concepts of energy and force, electrostatics), chemistry (basic thermodynamics and quantum mechanics) and mathematics (algebra and elementary calculus). The mathematical and theoretical developments, in particular, have been kept at a simple, unsophisticated level throughout. Vectors are omitted altogether. Most equations are derived from first principles, followed by examples of how they apply to specific situations. More complicated equations are stated, but are again carefully explained and demonstrated.

In a book such as this, of modest size yet covering such a wide spectrum, it has not been possible to treat each topic exhaustively or rigorously, and specialists may find their particular subject discussed somewhat superficially.

The text is divided into three parts, the first dealing with the interactions between atoms and molecules, the second with the interactions between 'hard' particles and surfaces, and the third with 'soft' molecular aggregates in solution such as micelles (aggregates of surfactant molecules) and biological membranes (aggregates of lipids and proteins). While the fundamental forces are, of course, the same in each of these categories, they manifest themselves in sufficiently different ways that, I believe, they are best treated in three parts.

The primary aim of the book is to provide a thorough grounding in the theories and concepts of intermolecular forces so that the reader will be able to appreciate which forces are important in whatever system he or she is dealing with and to apply these theories correctly to specific problems (research or otherwise). The book is intended for final-year undergraduate students, graduate students, and non-specialist research workers.

I am deeply grateful to the following people who have read the text and made valuable comments for improving its presentation: Derek Chan, David Gruen, Bertil Halle, Roger Horn, Stjepan Marcelja, John Mitchell, Håkan Wennerström and Lee White. My thanks also extend to Diana Wallace for typing the manuscript and to Tim Sawkins for his careful drawing of most of the figures. But above all, I am indebted to my wife, Karin, without whose constant support this book would not have been written.

UNITS AND SYMBOLS

Much of the published literature and equations on intermolecular and surface forces are based on the CGS system of units. In this book the *Système International* (SI) is used. In this system the basic units are the **kilogramme** (kg) for mass, the **metre** (m) for length, the **second** (s) for time, the **kelvin** (K) for temperature, the **ampère** (A) for electrical quantities, and the **mole** (mol) for quantity of mass. Some old units such as gramme (1 gm = 10^{-3} kg), centimetre (1 cm = 10^{-2} m), ångstrom (1 Å = 10^{-10} m) and degree centigrade (°C) are still commonly used although they are not part of the SI system. The SI system has many advantages over the CGS, not least when it comes to forces. For example, force is expressed in newtons (N) without reference to the acceleration due to the earth's gravitation, which is implicit in some formulae based on the CGS system.

DERIVED SI UNITS

Quantity	SI unit	Symbol	Definition of unit
Energy	Joule	J	$\text{kg m}^2\text{ s}^{-2}$
Force	Newton	N	$\text{kg m s}^{-2} = \text{J m}^{-1}$
Power	Watt	W	$\text{kg m s}^{-2} = \text{J s}^{-1}$
Pressure	Pascal	Pa	N m^{-2}
Electric charge	Coulomb	C	A s
Electric potential	Volt	V	$\text{J A}^{-1}\text{ s}^{-1} = \text{J C}^{-1}$
Electric field	Volt/metre		V m^{-1}
Frequency	Hertz	Hz	s^{-1}

Fraction	10^9	10^6	10^3	10^{-1}	10^{-2}	10^{-3}	10^{-6}	10^{-9}	10^{-12}
Prefix symbol	G	M	k	d	c	m	μ	n	p

FUNDAMENTAL CONSTANTS

Constant	Symbol	SI	CGS
Avogadro's constant	N_0	6.022×10^{23} mol^{-1}	6.022×10^{23} mol^{-1}
Boltzmann's constant	k	1.381×10^{-23} J K^{-1}	1.381×10^{-16} erg deg^{-1}
Molar gas constant	$R = N_0 k$	8.314 J K^{-1} mol^{-1}	8.314×10^7 erg mol^{-1} deg^{-1}
Electronic charge	$-e$	1.602×10^{-19} C	4.803×10^{-10} esu
Faraday constant	$F = N_0 e$	9.649×10^4 C mol^{-1}	9.649×10^4 C mol^{-1}
Planck's constant	$h (\hbar = 2\pi\hbar)$	6.626×10^{-34} J s	6.626×10^{-27} erg s
Permittivity of free space	ε_0	8.854×10^{-12} C^2 J^{-1} m^{-1}	1
Mass of $\frac{1}{12}$ of ^{12}C atom*	u	1.661×10^{-27} kg	1.661×10^{-24} g
Mass of hydrogen atom	m_H	1.673×10^{-27} kg	1.673×10^{-24} g
Mass of electron	m_e	9.109×10^{-31} kg	9.109×10^{-28} g
Gravitational constant	G	6.670×10^{-11} Nm^2 kg^{-2}	6.670×10^{-8} g^{-1} cm^3 s^{-2}
Speed of light in vacuum	c	2.998×10^8 m s^{-1}	2.998×10^{10} cm s^{-1}

* Atomic mass unit (also denoted by a.m.u. and a.u.).

CONVERSION FROM CGS TO SI

1 Å (ångstrom) $= 10^{-10}$ m $= 10^{-8}$ cm $= 10^{-4}$ μm $= 10^{-1}$ nm

1 litre $= 10^{-3}$ m$^3 = 1$ dm^3

1 erg $= 10^{-7}$ J

1 cal $= 4.184$ J

1 kcal mol$^{-1} = 4.184$ kJ mol^{-1}

1 kT $= 4.114 \times 10^{-14}$ erg $= 4.114 \times 10^{-21}$ J at 298 K ($\sim 25°$C)
$\quad = 4.045 \times 10^{-14}$ erg $= 4.045 \times 10^{-21}$ J at 293 K ($\sim 20°$C)

1 kT per molecule $= 0.592$ kcal mol$^{-1} = 2.478$ kJ mol^{-1} at 298 K

1 eV $= 1.602 \times 10^{-12}$ erg $= 1.602 \times 10^{-19}$ J

1 eV per molecule $= 23.06$ kcal mol$^{-1} = 96.48$ kJ mol^{-1}

1 cm^{-1} (wavenumber unit of energy) $= 1.986 \times 10^{-23}$ J

1 dyne $= 10^{-5}$ N

1 dyne cm$^{-1} = 1$ erg cm$^{-2} = 1$ mN m$^{-1} = 1$ mJ m^{-2} (unit of surface tension)

1 dyne cm$^{-2} = 10^{-1}$ Pa (N m^{-2})

1 atm $= 1.013 \times 10^6$ dyne cm$^{-2} = 1.013$ bar $= 1.013 \times 10^5$ Pa (N m^{-2})

1 torr $= 1$ mm Hg $= 1.316 \times 10^{-3}$ atm $= 133.3$ Pa (N m^{-2})

$0°$C $= 273.15$ K (triple point of water)

1 esu $= 3.336 \times 10^{-10}$ C

1 poise (P) $= 10$ gm cm^{-1} s$^{-1} = 10^{-1}$ kg m^{-1} s$^{-1} = 10^{-1}$ N s m^{-1}
$\quad$ (unit of viscosity)

1 stokes (St) $= 10^{-4}$ m^2 s^{-1} (unit of kinematic viscosity: viscosity/density)

Debye (D) $= 10^{-18}$ esu $= 3.336 \times 10^{-30}$ C m (unit of electric dipole moment)

CONVERSION FROM SI TO CGS

1 nm $= 10^{-9}$ m $= 10$ Å $= 10^{-7}$ cm

1 J $= 10^7$ erg $= 0.239$ cal $= 6.242 \times 10^{18}$ eV

$\quad = 5.034 \times 10^{22}$ cm$^{-1} = 7.243 \times 10^{22}$ K

1 kJ mol$^{-1} = 0.239$ kcal mol^{-1}

1 N $= 10^5$ dyne

1 Pa $= 1$ N m$^{-2} = 9.872 \times 10^{-6}$ atm $= 7.50 \times 10^{-3}$ torr $= 10$ dyne cm^{-2}

1 bar $= 10^5$ N m$^{-2} = 10^{-5}$ Pa $= 0.9868$ atm $= 750.06$ mm Hg

USEFUL QUANTITIES AND RELATIONS

Energy equivalent, mc^2, of one atomic mass unit (u) $= 1.492 \times 10^{-10}$ J

Mean volume occupied per molecule $= (MW)/(N_0 \times \text{density})$

Mass of any atom or molecule $= (MW) \times (1.661 \times 10^{-27})$ kg

Standard volume of ideal gas $= 22.414$ m^3 mol^{-1} (1 mol^{-1})

$kT/e = RT/F = -25.69$ mV at 298 K

1 C m$^{-2} = 1$ unit charge per 0.16 nm^2 (16Å^2)

κ^{-1} (Debye length) $= 0.304/\sqrt{M}$ nm for 1:1 electrolyte at 298 K, where

$\quad$ 1 M $= 1$ mol dm$^{-3} \equiv 6.022 \times 10^{26}$ molecules per m^3

Mass of the earth $= 5.976 \times 10^{24}$ kg

Density of earth (mean) $= 5.518 \times 10^3$ kg m^{-3}

Values of gravitational acceleration, g:

$\quad$ Equator (9.780 m s^{-2}), north and south poles (9.832 m s^{-2}),

$\quad$ New York (9.801 m s^{-2}), London (9.812 m s^{-2}).

COMMON SYMBOLS

A	Hamaker constant (J), area (m^2), Helmholtz free energy
a	Atomic or molecular radius (m), headgroup area (m^2)
a, b	Constants in equations of state
a_0	Bohr radius (0.053 nm)
C	Interaction constant (J m^6), aqueous solute concentration in mole fraction units (mol dm^{-3}/55.5)

D	Distance between two surfaces (m)
Da	Dalton unit of molecular weight (same as MW)
E	Electric field strength (V m^{-1})
F	Force (N)
G	Gibbs free energy
H	Enthalpy
I	Ionization potential (J)
i	$\sqrt{-1}$
K	Elastic modulus (N m^{-2}), spring constant (Nm^{-1})
K_a	Reaction constant, association constant (M^{-1})
K_d	Dissociation constant ($= 1/K_d$)
k_a	Area compressibility modulus (J m^{-2})
k_b	Bending modulus (J)
L	Latent heat (J mol^{-1}), thickness of polymer brush layer (m)
l	Length (m), unit segment length in polymer chain (m)
l_c	Critical hydrocarbon chain length (m)
M	Concentration (mol dm^{-3}), molecular weight, mean aggregation number
MW	Molecular weight (Da)
m	Mass (kg)
n	Refractive index, number of segments in a polymer chain
P	Pressure (Nm^{-2})
pK	$-\log_{10}$[concentration of H$^+$ ions in M]
Q, q	Charge (C)
R	Radius (m)
R_g	Radius of gyration of polymer (m)
R_F	Flory radius of polymer (m)
r	Interatomic distance (m), radius (m)
S	Entropy
s	Mean distance between polymer anchoring sites (m)
T	Temperature (K)
T_M, T_B	Melting or boiling points (K or °C)
T_c	Lipid chain melting temperature
u	Dipole moment (C m)
U	Molar cohesive energy (J mol^{-1}), internal energy (J mol^{-1})
V, v	Molar volume (m^3), volume (m^3)
W, w	Interaction free energy (J), pair potential (J)
X	Dimensionless concentration (e.g., mole fraction)
Y	Young's modulus (N m^{-2})
z	Valency
α	Polarizability (C^2 m^2 J^{-1}), interaction energy parameter (J or J m^{-1})

γ	Surface tension (N m^{-1}), surface energy (J m^{-2}), tanh$(e\psi_0/4kT) \approx$ tanh$[\psi_0(\text{mV})/103]$ at 298 K
γ_i, γ_{AB}	Interfacial energy (J m^{-2})
Γ	Surface coverage (number per m^2)
δ	Stern layer thickness (m)
ε	Relative permittivity, static dielectric constant
$\in$	Energy (J or J m^{-1})
θ	Angle (deg or rad), theta temperature of solvent (K)
κ	Inverse Debye length (m^{-1})
λ, λ_0	Characteristic exponential decay length, wavelength (m)
ξ	Correlation length (m)
μ	Chemical potential
μ^i, μ^0	Standard part of chemical potential due to interactions
ν, ν_I	Frequency (s^{-1} or Hz), ionization frequency (s^{-1})
ρ	Number density (m^{-3}) or mass density (kg m^{-3})
σ	Atomic or molecular diameter (m) surface charge density (C m^{-2}), standard deviation
τ	Characteristic relaxation time (s)
ϕ	Angle (deg or rad)
ψ	Electrostatic potential (V)
ψ_0	Electrostatic surface potential (V)
Π	Two dimensional surface pressure (N m^{-1})
$\approx, \sim$	Approximately equal to, roughly
$>, <$	Greater than, less than
$\gtrsim$	Slightly greater than
$\geqslant$	Greater than or equal to
$\equiv$	Equivalent to
$\propto$	Proportional to
Δ	Change or difference in
$\gg, \ll$	Much greater than, less than
$\langle X \rangle$	Average or mean of X
$[X]$	Concentration of X
$\rightarrow$	Approaches
$\square$	Start or end of Worked Example

PART ONE

THE FORCES BETWEEN ATOMS
AND MOLECULES:
PRINCIPLES AND CONCEPTS

CHAPTER 1

HISTORICAL PERSPECTIVE

1.1 THE FOUR FORCES OF NATURE

It is now well established that there are four distinct forces in nature. Two of these are the *strong* and *weak interactions* that act between neutrons, protons, electrons and other elementary particles. These two forces have a very short range of action, less than 10^{-5} nm, and belong to the domain of nuclear and high-energy physics. The other two forces are the *electromagnetic* and *gravitational interactions* that act between atoms and molecules (as well as between elementary particles). These forces are effective over a much larger range of distances, from subatomic to practically infinite distances, and are consequently the forces that govern the behaviour of everyday things (Fig. 1.1). For example, electromagnetic forces—the source of all intermolecular interactions—determine the properties of solids, liquids and gases, the behaviour of particles in solution, chemical reactions and the organization of biological structures. Gravitational forces account for tidal motion and many cosmological phenomena, and when acting together with intermolecular forces, determine such phenomena as the height that a liquid will rise in small capillaries and the maximum size that animals and trees can attain (Thompson, 1968).

This book is mainly concerned with intermolecular forces. Let us enter the subject by briefly reviewing its historical developments from the ancient Greeks to the present day.

1.2 GREEK AND MEDIEVAL NOTIONS OF INTERMOLECULAR FORCES

The Greeks found that they needed only two fundamental forces to account for all natural phenomena. One was Love, the other was Hate. The first brought things together while the second caused them to part. The idea was

3

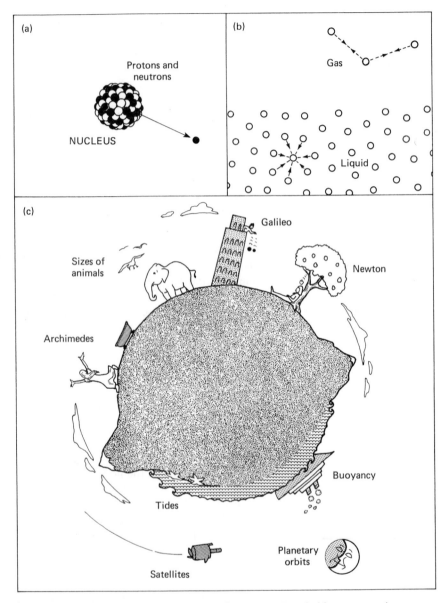

Fig. 1.1. The forces of nature. (a) Strong nuclear interactions hold protons and neutrons together in atomic nuclei. Weak interactions are involved in electron emission (β decay). (b) Electrostatic (intermolecular) forces determine the cohesive forces that hold atoms and molecules together in solids and liquids. (c) Gravitational forces affect tides, falling bodies and satellites. Gravitational and intermolecular forces acting together determine the maximum possible sizes of mountains, trees and animals.

first proposed by Empedocles around 450 B.C., was much 'improved' by Aristotle, and formed the basis of chemical theory for 2000 years.

The ancients appear to have been particularly inspired by certain mysterious forces, or influences, that sometimes appeared between various forms of matter (forces that we would now label as magnetic or electrostatic). They were intrigued by the 'action-at-a-distance' property displayed by these forces, as well as by gravitational forces, and were moved to reflect upon their virtues. What they lacked in concrete experimental facts they more than made up for by the abundant resources of their imagination (see Problem 1.1). Thus, magnetic forces could cure diseases, though they could also cause melancholy and thievery. Magnets could be used to find gold, and they were effective as love potions and for testing the chastity of women. Unfortunately, some magnetic substances lost their powers if rubbed with garlic (but they usually recovered when treated with goat's blood). Electric phenomena were endowed with attributes no less spectacular, manifesting themselves as visible sparks in addition to a miscellany of attractive or repulsive influences that appeared when different bodies were rubbed together. All these wondrous practices, and much else, were enjoyed by our forebears until well into the seventeenth century.

1.3   EARLY SCIENTIFIC PERIOD: CONTRASTS WITH GRAVITATIONAL FORCES

In the seventeenth century our subject entered its first phase of scientific scrutiny. Newton considered how the forces between molecules could be linked to the physical properties of matter, and later a number of eighteenth century researchers investigated the phenomenon of the capillary rise of liquids in glass tubes. In 1808 Clairaut suggested that capillarity could be explained if the attraction between the liquid and glass molecules was different from the attraction of the liquid molecules for themselves. It was also noticed that the height of rise of a liquid column does not depend on the capillary wall thickness, which led to the conclusion that these forces must be of very short range (or, in the language of the time, extended over 'insensible' distances).

During the nineteenth century it was believed that one simple universal force law (similar to Newton's law for the gravitational force) would eventually be found to account for all intermolecular attractions. To this end a number of interaction potentials were proposed that invariably contained the masses of the molecules, attesting to the belief at the time that these forces are related to gravitational forces. Thus, typical *interaction potentials* of two molecules were of the form $w(r) = -Cm_1m_2/r^n$, which is related to the *force law* $F(r)$ between them by

$$F(r) = -dw(r)/dr = -nCm_1m_2/r^{n+1}, \tag{1.1}$$

where m_1, m_2 are the molecular masses, r their separation, C a constant, and n some integer believed to be 4 or 5, which may be compared with $n = 1$ for the gravitational interaction:

$$w(r) = -Gm_1m_2/r, \qquad G = 6.67 \times 10^{-11}\,\text{N m}^2\,\text{kg}^{-2}. \tag{1.2}$$

It is instructive to see how the power-law index n was so chosen. It arose from an appreciation of the fact that if intermolecular forces are not to extend over large distances, the value of n must be greater than 3. Why this is so can be simply established as follows: suppose the attractive potential between two molecules or particles to be of the general form $w = -C/r^n$, where n is an integer. Now consider a region of space where the number density of these molecules is ρ. This region can be a solid, a liquid, a gas, or even a region in outer space extending over astronomical distances. Let us add all the interaction energies of one particular molecule with all the other molecules in the system. The number of molecules in a region of space between r and $(r + dr)$ away will be $\rho 4\pi r^2\,dr$ (since $4\pi r^2\,dr$ is the volume of a spherical shell of radius r and thickness dr, i.e., of area $4\pi r^2$ and thickness dr). The total interaction energy of one molecule with all the other molecules in the system will therefore be given by

$$\text{total energy} = \int_\sigma^L w(r)\rho 4\pi r^2\,dr = -4\pi C\rho \int_\sigma^L r^{2-n}\,dr$$

$$= \frac{-4\pi C\rho}{(n-3)\sigma^{n-3}}\left[1 - \left(\frac{\sigma}{L}\right)^{n-3}\right] \tag{1.3}$$

$$= -4\pi C\rho/(n-3)\sigma^{n-3} \qquad \text{for } n > 3 \quad \text{and} \quad L \gg \sigma, \tag{1.4}$$

where σ is the diameter of the molecules and L is the size of the system (e.g., the dimensions of a solid or the size of the box containing a gas). We can see that since σ must be smaller than L (i.e., $\sigma/L < 1$), large distance contributions to the interaction will disappear only for values of n greater than 3 (i.e., for $n = 4, 5, 6, \ldots$). But for n smaller than 3, the second term in Eq. (1.3) will be greater, and the contribution from more distant molecules will dominate over that of nearby molecules. In such cases the size of the system must be taken into account, as occurs for the gravitational force where

$n = 1$ and where distant planets, stars, and even galaxies are still strongly interacting with each other (see Problem 1.2).

In later chapters we shall see that theoretical derivations of intermolecular force potentials do indeed predict that n always exceeds 3 asymptotically, and it is for this reason that the bulk properties of solids, liquids and gases do not depend on the volume of material or on the size of the container (unless these are extremely small) but only on the forces between molecules in close proximity to each other. Important long-range intermolecular forces also exist, especially between macroscopic particles and surfaces, but their effective range of action rarely exceeds 100 nm.

Returning to the latter part of the nineteenth century, hopes for an all-embracing force law dwindled as it became increasingly apparent that no suitable candidate would be forthcoming to explain the multitude of

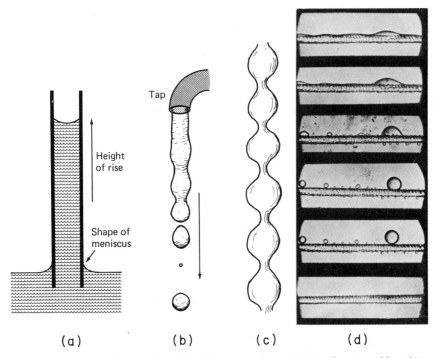

Fig. 1.2. Long-range effects produced by short-range forces. (a) Capillary rise of liquids in narrow channels. (b) Shape of water filament flowing out from a tap. (c) Unduloid shape of a spider's web, a 'frozen' intermediate state of (b). (d) Action of detergent molecules in removing oily dirt from a fabric. Top to bottom: progressive addition of detergent diminishes the contact area of oil droplets on fibre until they finally detach. ((d) from Adam and Stevenson, 1953.)

phenomena. However, by this time the modern concept of surface tension forces was firmly established, as was the recognition that these forces are the same as those that hold molecules together in solids and liquids, and that in both cases they arise from interactions acting over very short distances. In addition, it was shown how these very short-range surface tension forces can account for such phenomena as capillarity, the shapes of macroscopic liquid droplets on surfaces, the contact angle between coalescing soap bubbles, and the breakup of a jet of water into spherical droplets (Fig. 1.2). Thus it was established that very short-range forces can lead to very long-range (i.e., macroscopic) effects. But further developments were to be forthcoming from quite different quarters: from work on gases rather than liquids.

1.4 FIRST SUCCESSFUL PHENOMENOLOGICAL THEORIES

In an attempt to explain why real gases did not obey the ideal gas law ($PV = RT$, where P is the pressure, V the molar volume, R the gas constant, and T the temperature), the Dutch physicist J. D. van der Waals considered the effects of attractive forces between molecules (at a time when the very existence of molecules as we know them today was still being hotly debated). In 1873 he arrived at his famous equation of state for gases and liquids,

$$(P + a/V^2)(V - b) = RT, \tag{1.5}$$

in which he subtracted the term b from the volume to account for the finite size of molecules and added the term a/V^2 to the pressure to account for the attractive intermolecular forces now known as *van der Waals forces*.

By the early twentieth century it was recognized that intermolecular forces are not of a simple nature, and the pursuit of one basic force law gave way to a less ambitious search for semiempirical expressions that could account for specific phenomena. In this vein Mie, in 1903, proposed an interaction 'pair potential' of the form

$$w(r) = -A/r^n + B/r^m, \tag{1.6}$$

which for the first time included a repulsive term as well as an attractive term. This was the first of a number of similar laws that successfully accounted for a wide range of phenomena, and it is still used today, as is the van der Waals equation of state. Later we shall see how the parameters in potentials such as the Mie potential can be related to the constants a and b in the van der Waals equation of state. Lamentably, it was soon found that many different potentials with a wide range of (adjustable) parameters would satisfactorily account for the same experimental data, such as the elasticity of solids or the $P-V-T$ behaviour of gases. Thus, while such empirical equations were useful, the nature and origin of the forces themselves remained a mystery.

☐   ● **WORKED EXAMPLE** ●

Question: The *Lennard-Jones potential*

$$w(r) = -A/r^6 + B/r^{12} \tag{1.7}$$

is a special case of the Mie potential, Eq. (1.6). In this potential the attractive (negative) contribution is the van der Waals interaction potential which varies with the inverse-sixth power of the distance (Chapter 6). Make a sketch of how the energy $w(r)$ and force $F(r)$ vary with r. What does the Lennard-Jones potential predict for (i) the separation $r = r_e$ when the energy is at the minimum (equilibrium) value, w_{min}, (ii) the ratio of w_{min} to the purely attractive van der Waals component of the interaction potential at r_e, (iii) the ratio of r_e to r_0 defined by $w(r_0) = 0$, and (iv) the ratio of r_s to r_0, where r_s is the separation where the (adhesive) force is maximum, F_{max}?

For the interaction between two atoms the values of A and B are known to be $A = 10^{-77}$ J m^6 and $B = 10^{-134}$ J m^{12}. What is w_{min} for this interaction in units of kT at 298 K, and what is the maximum adhesion force F_{max} between the two atoms? Is this force measurable with a sensitive balance?

Answer:
● Figure 1.3 shows schematic plots of $w(r)$ and $F(r)$.
● $w(r)$ is minimum when $dw/dr = 0$. This occurs at $r = r_e = (2B/A)^{1/6}$.
● Substituting r_e into Eq. (1.7) gives $w(r_e) = w_{min} = -A^2/4B = -A/2r_e^6$.
● $w_{min}(r_e)/w_{VDW}(r_e) = (-A/2r_e^6)/(-A/r_e^6) = \frac{1}{2}$.
● Since $w(r) = 0$ at $r = r_0 = (B/A)^{1/6}$, we obtain $r_e/r_0 = 2^{1/6} = 1.12$.
● The force is given by $F = -dw/dr$, and F_{max} occurs at $d^2w/dr^2 = 0$, i.e. when $r = r_s = (26B/7A)^{1/6}$. Thus $r_s/r_0 = (26/7)^{1/6} = 1.24$.
● $w_{min} = -A^2/4B = -2.5 \times 10^{-21}$ J $\rightarrow 2.5 \times 10^{-21}/4.1 \times 10^{-21} = 0.61$ kT at 298 K.
● $F_{max} = -dw/dr = -6A/r^7 + 12B/r^{13}$ at $r = (26B/7A)^{1/6} = 0.3935$ nm. Thus $F_{max} = -(126A^2/169B)/(26B/7A)^{1/6} = -1.89 \times 10^{-11}$ N (attractive).
● Conventional all-purpose laboratory balances can measure down to about 10^{-9} N. To measure weaker forces one needs specialized techniques. The

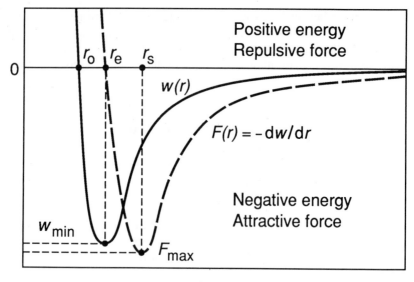

Fig. 1.3.

Atomic Force Microscope (AFM) can measure forces down to 10^{-11} N using piezoelectric, capacitance or optical fibre technology for determining spring deflections, while forces as small as 10^{-13} N (10 pg) have been measured between a colloidal particle in water and a surface using optical techniques (see Section 10.7).   ☐

By the beginning of this century our subject had reached the end of its first scientific phase, coinciding (not unexpectedly) with the end of the classical era of physics and chemistry. But a number of important conceptual changes had also occurred. The period that had started with Newton and ended with van der Waals, Boltzmann, Maxwell and Gibbs saw the abandonment of the purely mechanistic view of intermolecular forces and the adoption of thermodynamic and probabilistic concepts such as free energy and entropy. It was now appreciated that heat is also a form of energy and that the thermal energy and entropy associated with molecules are also involved in determining their collective properties. In particular, it rapidly became apparent that there is a big gap between knowing the force or pair potential between two isolated molecules and understanding how an ensemble of such molecules will behave. For example, the mere knowledge that air molecules attract each other does not mean that they will condense into a liquid or a solid at any given temperature or pressure. Even today there is no ready recipe for deriving the properties of condensed phases from the intermolecular pair potentials, and vice versa.

1.5   MODERN VIEW OF THE ORIGIN OF INTERMOLECULAR FORCES

Only with the elucidation of the electronic structure of atoms and molecules and the development of the quantum theory in the 1920s was it possible to understand the origin of intermolecular forces and derive expressions for their interaction potentials. It was soon established that all intermolecular forces are essentially electrostatic in origin. This is encapsulated in the *Hellman–Feynman theorem*, which states that once the spatial distribution of the electron clouds has been determined by solving the Schrödinger equation, the intermolecular forces may be calculated on the basis of straightforward classical electrostatics. This theorem greatly simplified notions of the nature of intermolecular forces.[1]

Thus, for two charges, we have the familiar inverse-square Coulomb force, while for moving charges we have electromagnetic forces, and for the complex fluctuating charge distributions occurring in and around atoms, we obtain the various interatomic and intermolecular bonding forces familiar to physics, chemistry and biology.

This seems marvellously simple. Unfortunately, exact solutions of the Schrödinger equation are not easy to come by. In fact, it is even too difficult to solve (exactly) something as simple as two hydrogen atoms interacting in vacuum. For this reason, it has been found useful to classify intermolecular interactions into a number of seemingly different categories even though they all have the same fundamental origin.[2] Thus, such commonly encountered terms as ionic bonds, metallic bonds, van der Waals forces, hydrophobic interactions, hydrogen bonding and solvation forces are a result of this classification, often accompanied by further divisions into strong and weak interactions and short-range and long-range forces. Such distinctions can be very useful, but they can also lead to confusion, for example, when the same interaction is 'counted twice' or when two normally distinct interactions are strongly coupled.

1.6   RECENT TRENDS

Today, as more and more information is accumulating on the properties of diverse systems at the molecular level, there is a natural desire to understand

---

[1] Later we shall encounter a very useful analogous theorem—the 'contact value theorem'—which gives the force between two surfaces once the density distribution of molecules, ions or particles in the space between them is known.

[2] Some physicists believe that some of the four fundamental forces of nature are likewise related.

these phenomena in terms of the operative forces. Until recently there were three main areas of activity. The first was largely devoted to the forces acting between simple atoms and molecules in gases, where various quantum mechanical and statistical mechanical calculations are able to account for many of their physical properties (Hirschfelder *et al.*, 1954). The second area was concerned with the chemical bonding of ions, atoms and molecules in solids (Pauling, 1960), while the third dealt with the long-range interactions between surfaces and small 'colloidal' particles suspended in liquids (Verwey and Overbeek, 1948)—an area that is traditionally referred to as *colloid science*.

More recently, the scope of endeavour has broadened to include liquid structure, surface and thin-film phenomena, 'complex fluids' such as surfactant and polymer 'self-assembling' systems, material properties, biological macromolecules and the interactions of biological structures (all of which are considered in this book). Not only are static (equilibrium) forces being investigated but also dynamic (e.g., viscous and time-dependent) forces, and modern theoretical tools now routinely include Monte Carlo, molecular dynamics and other *computer simulation* techniques.

The subject has become so spread out that a tendency has developed for different disciplines to adopt their own concepts and terminology,[3] and even to emphasize quite different aspects of interactions that are essentially the same. For example, in chemistry and biology, emphasis is placed almost entirely on the *short-range* force fields around atoms and molecules, rarely extending more than one or two atomic distances. The language of the present-day molecular biologist is full of terms such as molecular packing, specific binding sites, lock and key mechanisms, etc., all of which are essentially short range. In the different, though closely related area of colloid science, the emphasis is quite often on the *long-range* forces, which may determine whether two surfaces or particles are able to get close enough in the first place before they can interact via the types of short-range forces mentioned above. In this discipline one is more likely to hear about electric double-layer forces, van der Waals forces, steric polymer interactions, etc., all of which are essentially long range.

The situation is not quite as silly as it sounds. Often the important interactions in chemistry and biochemistry *are* the short-range ones, while those in colloidal systems are the long-range ones. But such contemporary issues no longer form part of our historical perspective. Their understanding requires some knowledge of the strength, range, and mode of action of

---

[3] Even the words *force* and *interaction* have become ambiguous. In this book I use *force* when specifically referring to a force, while *interaction* covers all the effects that two bodies may have on each other including the torques, induced shape-changes, molecular rearrangements *and* the forces between them.

different types of forces and thus leads us naturally into the subject matter of this book.

PROBLEMS AND DISCUSSION TOPICS

1.1 *Aristotle*: Now, Plato, let us lay the foundations of the physics of motion. We agree that the motion of any natural body is caused by its need for fulfilment?

Plato: Yes, my dear Aristotle.

Aristotle: And that each body is naturally carried to its appropriate place, which may be up or down?

Galileo: But by what impetus is this movement accomplished?

Aristotle: Ah, let me complete the theory, Simplisticus. Every body experiences a natural impulse commensurate with its virtue of weight or lightness. So two bodies which differ from one another owing to excess of weight or of lightness, but are alike in other respects, move faster over an equal distance akin to the ratio which their magnitudes bear to one another.

Galileo: I see. Actually, my name is Galileo. So a heavier body falls faster than a lighter body?

Aristotle: Of course. That is the obvious conclusion.

Galileo: But then if a heavy body falls faster than a light body, I infer from your theory that an even heavier body falls more slowly!

Aristotle: How is that?

Galileo: For if I join the two bodies together by a string, the lighter will slow down the heavier while the heavier will speed up the lighter. They will therefore fall at a rate intermediate between the two bodies falling alone. But, since I tied them together, the two bodies have been turned into a single mass which is heavier than either. So surely it should fall faster than either body. We are therefore faced with a paradox. It appears to me that the only way to resolve this paradox is to conclude that all bodies must fall at the same rate.

Aristotle: Your paradox is indeed intriguing.

Galileo: How about we do an experiment with two balls to test it?

Aristotle: Eh?

Question: Is either of them right? If you claim Galileo to be right explain how he could arrive at the correct answer when none of the postulates are based on any experimental observation? (This problem furnishes a good example of Rothchild's Rule: '*For every phenomenon, however complex, someone will eventually come up with a simple and elegant theory. This theory will be wrong.*' For further reading on early ideas on gravitation see Aristotle's

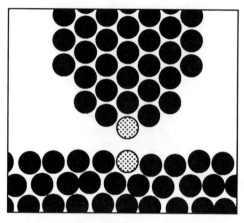

Fig. 1.4.

Physics, Book IV; Galileo Galilei, *Discorsi e dimostrazioni matematiche, intorno à due nuove scienze*, Leyden, The Netherlands (1638); Jammer (1957); Hesse (1961).)

1.2 Consider the universe as composed of particles, stars and galaxies distributed randomly but uniformly within a spherical region of space of average mass density ρ and radius R. The particles interact via the inverse square gravitational force law given by Eq. (1.1). One particle of mass m is at a finite distance r from the centre. What is the force acting on this particle when (i) $r \gg R$, (ii) $r \ll R$ and (iii) $R = \infty$ (infinite universe)? What are the implications of your results for the effect of faraway particles on neighbouring particles when all interact via an inverse square force? (For a related phenomenon to do with light find out about *Olbers' Paradox*.)

1.3 All the atoms in a particular solid interact via a Lennard–Jones potential, Eq. (1.7), with interaction constants $A = 10^{-77} \, \mathrm{J \, m^6}$ and $B = 10^{-134} \, \mathrm{J \, m^{12}}$ (as in the worked example in Section 1.4). Consider the approach of two such atoms where, as illustrated in Fig. 1.4, the lower is part of a solid surface and the other is at the end of a fine tip which is slowly brought vertically down. We may model this system as if the top atom is suspended from the end of a spring of effective stiffness K. If $K = 0.1 \, \mathrm{N \, m^{-1}}$, calculate the value of r at which an instability occurs and the tip 'jumps' into contact with the surface. Will there be another instability, and an outward 'jump', on separating the surfaces? In reality, will the top atom or the whole tip tend to move sideways during the approach and separation process?

(*Hint*: to understand this problem fully it is instructive to be able to solve it graphically as well as numerically. First, plot the force $F(r)$ against r (suggested range: $0.2 < r < 1.2$ nm). Next, find at what point (or points) on

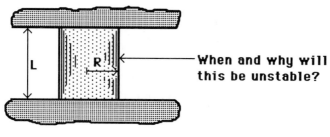

Fig. 1.5.

the curve the slope is $+0.1\ \mathrm{N\,m^{-1}}$. Draw a line through this point having this slope, and find where it cuts the curve again at another point. If you think about it, these two points on the line give the start and end points of a 'jump'. This problem can also be solved by considering the full energy–distance plot.)

Answer to numerical part: there is an instability and an inward jump from $r = 0.48$ nm, and another instability and an outward jump on separation from $r = 0.41$ nm. From your graphical solution you should also find the end points of these two jumps. Such mechanical instabilities occur when measuring intermolecular forces using an atomic force microscope or the forces between macroscopic surfaces using a surface forces apparatus, both of which are described in Section 10.7. They are also important in determining the displacements of atoms from their equilibrium lattice positions when two 'incommensurate' crystal lattices are in contact.

1.4   As was shown in Fig. 1.2, a long cylindrical filament of liquid is inherently unstable and tends to become unduloid and then break up into one or more spheres. For a liquid cylinder of radius R (Fig. 1.5) show that the length of tube L beyond which it becomes *mechanically* unstable is given by $L = 2\pi R$. Is this value the same as that for which the cylinder becomes *thermodynamically* unstable? Discuss the differences between the two. Would you expect similar instabilities for a liquid sheet or for a thin liquid film adsorbed on a surface?

CHAPTER 2

SOME THERMODYNAMIC ASPECTS OF INTERMOLECULAR FORCES

2.1 INTERACTION ENERGIES OF MOLECULES IN FREE SPACE AND IN A MEDIUM

While this book is not primarily concerned with thermodynamics, it is nevertheless appropriate to start by considering some fundamental thermodynamic principles without which a mere knowledge of interaction forces will not always be very meaningful. In this chapter we shall introduce a number of simple but important thermodynamic relations and then illustrate how these, when taken together with the strengths of intermolecular forces, determine the properties of a system of molecules.

At the most basic molecular level we have the interaction potential $w(r)$ between two molecules or particles. This is usually known as the *pair potential* or, when the interaction takes place in a solvent medium, the *potential of mean force.* The interaction potential $w(r)$ is related to the force between two molecules or particles by $F = -dw(r)/dr$. Since the derivative of $w(r)$ with respect to distance r gives the force, and hence the work that can be done by the force, $w(r)$ is often referred to as the *free energy* or *available energy.*

In considering the forces between two molecules or particles in liquids, several effects are involved that do not arise when the interaction occurs in free space. This is because an interaction in a medium always involves many solvent molecules, i.e., it is essentially a *many-body interaction.* Some of these effects are illustrated in Fig. 2.1 and will now be described.

(i) For two solute molecules in a solvent, their pair potential $w(r)$ includes not only the direct solute–solute interaction energy but also any changes in the solute–solvent and solvent–solvent interaction energies as the two solute molecules approach each other. A dissolved solute molecule can approach another only by displacing solvent molecules from its path (Fig. 2.1a). The net force therefore also depends on the attraction between the solute molecules and the solvent molecules. Thus, while two molecules may attract each other

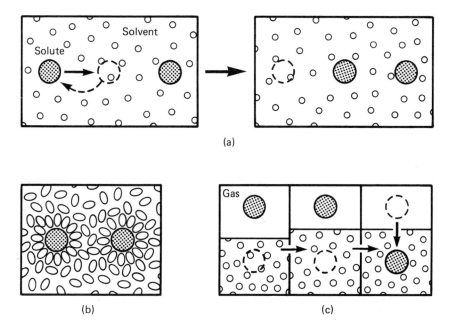

Fig. 2.1. Some solvent effects involved in the interactions of dissolved solute molecules or particles. (a) Displacement of solvent by two approaching solute molecules. (b) Solvation (reordering) of solvent molecules by solute. (c) Cavity formation by solvent prior to solute insertion.

in free space, they may repel each other in a medium if the work that must be done to displace the solvent exceeds that gained by the approaching solute molecules.

(ii) Solute molecules often perturb the local ordering or 'structure' of solvent molecules (Fig. 2.1b). If the free energy associated with this perturbation varies with the distance between the two dissolved molecules, it produces an additional 'solvation' or 'structural' force between them.

(iii) Solute–solvent interactions can change the properties of dissolved molecules, such as their dipole moment and charge (degree of ionization). The properties of dissolved molecules may therefore be different in different media.

(iv) Finally, when an individual molecule is introduced into a condensed medium, we must not forget the cavity energy expended by the medium when it forms the cavity to accommodate the guest molecule (Fig. 2.1c).

These effects are obviously interrelated and are collectively referred to as *solvent effects* or *medium effects.* They manifest themselves to different degrees depending on the nature and strength of solute–solute, solute–solvent and solvent–solvent interactions. In later chapters we shall investigate the various functional forms of $w(r)$ for different types of forces and examine how solvent effects influence intermolecular and interparticle interaction potentials. In this chapter, we shall address ourselves to the thermodynamic implications arising from the existence of interaction potentials without, at this stage, inquiring as to their origins.

When an individual molecule is in a medium (gas or liquid), it has what may be called a 'cohesive energy' or 'self-energy', μ^i, associated with it. This is given by the sum of its interactions with all the surrounding molecules (which includes any change in the energy of the solvent brought about by the presence of the solute molecule). In many cases one needs to know the value of μ^i of an isolated molecule in a medium rather than its pair potential $w(r)$ with another individual molecule. How are μ^i and $w(r)$ related? Let us first consider a molecule in the gas phase, where $w(r)$ is usually a simple power law of the form

$$w(r) = -C/r^n \qquad \text{for } r > \sigma \quad (\text{where } n > 3),$$

$$= \infty \qquad \text{for } r < \sigma, \tag{2.1}$$

where σ is the so-called *hard sphere diameter* of the molecules. We may now calculate μ^i by summing all the pair potentials $w(r)$ over all of space (as was done in Section 1.3) and obtain

$$\mu^i_{\text{gas}} = \int_\sigma^\infty w(r)\rho 4\pi r^2 \, dr = -4\pi C\rho/(n-3)\sigma^{n-3}. \tag{2.2}$$

This result will be used later for deriving the van der Waals equation of state.

When a molecule is introduced from vapour into a condensed phase, μ^i must also include the cavity energy (Fig. 2.1c). For example, in a liquid or solid each molecule can have up to 12 other molecules in contact with it (known as *close packing*). If a molecule is introduced into its own liquid medium, then 12 liquid molecules must first separate from each other to form the hole. This costs $-6w(\sigma)$ in breaking the six 'bonds' holding the 12 molecules together, where $w(\sigma)$ is the pair energy of two molecules in contact at $r = \sigma$. On introducing the guest molecule, 12 new 'bonds' are formed costing $+12w(\sigma)$. The net energy change is therefore

$$\mu^i_{\text{liq}} \approx 6w(\sigma), \tag{2.3}$$

which is *half* the total interaction energy of the molecule with its 12 nearest neighbours. Thus, the molar cohesive energy of a simple liquid (or solid) may be expected to be

$$U = -N_0 \mu_{\text{liq}}^i \approx -6N_0 w(\sigma). \tag{2.4}$$

It is interesting that a derivation similar to Eq. (2.2) also predicts a similar value for U. In a pure liquid (or solid) the number density of molecules ρ would be equal to

$$\rho = 1/(\text{molecular volume}) = 1/[(4\pi/3)(\sigma/2)^3], \tag{2.5}$$

where $\sigma/2$ is the molecular radius, and so we find that

$$\mu_{\text{liq}}^i = \frac{1}{2} \int_\sigma^\infty w(r)\rho 4\pi r^2 \, dr = \frac{-12C}{(n-3)\sigma^n} \approx \frac{12}{(n-3)} w(\sigma), \tag{2.6}$$

which for the van der Waals interaction where $n = 6$ (Chapter 6), gives

$$\mu_{\text{liq}}^i \approx 4w(\sigma) \quad \text{or} \quad U \approx -4N_0 w(\sigma). \tag{2.7}$$

As a rule of thumb we may therefore expect that the cohesive energy of a molecule in a pure liquid or solid will be somewhere between four and six times the pair energy, the higher value being applicable to simple spherical molecules that condense into close-packed structures.

An *accurate* calculation of a molecule's free energy μ^i in a liquid from the pair potential $w(r)$ is extremely difficult. The mean number of molecules surrounding any particular molecule is not known in advance. It can be as high as 12 (for simple molecules such as the inert gases) and as low as four (for water). Further, the density ρ of neighbouring molecules is not uniform locally, but rather it depends on the distance r from the reference molecule (cf. Fig. 2.1b). Thus, ρ in Eq. (2.2) should really be a function of r, i.e., $\rho(r)$. This is known as the *density distribution function*, which can be measured (see Chapters 7 and 13) but which can only be approximately determined *a priori*.

For a *solute* molecule dissolved in a *solvent* medium, i.e., when it is surrounded by different molecules, the calculation of its free energy in the medium becomes more difficult. Again, if we consider the simplest case of a solute molecule (s) surrounded by 12 solvent molecules (m) of similar size, then the change in cohesive energy on transferring the solute molecule from free space into the medium will be

$$\mu_{\text{liq}}^i \approx -[6w_{\text{mm}}(\sigma) - 12w_{\text{sm}}(\sigma)], \tag{2.8}$$

since six solvent pairs must first be separated before the solute molecule can enter the medium and interact with the 12 solvent molecules. Note that if the solvent and solute molecules are the same, then $w_{mm}(\sigma) = w_{sm}(\sigma)$ and the above expression reduces to the earlier result, $\mu^i \approx 6w_{mm}(\sigma)$, as expected.

Finally, ⌈it is important to note that the effective pair potential between two dissolved solute molecules in a medium is just the change in the sum of their free energies μ^i as they approach each other.⌉

2.2 THE BOLTZMANN DISTRIBUTION

If the molecular interaction energy of a particular type of molecule or particle has different values μ_1^i and μ_2^i in two regions of a system (e.g., a liquid in equilibrium with its vapour, or two coexisting phases), then at equilibrium the concentrations X_1 and X_2 of these molecules in the two regions is given by the well-known *Boltzmann distribution*

$$X_1 = X_2 \exp[-(\mu_1^i - \mu_2^i)/kT], \tag{2.9}$$

which may also be written as

$$\mu_1^i + kT \log X_1 = \mu_2^i + kT \log X_2, \tag{2.10}$$

where log means $\log_e$ or ln. Strictly, Eqs (2.9) and (2.10) are exact only when molecules mix ideally in both regions, that is, for a dilute system.

If there are many different regions or states in a system, each with different energies μ_n^i, then the condition of equilibrium is simply an extension of the above equation to all the states, viz.,

$$\mu_n^i + kT \log X_n = \text{constant} \qquad \text{for all states } n = 1, 2, 3, \ldots$$

$$= \mu. \tag{2.11}$$

In other words, there will be a flow of molecules between all the different states of the system until Eq. (2.11) is satisfied, that is, equilibrium is reached when the value of $\mu_n^i + kT \log X_n$ is uniform throughout. The quantity μ is known as the *chemical potential*, and it gives the total free energy per molecule; it includes the interaction energy as well as the contribution associated with its thermal energy. The $k \log X_n$ factor gives the entropy of confining the molecules and is known by a variety of names: the ideal gas entropy, the configurational entropy, the entropy of confinement, the ideal solution entropy, the translational entropy, the entropy of dilution and the entropy of mixing. The dimensionless concentrations X_n are usually expressed as mole fractions or volume fractions. For a pure solid or liquid, $X_1 = 1$ so that $\log X_1 = 0$.

2.3 The distribution of molecules and particles in systems at equilibrium

The requirement of equality of the chemical potentials, as expressed in Eq. (2.11), provides a very general and useful starting point for formulating conditions of equilibrium within a molecular framework (Castellan, 1972) and may be applied to both simple and complex multicomponent systems, e.g., the properties of 'self-assembling' molecular aggregates such as surfactant micelles and lipid bilayers, discussed in Part III. In the remainder of this chapter we shall consider some simpler cases.

Suppose we wish to calculate how the number density ρ of molecules in the earth's atmosphere varies with altitude z. If we only consider gravitational forces, we may write

$$\mu_z^i + kT \log \rho_z = \mu_0^i + kT \log \rho_0,$$

so that

$$\rho_z = \rho_0 \exp[-(\mu_z^i - \mu_0^i)/kT]. \tag{2.12}$$

Since $(\mu_z^i - \mu_0^i) = mgz$, where m is the molecular mass and g the gravitational acceleration, we immediately obtain

$$\rho_z = \rho_0 \exp(-mgz/kT) \tag{2.13}$$

which gives the density at height z in terms of the density at ground level ρ_0. This is the familiar gravitational or barometric distribution law.

Similarly, for charged molecules or ions, each carrying a charge e, if ψ_1 and ψ_2 are the electric potentials (in units of volts) in two regions of a systems, then $(\mu_2^i - \mu_1^i) = e(\psi_2 - \psi_1)$, and we obtain

$$\rho_2 = \rho_1 \exp[-e(\psi_2 - \psi_1)/kT], \tag{2.14}$$

which is known as the *Nernst equation*.

In the above examples the interaction energy did not arise from local intermolecular interactions, but from interactions with an externally applied gravitational or electric field. Let us now consider a two-phase system where $\mu_1^i - \mu_2^i$ is the difference in energy due to the different intermolecular interactions in the two phases. If one of the phases ($n = 1$) is a pure solid or liquid ($\log X_1 = 0$), we have

$$\mu_1^i = \mu_2^i + kT \log X_2,$$

thus

$$X_2 = X_1 \exp[-(\mu_2^i - \mu_1^i)/kT] = X_1 \exp(-\Delta\mu^i/kT). \qquad (2.15)$$

Here, for example, μ_2^i could be the energy of molecules in solution relative to their energy in the solid, μ_1^i, whence X_2 is their *solubility*.

In each of the above three examples it was assumed that only one type of interaction contributes to the chemical potential. More generally, if the two regions or phases are composed of different chemical species, are at different heights, Δz, and have a potential difference of $\Delta\psi$ volts between them, then when all the three effects of intermolecular, gravitational and electrostatic interactions are included, the distribution equation for the solute molecules or particles becomes

$$X_2 = X_1 \exp[-(\Delta\mu^i + mg\Delta z + e\Delta\psi)/kT]. \qquad (2.16)$$

□   ● WORKED EXAMPLE ●

Question: Consider two immiscible liquids such as water and oil. If a spherical oil molecule of radius r is taken out of the oil phase and placed in the water phase the unfavourable energy of this transfer is proportional to the area of the solute (oil) molecule newly exposed to the solvent (water) multiplied by the *interfacial energy*, γ_i, of the oil–water interface (see Chapter 15). The interfacial energy of the bulk cyclohexane–water interface is 50 mJ m^{-2}, and the radius of a cyclohexane molecule is 0.28 nm. Estimate the solubility of cyclohexane in water at 25°C in units of mol dm^{-3} (mol l^{-1} or M) and comment on your result.

Answer:
● $\Delta\mu^i/kT = 4\pi r^2\gamma_i/kT = 4 \times 3.142 \times (0.28 \times 10^{-9})^2 \times 50 \times 10^{-3}/4.12 \times 10^{-21} = 12.0$. Thus $X_2/X_1 = e^{-12} = 6 \times 10^{-6}$ in mole fraction units. Since this is very small, phase 2 is almost pure water (55.5 mol l^{-1}) and we may assume phase 1 to be almost pure cyclohexane so that $X_1 \approx 1$.

Thus the solubility of cyclohexane in water is calculated to be $X_2 \approx 55.5 \times 6 \times 10^{-6} = 3.4 \times 10^{-4}$ M.

• The literature value is 7×10^{-4} M, about twice the calculated value. Note, however, that if the transfer energy $\Delta\mu^i$ were only 6% less, i.e. $\Delta\mu^i/kT = 11.3$ instead of 12.0, the correct value would have been obtained. Thus, if this problem had been posed in reverse, i.e., if we had been given the solubility and asked to estimate the molecular radius, our result would have differed from the correct value by only 3% (instead of differing by 50%). This example shows that macroscopic values often work surprisingly well at the molecular level, but it also shows that interaction energies have to be calculated very accurately if they are to predict measurable quantities with reasonable accuracy. □

2.4 The van der Waals equation of state

We proceed with our consideration of the role of intermolecular interactions, starting with an analysis of a vapour in equilibrium with liquid. If the gas molecules interact through an attractive pair potential $w(r) = -C/r^n$, Eq. (2.2) may be expressed as

$$\mu^i_2 = \mu^i_{gas} = -4\pi C\rho/(n-3)\sigma^{n-3} = -A\rho, \qquad (2.17)$$

where $A = 4\pi C/(n-3)\sigma^{n-3} = $ constant. For molecules of finite size, we may also write

$$X_2 = 1/(v-B) = \rho/(1-B\rho) \qquad (2.18)$$

for the effective density of the nonideal gas molecules, where v is the gaseous volume occupied per molecule and $B = 4\pi\sigma^3/3$ is the excluded volume since σ is the closest distance that one molecule can approach another. We therefore have for the chemical potential of a gas:[1]

$$\mu = -A\rho + kT \log[\rho/(1-B\rho)]. \qquad (2.19)$$

---

[1] The complete expression for the chemical potential also includes additional purely temperature-dependent terms. One of these is the translational *kinetic energy* $\frac{3}{2}kT$ per molecule. Since the kinetic energy depends only on T, this term does not contribute to the pressure since it drops out of the derivation of the van der Waals equation when we calculate $(\partial\mu/\partial\rho)_T$ in Eq. (2.20). Another purely temperature-dependent term is $kT \log \lambda_T^3$, where $\lambda_T = (h^2/2\pi mkT)^{1/2}$ is known as the *de Broglie wavelength* or *thermal wavelength,* and where m is the molecular weight of the particles or molecules. λ_T^3 has units of volume, which ensures that the density $\lambda_T^3\rho$ is dimensionless. Boltzmann statistics apply whenever $\lambda_T^3\rho \ll 1$.

Now the pressure P is related to μ via the well-known thermodynamic relation

$$(\partial\mu/\partial P)_T = v = 1/\rho \qquad \text{or} \qquad (\partial P/\partial\rho)_T = \rho(\partial\mu/\partial\rho)_T. \qquad (2.20)$$

Thus we find

$$P = \int_0^\rho \rho\left(\frac{\partial\mu}{\partial\rho}\right)_T d\rho = \int_0^\rho \left[-A\rho + \frac{kT}{(1-B\rho)}\right] d\rho$$

$$= -\tfrac{1}{2}A\rho^2 - \frac{kT}{B}\log(1-B\rho)$$

and for $B\rho < 1$, we can expand the $\log_e$ term as

$$\log(1-B\rho) = -B\rho - \tfrac{1}{2}(B\rho)^2 + \ldots \approx -B\rho(1+\tfrac{1}{2}B\rho) \approx$$
$$-B\rho/(1-\tfrac{1}{2}B\rho) \approx -B/(v-\tfrac{1}{2}B),$$

so that

$$P = -\frac{\tfrac{1}{2}A}{v^2} + \frac{kT}{v-\tfrac{1}{2}B} \qquad \text{or} \qquad \left(P + \frac{a}{v^2}\right)(v-b) = kT, \qquad (2.21)$$

which is the van der Waals equation of state in terms of the molecular parameters

$$a = \tfrac{1}{2}A = 2\pi C/(n-3)\sigma^{n-3} \qquad \text{and} \qquad b = \tfrac{1}{2}B = 2\pi\sigma^3/3. \qquad (2.22)$$

Note that b depends only on the molecular size, σ, and thus on the stabilizing repulsive contribution to the total pair potential. Thus, conceptually, the constants a and b can be thought of as accounting for the attractive and repulsive forces between the molecules. In Chapters 6 and 7 we shall see how a and b can be related to other properties of molecules. The van der Waals equation is neither rigorous nor exact, but merely one of many equations that have been found useful for describing the properties of gases (PVT data) and gas–liquid phase transitions.

2.5   THE CRITERION OF THE THERMAL ENERGY kT FOR GAUGING THE STRENGTH OF AN INTERACTION

As we have seen, the fundamental significance of the thermal energy kT has to do with the partitioning of molecules among the different energy levels of a system. The magnitude of kT is also often used as a rough indicator of the strength of an interaction, the idea being that if the interaction energy exceeds kT it will 'win out' over the opposing or disorganizing effects of thermal motion. However, molecular motion or disorder can appear in various ways, e.g., as translational or positional disorder or as orientational disorder, and it is important to recognize how it manifests itself.

Let us first consider how strong the intermolecular attraction must be if it is to condense molecules into a liquid at a particular temperature and pressure. This amounts to finding the relation between the interaction (or cohesive) energy and the boiling point. Now, at standard atmospheric temperature and pressure (STP, where $T = 273$ K, $P = 1$ atm) one mole of gas occupies a volume of $\sim 22\,400$ cm^3, while in the condensed state a typical value would be about 20 cm^3. Equating chemical potentials for gas and liquid molecules in equilibrium with each other gives

$$\mu^i_{gas} + kT \log X_{gas} = \mu^i_{liq} + kT \log X_{liq} \tag{2.23}$$

and since the magnitude of μ^i_{liq} greatly exceeds μ^i_{gas} we may write

$$\mu^i_{gas} - \mu^i_{liq} \approx -\mu^i_{liq} = kT \log(X_{liq}/X_{gas}) \approx kT \log(22\,400/20) \approx 7 \text{ kT}.$$

If the gas obeys the gas law $PV \approx RT$, then at any other temperature the log term becomes $\log(22\,400 \times T/20 \times 273)$. It is straightforward to verify that the log term is not very sensitive to temperature and that in the range $T = 100$ to 500 K the log term changes by only 13%. Over this range of temperature we therefore find

$$-\mu^i_{liq} \approx 7\,kT_B \qquad \text{or} \qquad -N_0\mu^i_{liq}/T_B \approx 7\,N_0 k = 7R, \tag{2.24}$$

where T_B is the boiling temperature. This is an important result. First, it shows that the boiling point of a liquid is simply proportional to the energy needed to take a molecule from liquid into vapour. For one mole of molecules, the energy of vaporization U_{vap} is given by Eq. (2.4) as $U_{vap} = -N_0\mu^i_{liq}$, while the enthalpy or latent heat of vaporization, L_{vap}, is related to U_{vap} by

$$L_{vap} = H_{vap} = U_{vap} + PV \approx U_{vap} + RT_B. \tag{2.25}$$

Thus Eq. (2.24) predicts

$$L_{vap}/T_B \approx (U_{vap}/T_B) + R \approx 7R + R = 8R \approx 70 \text{ J K}^{-1} \text{ mol}^{-1}. \tag{2.26}$$

We have derived, very crudely, the well-known empirical relationship, known

TABLE 2.1 Boiling points T_B and latent heats of vaporization L_{vap} of some common substances

Substance		T_B at 1 atm (K)	L_{vap} (kJ mol^{-1})	L_{vap}/T_B^a (J K^{-1} mol^{-1})
Neon	Ne	27	1.8	65
Nitrogen	N_2	77	5.6	72
Argon	A	88	6.5	74
Oxygen	O_2	90	6.8	76
Methane	CH_4	112	8.2	73
Hydrogen chloride	HCl	118	16.2	86
Ammonia	NH_3	140	23.4	97
Hydrogen fluoride	HF	293	32.6	111
Ethanol	C_2H_5OH	352	39.4	112
Benzene	C_6H_6	353	20.8	87
Water	H_2O	373	40.7	109
Acetic acid	CH_3COOH	391	24.2	62
Iodine	I_2	456	41.7	91
Sodium	Na	1156	91.2	79
Lithium	Li	1645	129	78

[a] For most 'normal' substances, L_{vap}/T_B falls in the range of 75 to 90 J K^{-1} mol^{-1} (Trouton's rule). Higher values can usually be traced to cooperative association of molecules in the *liquid* (e.g., HF, NH_3, C_2H_5OH and H_2O), while lower values to association, e.g., dimerization, in the *vapour*, as occurs for CH_3COOH (see Fig. 8.2d).

as *Trouton's rule*, which states that the latent heat of vaporization is related to the normal boiling point of a liquid (at 1 atm) by

$$L_{vap}/T_B \approx 80 \text{ J K}^{-1} \text{ mol}^{-1}, \qquad (2.27)$$

which corresponds to a cohesive energy μ_{liq}^i of about $9 kT$ per molecule.

Trouton's rule, Eq. (2.27), applies to a great variety of substances, as illustrated in Table 2.1, and shows that the boiling point of a substance provides a reasonably accurate indication of the strength of the cohesive forces or energies holding molecules together in condensed phases. For solids, L_{vap} and T_B are replaced by the heat of sublimation and the sublimation temperature, respectively.

Equation (2.27) also tells us that molecules will condense once their cohesive energy μ^i with all the other molecules in the condensed phase exceeds about $9 kT$. Since we previously saw (Eq. (2.3)) that $\mu^i \approx 6w(\sigma)$, we may further conclude that if the pair interaction energy of two molecules or particles in contact exceeds about $\frac{3}{2} kT$, then it is strong enough to condense them into a liquid or solid (see Table 6.1, Chapter 6). It is for this reason that the thermal energy $\sim \frac{3}{2}kT$ can be used as a standard reference for gauging the cohesive strength of an interaction potential, though it is essential to note that this indicator, and Trouton's rule, are valid only because of the particular value of the atmospheric pressure on the earth's surface. This pressure determines that a gas molecule will occupy a volume of about 4×10^{-20} cm^3 at or near STP, which is needed for deriving Eqs (2.24)–(2.26).

It is also worth mentioning that the notion that molecules go into the vapour phase because of the kinetic energy $\frac{3}{2}kT$ they acquire is not correct. Their kinetic energy does not disappear in a liquid or solid; the molecule's motion is merely restricted to a narrower region of space around a potential-energy minimum. The average translational kinetic energy of a molecule is $\frac{3}{2}kT$ irrespective of whether it is in the gas, liquid or solid state. We shall not here consider the very complex energies associated with rotational and vibrational states of molecules in solids, liquids and gases, which would take us outside the scope of this book.

So far the Boltzmann distribution has been used to find the spatial or density distribution of molecules in different regions of a system. The Boltzmann distribution can also be used to determine the *orientational distribution* of molecules. For example, if the pair potential also depends on the mutual orientation of two anisotropic molecules, i.e., if $w(r)$ is also angle-dependent so that it may be written as $w(r, \theta)$, then the angular distribution of two molecules at a fixed distance r apart will be

$$X(\theta_2) = X(\theta_1) \exp\left(-\frac{w(r, \theta_2) - w(r, \theta_1)}{kT} \right), \qquad (2.28)$$

and here again the factor kT appears as a convenient energy unit, but this time it appears for the strength of orientation-dependent interactions needed to align molecules mutually (e.g., solvent molecules around a dissolved solute molecule, discussed in Chapters 4 and 5).

2.6 · CLASSIFICATION OF FORCES

Intermolecular forces can be loosely classified into three categories. First, there are those that are *purely electrostatic* in origin arising from the *Coulomb force* between charges. The interactions between charges, permanent dipoles, quadrupoles, etc., fall into this category. Second, there are *polarization forces* that arise from the dipole moments *induced* in atoms and molecules by the electric fields of nearby charges and permanent dipoles. All interactions in

Type of interaction		Interaction energy $w(r)$

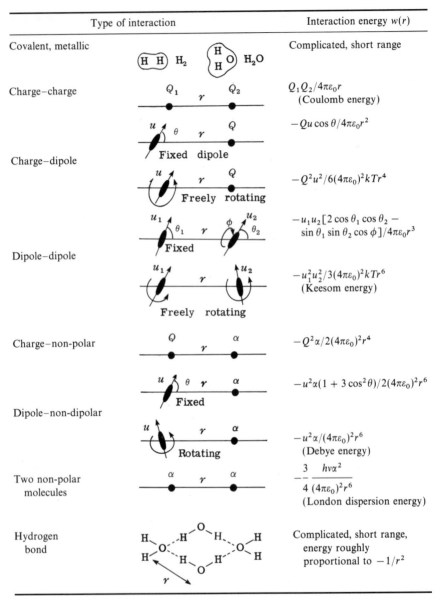

Covalent, metallic	Complicated, short range
Charge–charge	$Q_1 Q_2 / 4\pi\varepsilon_0 r$ (Coulomb energy)
Charge–dipole	$-Qu\cos\theta / 4\pi\varepsilon_0 r^2$
	$-Q^2 u^2 / 6(4\pi\varepsilon_0)^2 k T r^4$
Dipole–dipole	$-u_1 u_2 [2\cos\theta_1 \cos\theta_2 - \sin\theta_1 \sin\theta_2 \cos\phi] / 4\pi\varepsilon_0 r^3$
	$-u_1^2 u_2^2 / 3(4\pi\varepsilon_0)^2 k T r^6$ (Keesom energy)
Charge–non-polar	$-Q^2\alpha / 2(4\pi\varepsilon_0)^2 r^4$
Dipole–non-dipolar	$-u^2\alpha(1 + 3\cos^2\theta) / 2(4\pi\varepsilon_0)^2 r^6$
	$-u^2\alpha / (4\pi\varepsilon_0)^2 r^6$ (Debye energy)
Two non-polar molecules	$-\dfrac{3}{4}\dfrac{h\nu\alpha^2}{(4\pi\varepsilon_0)^2 r^6}$ (London dispersion energy)
Hydrogen bond	Complicated, short range, energy roughly proportional to $-1/r^2$

Fig. 2.2. Common types of interactions between atoms, ions and molecules in vacuum. $w(r)$ is the interaction free energy (in J); Q, electric charge (C); u, electric dipole moment (C m); α, electric polarizability ($C^2 m^2 J^{-1}$); r, distance between interacting atoms or molecules (m); k, Boltzmann constant ($1.381 \times 10^{-23} J K^{-1}$); T, absolute temperature (K); h, Planck's constant ($6.626 \times 10^{-34} J s$); ν, electronic absorption (ionization) frequency (s^{-1}); ε_0, dielectric permittivity of free space ($8.854 \times 10^{-12} C^2 J^{-1} m^{-1}$). The force is obtained by differentiating the energy $w(r)$ with respect to distance r.

a solvent medium involve polarization effects. Third, there are forces that are *quantum mechanical* in nature. Such forces give rise to *covalent* or *chemical bonding* (including charge-transfer interactions) and to the repulsive *steric* or *exchange interactions* (due to the Pauli exclusion principle) that balance the attractive forces at very short distances.

These three categories should be considered as neither rigid nor exhaustive: for certain types of forces, e.g., van der Waals forces, an unambiguous classification is not possible, while some intermolecular interactions (e.g., magnetic forces) will not even be mentioned, since for the systems we shall consider, they are always very weak. A commonly encountered and even more artificial classification of forces into short-range forces and long-range forces will not be adhered to here except in an intuitive sense whereby short-range forces refer to those interactions occurring at or very near atomic or molecular contacts.

Falling into the above categories are a number of fairly distinct interactions whose pair potentials in vacuum are given in Fig. 2.2. In the following chapters these will be considered in turn, and in the process we shall introduce important conceptual aspects of intermolecular forces especially for interactions occurring in a medium: in a condensed medium, whether liquid or solid, the force between two molecules depends on whether they are immobilized or free to rotate (see Fig. 2.2) and is rarely given by simply dividing the vacuum interaction by the medium's dielectric constant, ε.

PROBLEMS AND DISCUSSION TOPICS

2.1 For molecules constrained to interact on a surface (as occurs on adsorption and in surface monolayers) there is a 'two-dimensional' van der Waals equation of state, analogous to the three-dimensional one (Eq. (2.21)). This may be written as

$$(\Pi + a/A^2)(A - b) = kT, \tag{2.29}$$

where Π is the externally applied surface pressure (in units of $N\,m^{-1}$), A the mean area occupied per molecule, and a and b are constants. Derive this equation for molecules of diameter σ interacting with a van der Waals-type interaction pair potential given by $w(r) = -C/r^6$, and find the relation between the constants a, b and C, σ.

Can this approach, which predicts the existence of a gas–liquid transition, be extended to one-dimension? (*Hint*: carefully check the initial assumptions of this approach.)

2.2   Why do the attractive forces between the molecules of a gas affect the pressure of the gas while the attractive forces between the gas molecules and the wall molecules do not?

2.3   Why is there no simple rule, such as Trouton's Rule, relating the latent heat of melting to the melting temperature?

2.4   In Section 1.3 it is argued that long-range effects could arise for certain types of intermolecular potentials. Can you think of any examples (e.g., dipolar molecules, two-dimensional structures, metals, liquid crystals), explaining why and how the long-range effects manifest themselves. (Use Fig. 2.2 and other literature sources.)

CHAPTER 3

STRONG INTERMOLECULAR FORCES: COVALENT AND COULOMB INTERACTIONS

3.1 COVALENT OR CHEMICAL BONDING FORCES

When two or more atoms come together to form a molecule, as when two hydrogen atoms and one oxygen atom combine to form a water molecule, the forces that tightly bind the atoms together within the molecule are called *covalent forces*, and the interatomic bonds formed are called *covalent bonds*. Closely allied to covalent bonds are metallic bonds. In both cases the bonds are characterized by the electrons being shared between two or more atoms so that the discrete nature of the atoms is lost.

Depending on the position an atom (or element) occupies in the periodic table, it can participate in a certain number of covalent bonds with other atoms. This number or stoichiometry is known as the atomic *valency*; for example, it is zero for the inert gases (e.g., argon) which cannot normally form covalent bonds with other atoms, one for hydrogen, two for oxygen, three for nitrogen and four for carbon. A further characteristic of covalent bonds is their *directionality*, that is, they are directed or oriented at well-defined angles relative to each other. Thus, for multivalent atoms, their covalent bonds determine the way they will coordinate themselves in molecules or in crystalline solids to form an ordered three-dimensional lattice. For example, they determine the way carbon atoms arrange themselves to form the perfectly ordered diamond structure.

Covalent forces are of short range, that is, they operate over very short distances of the order of interatomic separations (0.1–0.2 nm). Table 3.1 shows the strength of some common covalent bonds. As can be seen they are mainly in the range 100–300 kT per bond (200–800 kJ mol^{-1}), and they tend to decrease in strength with increasing bond length—a characteristic property of most intermolecular interactions.

TABLE 3.1 Strengths of covalent bonds

Bond type	Strength (kJ mol^{-1})	Bond type	Strength (kJ mol^{-1})
C≡N (HCN)	870	Si—O	370
C═O (HCHO)	690	C—C (C_2H_6)	360
C═C (C_2H_4)	600	C—O (CH_3OH)	340
O—H (H_2O)	460	N—O (NH_2OH)	200
C—H (CH_4)	430	F—F (F_2)	150

[a] The strength of a covalent bond depends on the type of other bonds nearby in the molecule. For example, the C—H bond strength can be as low as 360 kJ mol^{-1} (in H—CHO) and as high as 500 kJ mol^{-1} (in H—C≡N). Note that 1 kJ mol^{-1} corresponds to about 0.4 kT per bond at 298 K.

3.2 PHYSICAL AND CHEMICAL BONDS

The complex quantum mechanical interactions that give rise to covalent bonding will not be a major concern in this book, which is devoted more to the forces between unbonded *discrete* atoms and molecules. These are usually referred to as *physical forces* and they give rise to *physical bonds*, in contrast to chemical forces, which give rise to *chemical* or *covalent bonds*.

Physical bonds usually lack the specificity, stoichiometry and strong directionality of covalent bonds. They are therefore the ideal candidates for holding molecules together in liquids, since the molecules can move about and rotate while still remaining 'bonded' to each other. Strictly, physical 'bonds' should not be considered as bonds at all, for during covalent binding the electron charge distributions of the uniting atoms change completely and merge, whereas during physical binding they are merely perturbed, the atoms remaining as distinct entities. Nevertheless, physical binding forces can be as strong as covalent bonds, and even the weakest is strong enough to hold all but the smallest atoms and molecules together in solids and liquids at room temperature, as well as in colloidal and biological assemblies. These properties, coupled with the long-range nature of physical forces, makes them the regulating forces in all phenomena that do not involve chemical reactions.

3.3 COULOMB FORCES OR CHARGE–CHARGE INTERACTIONS

The inverse-square Coulomb force between two charged atoms, or ions, is by far the strongest of the physical forces we shall be considering—stronger even than most chemical binding forces.

The free energy for the Coulomb interaction between two charges Q_1 and Q_2 is given by

$$w(r) = \frac{Q_1 Q_2}{4\pi\varepsilon_0 \varepsilon r} = \frac{z_1 z_2 e^2}{4\pi\varepsilon_0 \varepsilon r}, \tag{3.1}$$

where ε is the *relative permittivity* or *dielectric constant* of the medium and r the distance between the two charges. The expression on the right is commonly used for ionic interactions, where the magnitude and sign of each ionic charge is given in terms of the elementary charge ($e = 1.602 \times 10^{-19}$ C) multiplied by the ionic valency z. For example, $z = +1$ for monovalent *cations* such as Na^+, $z = -1$ for monovalent *anions* such as Cl^-, $z = +2$ for divalent cations such as Ca^{2+}, etc.

The Coulomb force F is given by

$$F = -\frac{dw(r)}{dr} = \frac{Q_1 Q_2}{4\pi\varepsilon_0 \varepsilon r^2} = \frac{z_1 z_2 e^2}{4\pi\varepsilon_0 \varepsilon r^2}. \tag{3.2}$$

For like charges, both w and F are positive and the force is repulsive, while for unlike charges they are negative and the force is attractive.

For completeness, we may also note that the electric field E at a distance r away from a charge Q_1 is defined by

$$E_1 = \frac{Q_1}{4\pi\varepsilon_0 \varepsilon r^2} \quad \text{V m}^{-1}. \tag{3.3}$$

This field, when acting on a second charge Q_2 at r, gives rise to a force

$$F = Q_2 E_1 = \frac{Q_1 Q_2}{4\pi\varepsilon_0 \varepsilon r^2}$$

which is the same as Eq. (3.2).

Let us put the strength of the Coulomb interaction into perspective. For two isolated ions (e.g., Na^+ and Cl^-) in contact, r is now the sum of the two ionic radii (0.276 nm), and the binding energy is

$$w(r) = \frac{-(1.602 \times 10^{-19})^2}{4\pi(8.854 \times 10^{-12})(0.276 \times 10^{-9})} = -8.4 \times 10^{-19} \text{ J}.$$

In terms of the thermal energy $kT = (1.38 \times 10^{-23})(300) = 4.1 \times 10^{-21}$ J at 300 K, this energy turns out to be of order $200\ kT$ per ion pair in vacuum, viz. similar to the energies of covalent bonds (see Table 3.1). Only at a separation r greater than about 56 nm will the Coulomb energy fall below kT. We have thus established that the Coulomb interaction is very strong and of long range.

☐ ● **WORKED EXAMPLE** ●

Question: Show (i) that the Coulomb force on a charge e near a flat surface of uniform charge density σ is independent of the distance of the charge from the surface, and (ii) that the force on a charge e near a sphere containing a net charge Q uniformly distributed on its surface is the same as if all the charge Q were concentrated at the centre of the sphere.

Answer:

- (i) Consider a circular strip of radius r and width dr on the charged surface. Position the charge e on the axis passing through the centre of the circle at a distance z from the surface. The distance between the charge and the circular strip is therefore $\sqrt{z^2 + r^2}$. If θ is the angle at the charge between the axis and the circle then the normal field at z due to the circular strip of charge $2\pi r \, dr\sigma$ is

$$E_z = \sigma \int_0^\infty \frac{2\pi r \, dr \cos\theta}{4\pi\varepsilon_0 [z^2 + r^2]} = \frac{z\sigma}{2\varepsilon_0} \int_0^\infty \frac{r \, dr}{[z^2 + r^2]^{3/2}} = \frac{\sigma}{2\varepsilon_0}.$$

Thus the force on the charge $e\sigma/2\varepsilon_0$ is independent of its distance from the surface, z. Note that in practice, to ensure charge neutrality, there must be a countercharge somewhere else, and this may affect the net field experienced by the charge e. For example, let there be a surface of opposite charge density $-\sigma$ parallel to the first surface (as in a capacitor). Then the total field will be σ/ε_0 *between* the two surfaces (where the fields augment each other) and zero *outside* the surfaces (where they cancel). Likewise, between the two surfaces the net force on the charge is $e\sigma/\varepsilon_0$ independent of its distance from any of the surfaces. If the medium has dielectric constant ε, then ε_0 becomes replaced by $\varepsilon\varepsilon_0$ in the above equations.

- (ii) This part will be solved by calculating the electric potential at the charge e. Let the centre of the sphere (of radius R) be at $x = 0$ and position the charge on the x-axis at $x = z$, where $z > R$. Let $\sigma = Q/4\pi R^2$ be the surface charge density of the sphere. Consider a circular strip on the sphere centred at $x = R\cos\theta$ of radius $R\sin\theta$ and width $R \, d\theta$. The distance between the charge and the circular strip is therefore

$\sqrt{(R \sin \theta)^2 + (z - R \cos \theta)^2} = \sqrt{z^2 - 2zR \cos \theta + R^2}$, so that the potential at $x = z$ is

$$\psi_z = \int_0^\pi \frac{2\pi R^2 \sigma \sin \theta \, d\theta}{4\pi\varepsilon_0 [z^2 - 2zR \cos \theta + R^2]^{1/2}} = \frac{Q}{4\pi\varepsilon_0 z}.$$

Thus, the electric field $E = d\psi/dz$ and force $F = eE$ is the same as for a point charge Q located at the centre of the sphere, and therefore independent of the radius R of the sphere.

(Comment: Either of the above results could have been obtained by summing the potential and then differentiating with respect to z, as in (ii), or by summing the resolved force, as in (i). The above two examples are illustrations of *Gauss' Law*, which also applies to other $1/r^2$ forces such as the gravitational force. Gauss' Law is very useful for determining electrostatic fields and forces.) ☐

3.4   IONIC CRYSTALS

Coulomb forces (also known as *ionic forces*) hold the sodium and chloride ions together in the rigid salt lattice composed of alternate sodium and chloride ions, and the bond they give rise to is often referred to as the *ionic bond*. However, the above calculation of the binding energy of an *isolated* pair of Na^+–Cl^- ions is too simplistic for estimating the mean energy of an ionic bond in a lattice. As was shown above, Coulomb forces are of very long range, which is manifested in the $1/r$ distance dependence of $w(r)$. For an accurate determination of the lattice energy the Coulomb energy of an ion with all the other ions in the lattice has to be summed, and not only with its nearest neighbours. Thus, in the NaCl crystal lattice each Na^+ has six nearest neighbour Cl^- ions at $r = 0.276$ nm, 12 next-nearest neighbour Na^+ ions at $\sqrt{2}r$, eight more Cl^- ions at $\sqrt{3}r$, etc.

The total interaction energy for a pair of Na^+–Cl^- ions in the lattice is therefore

$$\mu^i = -\frac{e^2}{4\pi\varepsilon_0 r} \left[6 - \frac{12}{\sqrt{2}} + \frac{8}{\sqrt{3}} - \frac{6}{2} + \ldots \right]$$

$$= -\frac{e^2}{4\pi\varepsilon_0 r} [6 - 8.485 + 4.619 - 3.000 + \ldots]$$

$$= -1.748 \frac{e^2}{4\pi\varepsilon_0 r} = -1.46 \times 10^{-18} \text{ J.} \tag{3.4}$$

The constant 1.748 is known as the *Madelung constant* and has different values for other crystal structures, varying between 1.638 and 1.763 for crystals composed of monovalent ions such as NaCl and CsCl, rising to about 5 for monovalent–divalent ion pairs such as CaF_2, and higher for multivalent ions such as SiO_2 and TiO_2. However, it is clear that the net (binding) energy is negative and that it is of the same order as for isolated ion pairs. To obtain the theoretical *molar lattice energy* or *cohesive energy* U of a NaCl crystal, we must multiply the above by Avogadro's number N_0, thus

$$U = -N_0\mu^i = (6.02 \times 10^{23})(1.46 \times 10^{-18}) = 880 \text{ kJ mol}^{-1}. \qquad (3.5)$$

This is about 15% higher than the measured value due to our neglect of the repulsive forces at contact which lower the final binding energy. These very short-range repulsive forces will be discussed in Chapter 7. The above value for the molar lattice energy of NaCl is fairly representative of other alkali halide energies, which range from about 600 to 1000 kJ mol^{-1} going from RbI to LiF, i.e., increasing with decreasing ionic size.

3.5 REFERENCE STATES

It is important to always have the right reference state in mind when considering intermolecular interactions. Any value for the free energy is not very meaningful unless referred to some state with which it is being compared. Thus, when ions or molecules come together to form a condensed phase from the gaseous state, the reference state is at $r = \infty$, and the interaction occurs in vacuum ($\varepsilon = 1$). It is for this reason that Eq. (3.4) for the lattice energy of ionic crystals does not contain the dielectric constant of the medium (e.g., $\varepsilon \approx 6$ for NaCl). On the other hand, if two ions are interacting in a condensed liquid medium, the reference state is also at $r = \infty$, but the dielectric constant now appears in the denominator of any expression for the Coulomb interaction since the interaction is now occurring entirely within the solvent medium.

3.6 RANGE OF COULOMB FORCES

One very important aspect of Coulomb forces concerns the range of the interaction. The inverse-square distance dependence of the Coulomb force, the same as for the gravitational force, appears to make it very long ranged, in apparent contradiction with the statement made earlier that all intermolecular force laws must fall with distance faster than $1/r^4$ ($1/r^3$ for the energy). Since positive ions always have negative ions nearby, whether they are in a lattice or in solution, the electric field becomes *screened* and decays more rapidly away from them than from a truly isolated ion. In Chapter 12 we shall see that at large distances the decay is always exponential with distance, thus making all Coulomb interactions between ionic crystals, charged surfaces, and dissolved ions of much shorter range (though still of much longer range than covalent forces).

3.7   THE BORN ENERGY OF AN ION

When a single ion is in a vacuum or in a medium, even though it may not be interacting with other ions, it still has an electrostatic free energy—equal to the electrostatic work done in forming the ion—associated with it. In vacuum this energy is referred to as the *self-energy*, while in a medium it is referred to as the *Born* or *solvation energy* of an ion, and it is an important quantity since it determines among other things the extent to which ions will dissolve and partition in different solvents. Let us see how the Born energy arises.

Imagine the process of charging an atom or sphere of radius a by gradually increasing its charge from zero up to its full charge Q. At any stage of this process let the ionic charge be q, and let this be incremented by dq. The work done in bringing this additional charge from infinity to $r = a$ is therefore, from Eq. (3.1) putting $Q_1 = q$, $Q_2 = dq$, and $r = a$,

$$dw = \frac{q\,dq}{4\pi\varepsilon_0\varepsilon a},\tag{3.6}$$

so that the total free energy of charging the ion, the Born energy, is

$$\mu^i = \int dw = \int_0^Q \frac{q\,dq}{4\pi\varepsilon_0\varepsilon a} = \frac{Q^2}{8\pi\varepsilon_0\varepsilon a} = \frac{(ze)^2}{8\pi\varepsilon_0\varepsilon a}.\tag{3.7}$$

The Born energy gives the electrostatic free energy of an ion in a medium of dielectric constant ε. It is positive because the energy is unfavourable, viz. it is the energy of keeping a net charge Q distributed on the surface of a sphere against its own electrostatic repulsion. In Chapter 5 we shall see how the Born energy can also be obtained by summing the pair potentials of an ion with its surrounding solvent molecules.

From Eq. (3.7) we see that the change in free energy on transferring an ion from a medium of low dielectric constant ε_1 to one of high dielectric constant ε_2 is negative, i.e., it is energetically favourable, and equal to

$$\Delta\mu^i = -\frac{z^2 e^2}{8\pi\varepsilon_0 a}\left[\frac{1}{\varepsilon_1} - \frac{1}{\varepsilon_2}\right] \text{J}$$

$$= -\frac{28z^2}{a}\left[\frac{1}{\varepsilon_1} - \frac{1}{\varepsilon_2}\right] \text{kT per ion at 300 K} \tag{3.8}$$

or

$$\Delta G = N_0 \Delta\mu^i = -\frac{69z^2}{a}\left[\frac{1}{\varepsilon_1} - \frac{1}{\varepsilon_2}\right] \text{kJ mol}^{-1}, \tag{3.9}$$

where a is given in nanometres. Thus, if one mole of monovalent cations and anions are transferred from the gas phase ($\varepsilon = 1$) into water ($\varepsilon = 78$), the gain in the molar free energy will be, assuming $a = 0.14$ nm for both cations and anions,

$$\Delta G = -\frac{2 \times 69}{0.14}\left[1 - \frac{1}{78}\right] \approx -1000 \text{ kJ mol}^{-1}. \tag{3.10}$$

Equations (3.7)–(3.9) provide the basis for calculating the partitioning of ions between different solvents. Note that the Born energy does not include the energy expended by solvents in forming the cavities for accommodating the ions; these are generally small compared to the large Born energy.

3.8 SOLUBILITY OF IONS IN DIFFERENT SOLVENTS

Closely related to partitioning is the solubility of ions in different solvents. Both the Coulomb energy and the Born energy are useful for understanding why ionic crystals such as Na^+Cl^-, in spite of their very high lattice energies, dissociate in water and in other solvents with high dielectric constants. If we consider the Coulomb interaction, Eq. (3.1), we immediately see that the electrostatic attraction between ions is much reduced, by a factor ε, in a medium. This is a somewhat superficial approach to the problem since the Coulomb law is strictly not valid at very small interionic distances where the molecularity of the medium makes the continuum description (in terms of ε) break down. However, this approach does predict the right trends, so let us follow it up.

On the simplest level we may consider the difference in energy on going from the associated state to the dissociated state to be roughly given by Eq. (3.1). Thus, the free energy change in separating two monovalent ions such as Na^+ and Cl^- from contact in a solvent medium of dielectric constant ε is

$$\Delta\mu^i \approx \frac{+e^2}{4\pi\varepsilon_0\varepsilon(a_+ + a_-)},$$

where a_+ and a_- are the ionic radii of Na^+ and Cl^-. This energy is positive since the attractive Coulomb interaction will always favour association. However, some fraction of the ions will always dissociate due to their entropy of dilution (Chapter 2). The concentration X_s of ions forming a saturated solution in equilibrium with the solid will therefore be given by Eq. (2.15):

$$X_s = \exp(-\Delta\mu^i/kT) \approx \exp\left[-\frac{e^2}{4\pi\varepsilon_0\varepsilon kT}\frac{1}{(a_+ + a_-)}\right], \tag{3.11}$$

where the value of the dimensionless parameter X_s may be identified with the solubility of the electrolyte in water, or in any solvent, in mole-fraction units. Thus, for NaCl in water, $(a_+ + a_-) = 0.276$ nm, $\varepsilon = 78$ at $T = 298$ K, we expect very roughly

$$X_s \approx e^{-2.6} \approx 0.075,$$

which may be compared with the experimental value of 0.11 mol mol^{-1} (36 g l^{-1}).

While Eq. (3.11) is far too simplistic to account quantitatively for the solubilities of all electrolytes, it does predict the observed trends for monovalent salts reasonably well. For example, it predicts that the solubility of a salt in different solvents will be proportional to $e^{-const/\varepsilon}$, where ε is the solvent dielectric constant. Thus a plot of $\log X_s$ against $1/\varepsilon$ should yield a straight line. This is more or less borne out in practice, as shown in Fig. 3.1 for NaCl and the amino acid glycine in different solvents. The large solubilizing power of water to ions is therefore seen as arising quite simply from its high dielectric constant (Table 3.2) and not because of some special property of water.

From Eq. (3.11) we may also expect larger ions to be more soluble than smaller ions. This too is usually borne out in practice: alkali halide salts with large ionic radii, such as CsBr and KI, are generally much more soluble in

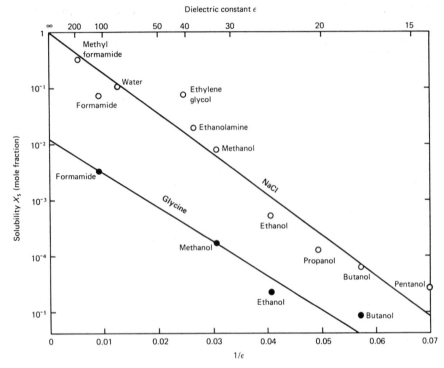

Fig. 3.1. Solubilities of sodium chloride $(NaCl = Na^+ + Cl^-)$ and glycine $(NH_2CH_2COOH = NH_3^+CH_2COO^-)$ in solvents of different static dielectric constants ε at 25°C. Solubility (in mole fraction units) is plotted as $\log X_s$ as a function of $1/\varepsilon$. For NaCl, the line passes through $X_s = 1$ at $\varepsilon = \infty$, which from Eq. (3.11) suggests that the interaction of $Na^+ + Cl^-$ with these solvents is purely Coulombic. For glycine, the line tends to a finite value $(X_s < 1)$ as ε tends to infinity, indicative of some additional type of solute–solute attraction—the van der Waals interaction (see Chapter 6 and Problem 4.1). Note that all the solvents are hydrogen-bonding liquids (Chapter 8). Non-hydrogen-bonding liquids are less effective as solvents for ionic species; for example, the solubility of NaCl in acetone $(\varepsilon = 20.7)$ is $X_s = 4 \times 10^{-7}$ while that of glycine in acetone is $X_s = 2 \times 10^{-6}$. Solubility data were taken from *GMELINS Handbuch*, Series 21, Vol. 7 for NaCl, and from the CRC *Handbook of Chemistry and Physics* for glycine.

various solvents than salts such as NaF and LiF; the latter has the smallest interionic distance and is the least soluble of the alkali halides in water.

☐   ● **WORKED EXAMPLE** ●

Question: From the data of Fig. 3.1, estimate the interionic spacing between a sodium and chloride ion when these are in contact. Compare your result with the known value of 0.28 nm in the NaCl crystal.

TABLE 3.2 Static dielectric constants[a] ε of some common liquids and solids at 25°C

Compound		ε	Compound	ε
Hydrogen bonding			*Polymers*	
Methyl			Nylon	3.7–4.2
formamide	HCONHCH$_3$	182.4	Fluorocarbons	2.1–3.6
Formamide	HCONH$_2$	109.5	Polycarbonate	3.0
Hydrogen			Polystyrene	2.4
fluoride	HF (at 0°C)	84	PTFE	2.0
Water	H$_2$O	78.5		
	D$_2$O	77.9	*Glasses*	
Formic acid	HCOOH (at 16°C)	58.5	Fused quartz SiO$_2$	3.8
Ethylene			Soda glass	7.0
glycol	C$_2$H$_4$(OH)$_2$	40.7	Borosilicate glass	4.5
Methanol	CH$_3$OH	32.6		
Ethanol	C$_2$H$_5$OH	24.3	*Crystalline solids*	
n-Propanol	C$_3$H$_7$OH	20.2	Diamond (carbon)	5.7
Ammonia	NH$_3$	16.9	Quartz SiO$_2$	4.5
Acetic acid	CH$_3$COOH	6.2	Micas	5.4–7.0
			Sodium chloride NaCl	6.0
			Alumina Al$_2$O$_3$	8.5
Non-hydrogen bonding				
Acetone	(CH$_3$)$_2$CO	20.7	*Miscellaneous*	
Chloroform	CHCl$_3$	4.8	Paraffin (liquid)	2.2
Benzene	C$_6$H$_6$	2.3	Paraffin wax (solid)	2.2
Carbon			Silicone oil	2.8
tetrachloride	CCl$_4$	2.2	Liquid helium (2–3 K)	1.055
Cyclohexane	C$_6$H$_{12}$	2.0	Water (liquid at 0°C)	87.9
Dodecane	C$_{12}$H$_{26}$	2.0	Water (ice at 0°C)	91.6–106.4
Hexane	C$_6$H$_{14}$	1.9	Air (dry)	1.00054

[a] The dielectric constant is a measure of the extent of reduction of electric fields and consequently of the reduced strengths of electrostatic interactions in a medium.

Answer:

- The slope of the NaCl line in Fig. 3.1 is $2.303 \times 5/0.069 = -167$. Using Eq. (3.11) this corresponds to an interionic spacing given by $(a_+ + a_-) = (1.602 \times 10^{-19})^2/4\pi(8.854 \times 10^{-12})(4.1 \times 10^{-21})(167) = 3.4 \times 10^{-10}$ m $= 0.34$ nm. This is about 0.06 nm larger than in the pure, dry crystal. Given the gross assumptions that were made in deriving Eq. (3.11) it is surprising that the calculated value differs from the correct value by only 20% (see Problem 3.5).   □

The solubility of an electrolyte is strictly determined by the difference in the free energy of the ions in the solid lattice from that in solution. In the above approach we did not consider the lattice energy specifically; but it is also possible to approach the problem in a way that takes the lattice energy into account, by splitting up the dissociation process into two well-defined stages. The first is the dissociation of the solid into isolated gaseous ions, and the second is the transfer of these ions into the solvent. In this approach (see, e.g., Dasent, 1970; Pass, 1973) the energy associated with the first stage is simply the positive lattice energy, while the second stage reduces this by the negative Born energies of transferring the ions from the gas phase ($\varepsilon = 1$) into the solvent medium of dielectric constant ε. However, for water, the theoretical Born energies turn out to be much too large, even larger than the lattice energies (compare Eq. (3.5) with Eq. (3.10)), and so to obtain agreement with measured solubility and other thermodynamic data it has been found necessary to 'correct' the crystal lattice radii of ions by increasing them by 0.02 to 0.09 nm when the ions are in water. The larger 'effective' sizes of ions in water arises from their *solvation* or *hydration shells* (discussed in Chapter 4).

While both the Coulomb and Born energy approaches usually predict the right trends, neither are quantitatively reliable, because they ignore the complex and sometimes specific interactions that can occur between dissolved ions and the solvent molecules in their immediate vicinity. This problem has already been referred to as the *solvent effect*, and it is particularly acute when the solvent is water and for multivalent ions.

3.9   SPECIFIC ION–SOLVENT EFFECTS

The solvent effect is at the heart of all problems concerning the interactions between molecules (and surfaces) in a liquid medium at small separations. Here the medium cannot always be treated as a structureless continuum defined solely in terms of its bulk properties such as its bulk dielectric constant, bulk density, etc., and where highly specific solvent effects can arise. This presents serious theoretical and conceptual problems because these effects necessarily involve the simultaneous interactions of many molecules.

There are two basic approaches to this problem. The first is to 'work down' from the continuum to the molecular level. In this approach one usually assumes that at least some of the bulk properties of the system, for example, the dielectric constant and density of the solvent, hold right down to molecular dimensions. In the second approach one starts from the specific interactions occurring at the atomic and molecular levels and then attempts to 'build up' the properties of the whole system. This dichotomy of approaches will be a recurring theme in theoretical treatments of various intermolecular forces. These two approaches are commonly referred to as *continuum* and *molecular* theories.

To understand the solvent effect for Coulomb interactions, we must investigate how a solvent affects the electric fields around dissolved ions. Ultimately, we shall have to consider the origin of the dielectric permittivity, ε, since this is what defines the 'solvent' in all equations for electrostatic interactions. Let us start with the continuum approach, and attempt to assess its limitations by examining in more detail the origin of the Born and Coulomb energies.

3.10   CONTINUUM APPROACH

The Born energy of an ion is strictly a measure of the free energy associated with the electric field around the ion. From electrostatic theory (Guggenheim, 1949, Ch. 12; Landau and Lifshitz, 1963, Ch. II) the free energy density of an electric field arising from a fixed charge is $\frac{1}{2}\varepsilon_0\varepsilon E^2$ per unit volume. By integrating the energy density over all of space we readily obtain Eq. (3.7) for the Born energy:

$$\mu^i = \frac{1}{2}\varepsilon_0\varepsilon \int E^2\,dV = \frac{1}{2}\varepsilon_0\varepsilon \int_a^\infty \frac{Q^2}{(4\pi\varepsilon_0\varepsilon r^2)^2}4\pi r^2\,dr = \frac{Q^2}{8\pi\varepsilon_0\varepsilon a}. \qquad (3.12)$$

The above derivation provides us with some important insights: First, the electrostatic self-energy is seen not to be concentrated on the ion itself, but rather it is spread out over the whole of space *around* the ion; thus, if in the above we integrate from $r = a$ to $r = R$, we find that the energy contained within a finite sphere of radius R around the ion is

$$\frac{-Q^2}{8\pi\varepsilon_0\varepsilon}\left[\frac{1}{a} - \frac{1}{R}\right]. \qquad (3.13)$$

For example, for an ion of radius 0.1 nm, 50% of its energy will be contained within a sphere of radius 0.2 nm and 90% within a radius of 1.0 nm, Thus, if the Born energy equation is to be applicable in a condensed medium, the value of the dielectric constant of the medium must already be equal to the bulk value at ~ 0.1 nm away from the ion, a distance that is smaller than even the smallest solvent molecule.

Second, it can be shown that the Coulomb interaction in a medium can also be derived from the change in the electric field energy, integrated over the whole of space, when two charges are brought together. The Coulomb interaction can therefore be considered as the change in the Born energies of two charges as they approach each other. This is conceptually important since it shows that the Coulomb interaction in a medium is not determined by the dielectric constant in the region *between* two charges but by its value in the region *surrounding* (as well as between) the charges. It is for this reason that the strength of the Coulomb interaction of two oppositely charged ions will be reduced in a solvent medium even if the ions still remain in contact, that is, even before there are any solvent molecules between them! This is yet another manifestation of the long-range nature of electrostatic interactions. And it clearly shows why the extremely strong ionic 'bond' is so easily disrupted in a medium of high dielectric constant such as water, in marked contrast to the short-range covalent bonds, which—though often weaker—are not generally broken by a solvent.

We have now established how a solvent medium affects the electrostatic Born and Coulomb energies of ions. We have seen that it is the value of ε of the locally surrounding medium that is important and that if the standard expressions for the Born and Coulomb energies (and other interaction energies that depend on ε) are to apply, the bulk value of ε must be attained already within the first shell of surrounding solvent molecules. In Section 4.8 we shall see that for water near monovalent ions, this is more or less expected on theoretical grounds. Experimentally, too, this is often the case, and we may recall how the Coulomb and Born energies are able to predict, at least semiquantitatively, the solubilities of monovalent ionic salts in different solvents. But we shall also encounter numerous instances where such continuum theories break down at small intermolecular distances.

3.11 MOLECULAR APPROACH: COMPUTER SIMULATIONS

In reality, two solute molecules do not see themselves surrounded by a homogeneous liquid medium, but rather by discrete molecules of a given size and shape, and, for water, possessing a large dipole moment. In recent years, much effort has been invested in developing molecular theories of liquids and of solute–solvent, solute–solute and surface–surface interactions in liquids. In the molecular approach, each molecule is individually modelled in terms of its known structure (including all the interatomic bond lengths, bond angles, atomic radii, charge distributions, etc.). A computer then works out how an 'ensemble' of such molecules will behave when they are allowed to interact according to some interaction potential, such as the Lennard–Jones potential, Eq. (1.7), for simple molecules or more complex potentials for more complex molecules (e.g., water). Such *computer simulations* or *computer experiments* are increasingly providing unique insights into the properties of different systems at the molecular level.

The two most popular types of computer simulations are the *Monte Carlo* (*MC*) and *Molecular Dynamics* (*MD*) techniques (Allen and Tildesley, 1987). In the *MC* technique a number of molecules (or ions or particles) are confined within a box or cell. One molecule, chosen at random, is then moved to a different position, also chosen at random (hence 'Monte Carlo'). The computer then determines whether to accept or reject this move depending on whether the total energy of the system has decreased or increased. This process is repeated many times until there is no further change in the energy and other computed properties of the system, at which point the system is deemed to have reached thermodynamic equilibrium.

In the *MD* technique the computer first calculates the force on each molecule arising from all the other molecules and then, by solving Newton's equations of motion, determines how the molecule moves in response to this force. This computation is done simultaneously and continuously for all the molecules in the box from which their trajectories can be followed in space and time. An *MD* simulation always gives the same final result as an *MC* simulation for the equilibrium state of a system, but *MD* is more expensive than *MC*. On the other hand, an *MD* simulation is usually more revealing because it provides information on how molecules actually move (hence 'molecular dynamics'). Thus, with *MD* one can also study time-dependent phenomena, non-equilibrium effects, fluid flow and other transport phenomena, which cannot be done with *MC*.

In a typical *MD* simulation of a system consisting of 1000 molecules, it will take about 10^{-5} s to compute the force on one molecule and follow its motion over some fraction of an ångstrom in response to this force. This move will correspond to 10^{-14} s of 'real' time. Thus to simulate 100 moves for each of the 1000 molecules (10^5 moves in all) will take $100 \times 1000 \times 10^{-5}$ s ~ 1 of computer time. Yet this simulates only 1 ps of real time—the time of a single bond vibration. To simulate a molecular rotation (~ 1 ns of real time) takes 10^3 s; to simulate slow molecular reorientations of polymers or the collision of two colloidal particles (~ 1 μs of real time) takes ~ 10 days, while 1 s of real time currently requires 30 000 years.

A properly executed computer simulation, whether *MC* or *MD*, is regarded as providing the *exact* solution to any well-defined problem, and all other (analytic) theories stand or fall depending on how well their predictions can be supported by computer simulation. Of course, the *correctness* of the results depends on using the correct interaction potentials in the first place, a matter that can only be established by comparison with real (as opposed to computer) experiments.

Computer simulations provide the most powerful theoretical tools available today for studying an almost endless variety of interesting phenomena (Ciccotti *et al.*, 1987), ranging from the way NaCl ionizes in water to the interactions of polymers and proteins to friction and lubrication. In future chapters we shall explore the results and implications of computer simulations after we have acquired a better overview of the various intermolecular potentials that they employ.

PROBLEMS AND DISCUSSION TOPICS

3.1   It is often argued that the Lennard–Jones potential $w(r) = B/r^{12} - A/r^6$ is applicable to chemical bonds as well as physical bonds, where the attractive van der Waals term remains unchanged and where only a difference in the repulsive coefficient, B, distinguishes between the two types of interactions.

Consider two atoms for which $A = 10^{-77}$ J m^6, and where their equilibrium separation is at $r_0 = 0.35$ nm and $r_0 = 0.15$ nm for the case of physical and chemical binding, respectively. Assuming the above hypothesis to be true, calculate the values of w_{min} in each case and then argue whether your result tends to support the above view.

3.2   How many negatively charged ions can be put in contact around one positively charged ion of the same radius such that the net Coulombic energy of the cluster is still negative (i.e. energetically favourable). Is your answer physically meaningful, and if so in what situations?

3.3   Calculate the total Coulomb interaction energy of a Na$^+$ ion with its 12 closest Cl$^-$ ions and 11 closest Na$^+$ ions in a NaCl lattice (cf. Section 3.4). Note that this cluster of 24 ions is overall electrically neutral. What does your answer suggest concerning the contribution of distant ions to the total lattice energy (or mean ionic bond energy) in the NaCl crystal? Check your conclusion by considering a neutral cluster of 26 ions.

3.4   Show that if a charged liquid drop evaporates in air, there will come a point at which it will become unstable and break up into smaller charged droplets. This is known as the *Rayleigh limit*. If the charge on the drop is $Q = 100e$ and if the liquid is water which has a surface tension of $\gamma = 72$ mN m^{-1} (numerically equal to the surface energy in units of mJ m^{-2}), at what drop radius R will this happen and what will be the nature of the fragmentation? (*Comment*: under realistic conditions this problem is much more subtle than appears at first sight, requiring consideration of the following: the vapour pressure of the water, whether the instability is mechanical or thermodynamic, alternative non-spherical droplet shapes and

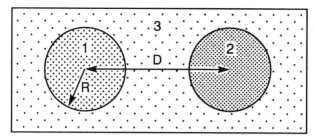

Fig. 3.2.

non-uniform charge distributions. You should also consider what prevents the charges from leaving the droplet in the first place.)

3.5 Look up the values for the solubility of KCl in some of the same solvents as those in Fig. 3.1. Use these to determine a value for the interionic spacing of K^+–Cl^-, and compare this with the value obtained for Na^+–Cl^- (see the worked example in Section 3.8). Is your result reasonable?

3.6 Two solid spheres, 1 and 2, of density $1.0 \, \text{g cm}^{-3}$ and $10.0 \, \text{g cm}^{-3}$ are in an inert liquid medium 3 of density $2.0 \, \text{g cm}^{-3}$ and dielectric constant 2.0 (Fig. 3.2). Sphere 1 carries a charge $+e$ and sphere 2 carries a charge $-e$. If the spheres have the same radius R calculate the value of R for there to be no long range force between them at any separation D.

3.7 Show that inside a uniformly charged sphere the electric field is everywhere zero. (Refer to Worked Example in Section 3.3.)

3.8 (i) In the language of computer simulations what are 'periodic boundary condition' and what is the meaning of 'ergodicity'?

(ii) How do the following factors limit the reliability of a simulation: (a) the use of a cell of finite size; (b) the use of a finite number of particles; (c) the finite time of a simulation; (d) the need to start the simulation with the particles in some arbitrary configuration.

(iii) Consider a colloidal dispersion of particles in a liquid where, due to surface-induced solvent structuring effects, the viscosity of the liquid is much higher near the surface of the particles than in the bulk. Thus, during a Brownian collision, two approaching particles would be slowed down and therefore experience an additional, effectively repulsive, hydrodynamic force. Would this effect modify the equilibrium mean distribution of the particles? What other possible effects would an increased viscosity near the surfaces have on any measurable properties of the colloidal dispersion?

CHAPTER 4

INTERACTIONS INVOLVING POLAR MOLECULES

4.1 WHAT ARE POLAR MOLECULES?

Most molecules carry no net charge, but many possess an *electric dipole*. For example, in the HCl molecule the chlorine atom tends to draw the hydrogen's electron towards itself, and this molecule therefore has a permanent dipole. Such molecules are called *polar* molecules. The dipoles of some molecules depend on their environment and can change substantially when they are transferred from one medium to another, especially when molecules become ionized in a solvent. For example, the amino acid molecule glycine contains an acidic group on one side and a basic group on the other. In water this molecule exists as a dipolar molecule in the following form:

Glycine   Glycine
                         in water

Such dipolar molecules in water are also referred to as *zwitterions*. Quite often the magnitude of the positive and negative charges on zwitterions are not the same, and these molecules therefore possess a net charge in addition to a dipole. Such molecules are then referred to as *dipolar ions*. It should already be apparent that the interactions and the solvent effects of polar molecules can be very complex.

The *dipole moment* of a polar molecule is defined as

$$u = ql,\qquad(4.1)$$

TABLE 4.1 Dipole moments of molecules, bonds, and molecular groups (in Debye units: $1 D = 3.336 \times 10^{-30}$ C m)[a]

Molecules			
Alkanes	0^{b}	H_2O	1.85
C_6H_6 (benzene)	0	CH_3OH, C_2H_5OH	1.7
CCl_4	0	Hexanol, octanol	1.7
CO_2	0	$C_6H_{11}OH$ (cyclohexanol)	1.7
CO	0.11	CH_3COOH (acetic acid)	1.7
$CHCl_3$ (chloroform)	1.06	C_2H_4O (ethylene oxide)	1.9
HCl	1.08	CH_3COCH_3 (acetone)	2.9
NH_3	1.47	$HCONH_2$ (formamide)	3.7
SO_2	1.62	C_6H_5OH (phenol)	1.5
CH_3Cl	1.87	$C_6H_5NH_2$ (aniline)	1.5
NaCl	8.5	C_6H_5Cl (chlorobenzene)	1.8
CsCl	10.4	$C_6H_5NO_2$ (nitrobenzene)	4.2

Bond moments					
$C\!-\!H^{+}$	0.4	$C\!-\!C$	0	$C^{+}\!-\!Cl$	1.5–1.7
$N\!-\!H^{+}$	1.31	$C\!=\!C$	0	$N^{+}\!-\!O$	0.3
$O\!-\!H^{+}$	1.51	$C^{+}\!-\!N$	0.22	$C^{+}\!=\!O$	2.3–2.7
$F\!-\!H^{+}$	1.94	$C^{+}\!-\!O$	0.74	$N^{+}\!=\!O$	2.0

Group moments					
$C\!-\!{}^{+}OH$	1.65	$C\!-\!{}^{+}CH_3$	0.4	$C\!-\!{}^{+}COOH$	1.7
$C\!-\!{}^{+}NH_2$	1.2–1.5	$C^{+}\!-\!NO_2$	3.1–3.8	$C\!-\!{}^{+}OCH_3$	1.3

[a] Data compiled from Wesson (1948), Smyth (1955), Davies (1965) and Landolt-Börnstein (1982).
[b] Depends on conformation (e.g., cyclopropane has a dipole moment).

where l is the distance between the two charges $+q$ and $-q$. Thus, for two electronic charges $q = \pm e$ separated by $l = 0.1$ nm, the dipole moment is $u = (1.602 \times 10^{-19})(10^{-10}) = 1.6 \times 10^{-29}$ C m = 4.8 D. The unit of dipole moment is the *Debye*, where 1 Debye = 1 D = 3.336×10^{-30} C m. Small polar molecules have moments of the order of 1 D, some of which are listed in Table 4.1. Permanent dipole moments only occur in asymmetric molecules and thus not in single atoms. For isolated molecules, they arise from the asymmetric displacements of electrons along the covalent bonds, and it is therefore not surprising that a characteristic dipole moment can be assigned to each type of covalent bond. Table 4.1 also lists some of these *bond moments*, which lie parallel to the axis of each bond. These values are approximate but very useful for estimating the dipole moments of molecules and especially of parts of molecules by vectorial summation of their bond moments. For example, the dipole moment of gaseous H_2O, where the H—O—H angle is

$\theta = 104.5°$, may be calculated from

$$u_{H_2O} = 2\cos(\tfrac{1}{2}\theta)u_{OH} = 2\cos(52.25°) \times 1.51 = 1.85 \text{ D}.$$

4.2   DIPOLE SELF-ENERGY

A dipole possesses an electrostatic self-energy μ^i that is analogous to the Born self-energy of an ion. The dipole self-energy is quite simply the sum of the Born energies of the two charges $\pm q$ (at infinity) minus the Coulomb energy of bringing the two charges together to form the dipole. Let us estimate this for two ions of equal radius a brought into contact to form a hypothetical dipolar molecule of length equal to the sum of the two ionic radii, $l = 2a$. We therefore have

$$\mu^i = \frac{1}{4\pi\varepsilon_0\varepsilon}\left[\frac{q^2}{2a} + \frac{q^2}{2a} - \frac{q^2}{r}\right] = \frac{q^2}{8\pi\varepsilon_0\varepsilon a} \quad \text{or} \quad \frac{u^2}{4\pi\varepsilon_0\varepsilon l^3}, \tag{4.2}$$

where $r = l = 2a$. The dipole self-energy is therefore seen to be of roughly the same magnitude as the Born energy of an individual ion, Eq. (3.7), and its dependence on the dielectric constant of the medium is also the same. We may therefore expect that the solubility of polar molecules in different solvents should likewise increase with their value of ε. While this is generally the case, as shown in Fig. 3.1 (Chapter 3) for glycine equation (4.2) for polar molecules is somewhat model dependent and not as useful as the Born equation for ions. First, the dipole moment u can vary from solvent to solvent (Davies, 1965). Second, since molecules are usually much bigger than ions there are additional large energy terms arising from non-electrostatic solute–solvent interactions, such as the van der Waals self-energy, which are not included in Eq. (4.2) (see Problem 4.1).

4.3   ION–DIPOLE INTERACTIONS

The second type of electrostatic pair interaction we shall consider is that between a charged atom and a polar molecule, for example, between Na^+ and H_2O. As an illustrative example we shall derive the interaction potential for this case from basic principles. Figure 4.1 shows a charge Q at a distance r from the centre of a polar molecule of dipole moment u subtending an angle θ to the line joining the two molecules. If the length of the dipole is l,

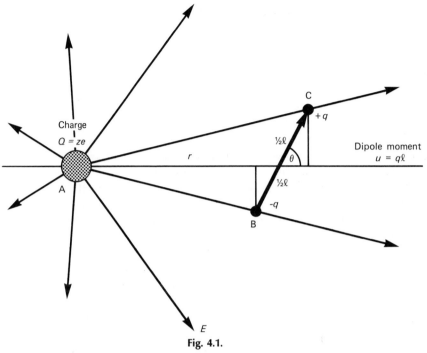

Fig. 4.1.

with charges $\pm q$ at each end, then the total interaction energy will be the sum of the Coulomb energies of Q with $-q$ at B and Q with $+q$ at C:

$$w(r) = -\frac{Qq}{4\pi\varepsilon_0\varepsilon}\left[\frac{1}{AB} - \frac{1}{AC}\right], \tag{4.3}$$

where

$$AB = [(r - \tfrac{1}{2}l\cos\theta)^2 + (\tfrac{1}{2}l\sin\theta)^2]^{1/2},$$
$$AC = [(r + \tfrac{1}{2}l\cos\theta)^2 + (\tfrac{1}{2}l\sin\theta)^2]^{1/2}. \tag{4.4}$$

At separations r exceeding the dipole's length l these distances can be written approximately as $AB \approx r - \tfrac{1}{2}l\cos\theta$, and $AC \approx r + \tfrac{1}{2}l\cos\theta$, and the interaction energy in this limit becomes

$$
\begin{aligned}
w(r) = w(r, \theta) &= -\frac{Qq}{4\pi\varepsilon_0\varepsilon}\left[\frac{1}{r - \tfrac{1}{2}l\cos\theta} - \frac{1}{r + \tfrac{1}{2}l\cos\theta}\right] \\
&= -\frac{Qq}{4\pi\varepsilon_0\varepsilon}\left[\frac{l\cos\theta}{r^2 - \tfrac{1}{4}l^2\cos^2\theta}\right] \\
&= -\frac{Qu\cos\theta}{4\pi\varepsilon_0\varepsilon r^2} = -\frac{(ze)u\cos\theta}{4\pi\varepsilon_0\varepsilon r^2}.
\end{aligned}
\tag{4.5}
$$

Note that since the electric field of the charge acting on the dipole is $E(r) = Q/4\pi\varepsilon_0\varepsilon r^2$, we see that in general the energy of a permanent dipole u in a field E may be written as

$$w(r, \theta) = -uE(r)\cos\theta. \tag{4.6}$$

Equation (4.5) gives the free energy for the interaction of a charge Q and a 'point dipole' u (for which $l = 0$) in a medium. Thus, when a cation is near a dipolar molecule maximum attraction (i.e., maximum negative energy) will occur when the dipole points away from the ion ($\theta = 0°$), while if the dipole points towards the ion ($\theta = 180°$) the interaction energy is positive and the force is repulsive. Figure 4.2 shows how the energy varies with distance for a monovalent cation ($z = +1$) interacting with a dipolar molecule of moment 1 D in vacuum. The solid curves are based on the exact solution calculated from Eqs. (4.3) and (4.4), while the dashed curves are for the point–dipole approximation, Eq. (4.5), which shows itself to be surprisingly accurate down to fairly small separations. Only at ion–dipole separations r below about $2l$ does the approximate equation deviate noticeably ($>10\%$) from that obtained using the exact formula. Thus, if the dipole moment arises from charges separated by less than about 0.1 nm, Eq. (4.5) will be valid at all physically realistic intermolecular separations. However, for greater dipole lengths—as occur in zwitterionic molecules—the deviations may be large, thereby requiring that the energy be calculated in terms of the separate Coulombic contributions. In such cases the interactions are always stronger than expected from Eq. (4.5), as can be inferred from Fig. 4.2.

It is also evident from Fig. 4.2 that the ion–dipole interaction is much stronger than kT at typical interatomic separations (0.2–0.4 nm). It is therefore strong enough to bind ions to polar molecules and mutually align them. Let us calculate the vacuum interaction between some common ions and water molecules. We shall assume that the water molecule may be treated as a simple spherical molecule of radius 0.14 nm with a point dipole of moment 1.85 D. This is a gross oversimplification: the distribution of charges in a water molecule is much more complex than for a simple dipole, as will be discussed later. But for our present purposes, we may ignore this complication. Thus, for the monovalent ion Na^+ ($z = 1$, $a = 0.095$ nm) near a water molecule ($a = 0.14$ nm, $u = 1.85$ D), the maximum interaction energy will be given by Eq. (4.5) as

$$w(r, \theta = 0°) = -\frac{(1.602 \times 10^{-19})(1.85 \times 3.336 \times 10^{-30})}{4\pi(8.854 \times 10^{-12})(0.235 \times 10^{-9})^2} = 1.6 \times 10^{-19} \text{ J}$$

$$= 39\,kT \quad \text{or} \quad 96\,\text{kJ mol}^{-1} \quad \text{at 300 K},$$

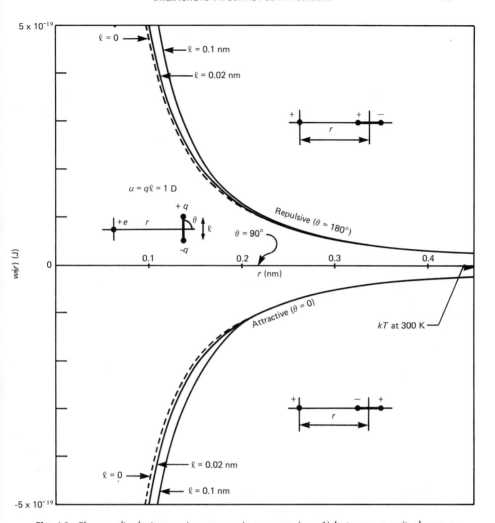

Fig. 4.2. Charge–dipole interaction energy in vacuum ($\varepsilon = 1$) between a unit charge e and a dipole of moment $u = ql = 1$ D oriented at different angles θ to the charge. Solid lines are exact solutions, Eqs (4.3) and (4.4), for finite-sized dipoles with $l = 0.02$ nm and $l = 0.10$ nm; dashed lines are exact solutions for $l = 0$, which correspond to the approximate point-dipole formula, Eq. (4.5). Note that for typical interatomic spacings ($r \approx 0.3$–0.4 nm) the strength of the pair interaction greatly exceeds the thermal energy kT at 300 K.

which compares surprisingly well with the experimental value of 100 kJ mol^{-1} (Saluja, 1976). For the smallest monovalent ion Li$^+$ ($a = 0.068$ nm), this rises to about $50\,kT$ or 125 kJ mol^{-1} (experimental value: 142 kJ mol^{-1}), while for the small divalent cations Mg^{2+} ($z = 2$, $a = 0.065$ nm) and Be^{2+}

($z = 2$, $a = 0.03$ nm), it will rise even further to about $100\,kT$ and $150\,kT$, respectively.

The strongly attractive interaction between ions and water is responsible for promoting the 'ionic nucleation' of rain drops in thunderclouds. In other types of clouds the nucleation of water occurs around uncharged particles, molecules or molecular groups that have a strong affinity for water. Such groups are referred to as being *hydrophilic, hygroscopic* or *deliquescent* and are discussed in Chapter 8.

4.4   IONS IN POLAR SOLVENTS

When ion–water interactions take place in bulk water the above energies will be reduced by a factor of about 80, the dielectric constant of water; but even then the strength of the interaction will exceed kT for small divalent and multivalent ions, and it is by no means negligible for small monovalent ions.

But before we proceed it is essential to understand what this interaction energy means. It cannot be the energy gained on bringing a water molecule up to an ion *in bulk water*, since this process must also involve removing one of the water molecules that was originally in contact with the ion and placing it where the guest water molecule was before it started off on its journey. That is, the whole process is no more than an *exchange* of two water molecules and cannot result in any change of free energy (note that this is a continuum approach; in fact, the free energy change is finite due to the finite size of molecules, as discussed in Chapter 7). However, within the continuum approach let us consider the ion–water interaction in bulk water as given by Eq. (4.5), where we note that it contains an orientation term $\cos\theta$. At large separations the water molecules would be randomly oriented relative to the ion, and if they remained randomly oriented right up to the ion, the interaction energy *would* be zero since the spatial average of $\cos\theta$ is zero. For an ion in a polar solvent, Eq. (4.5) therefore gives us an estimate of the free energy change brought about by orienting the polar solvent molecules around the ion, that is, the reference state of zero energy is for randomly oriented dipoles.

We have therefore established that the ion–dipole energies calculated for ions in water are comparable to or greater than kT and, from Eq. (2.28), reflect the strong aligning effect that small ions must have on their surrounding water molecules.

4.5 Strong ion–dipole interactions: hydrated ions

[For small or multivalent ions in highly polar solvents, the strong orientation dependence of their ion–dipole interaction will tend to orient the solvent molecules around them, favouring $\theta = 0°$ near cations and $\theta = 180°$ near anions. Thus, in water Li^+, Be^{2+}, Mg^{2+} and Al^{3+} ions have a number of water molecules orientationally bound to them.] Such ions are called *solvated ions* or *hydrated ions*, and the number of water molecules they bind—usually between 4 and 6—is known as their *hydration number* (Table 4.2). It shoud be noted, however, that these bound water molecules are not completely immobilized and that they do exchange with bulk water, albeit more slowly. The hydration number is more of a qualitative indicator of the degree to which ions bind water rather than an exact value.

Closely related to the hydration number is the effective radius or *hydrated*

TABLE 4.2 Hydrated radii and hydration numbers of ions in water (approximate)

Ion	Bare ion radius (nm)	Hydrated radius (nm)	Hydration number (± 1)	Lifetime (exchange rate) (s)
H_3O^+	—	0.28	3	—
Li^+	0.068	0.38	5–6	10^{-9}–10^{-8}
Na^+	0.095	0.36	4–5	10^{-9}
K^+	0.133	0.33	3–4	10^{-9}
Cs^+	0.169	0.33	1–2	10^{-10}–10^{-9}
Be^{2+}	0.031	0.46	4^a	10^{-3}–10^{-2}
Mg^{2+}	0.065	0.43	6^a	10^{-6}–10^{-5}
Ca^{2+}	0.099	0.41	6	10^{-8}
Al^{3+}	0.050	0.48	6^a	10^{-1}–1
OH^-	0.176	0.30	3	
F^-	0.136	0.35	2	
Cl^-	0.181	0.33	1	
Br^-	0.195	0.33	1	
I^-	0.216	0.33	0	
NO_3^-	0.264	0.34	0	
$N(CH_3)_4^+$	0.347	0.37	0	

[a] Number of water molecules forming a stoichiometric complex with the ion (e.g., $[Be(H_2O)_4]^{2+}$). The hydration number gives the number of water molecules in the primary shell, though the total number of water molecules affected can be much larger and depends on the method of measurement. Similarly, the hydrated radius depends on how it is measured. Different methods can yield radii that can be as much as 0.1 nm smaller or larger than those shown. Table compiled from data given by Nightingale (1959), Amis (1975), Saluja (1976), Bockris and Reddy (1970), Cotton and Wilkinson (1980).

radius of an ion in water, which is larger than its real radius (i.e., its crystal lattice radius) as shown in Table 4.2. Because smaller ions are more hydrated they tend to have larger hydrated radii than larger ions. Hydration numbers and radii can be deduced from measurements of the viscosity, diffusion, compressibility, conductivity, solubility, and various thermodynamic and spectroscopic properties of electrolyte solutions, the results rarely agreeing with one another (Amis, 1975; Saluja, 1976).

More insight into the nature of ion hydration can be gained by considering the average time that water molecules remain bound to ions. In the pure liquid at room temperature the water molecules tumble about with a mean reorientation time or *rotational correlation time* of about 10^{-11} s. This also gives an estimate of the lifetime of the water–water bonds formed in liquid water (the hydrogen bonds). But when the water molecules are near ions various techniques, such as oxygen nuclear magnetic resonance, show that the mean lifetimes or exchange rates of water molecules in the first hydration shell can be much longer, varying from 10^{-11} s to many hours (Cotton and Wilkinson, 1980; Hertz, 1973). For very weakly solvated ions (usually large monovalent ions) such as $N(CH_3)_4^+$, Cl^-, Br^- and I^-, these lifetimes are not much different from that for two water molecules, and they can even be shorter (referred to as *negative hydration*). Cations are generally more solvated than anions of the same valency since they are smaller—having lost rather than gained an electron. Thus, for K^+, Na^+ and Li^+, the residence times of water molecules in the primary hydration shells are about 10^{-9} s. Divalent cations are always more strongly solvated than monovalent cations, and for Ca^{2+} and Mg^{2+}, the bound water lifetimes are about 10^{-8} s and 10^{-6} s, respectively. Even longer lifetimes are observed for very small divalent cations such as Be^{2+} (10^{-3} s), while for trivalent cations such as Al^{3+}, La^{3+}, and Cr^{3+} these can be many seconds or hours. In such cases the binding is so strong that an ion–water complex is actually formed of fixed stoichiometry (see Table 4.2). In fact, these quasi-stable complexes begin to take on the appearance of (charged) molecules and are often designated as such, e.g. $[Mg(H_2O)_6]^{2+}$, $[Be(H_2O)_4]^{2+}$. In particular, protons (H^+) always associate with one water molecule, which goes by the name of the *hydronium ion* or *oxonium ion* (H_3O^+), while three water molecules are solvated around this ion to form $H_3O^+(H_2O)_3$. Likewise, the hydroxyl ion (OH^-) is believed to be solvated by three water molecules forming $OH^-(H_2O)_3$.

4.6   SOLVATION FORCES, STRUCTURAL FORCES, HYDRATION FORCES

The first shell of water molecules around a strongly solvated ion is usually referred to as the *primary hydration shell*. This is where the water molecules are most restricted in their motion. But the effect does not end there; it propagates out well beyond the first shell. Thus, in the second hydration shell the water molecules are freer to rotate and exchange with bulk water; in the third shell even more so, and so on.

Later we shall find that other types of interactions can also lead to a modified molecular ordering around solute molecules and surfaces and that the effect decays roughly exponentially with distance, extending a few molecular diameters. We may refer to this region of modified *solvent structure* as the *solvation zone* wherein the properties of the solvent (e.g., density, positional and orientational order, and mobility) are significantly different from the corresponding bulk values, as was shown schematically in Fig. 2.1b.

The existence of a solvation zone around dissolved ions, molecules or particle surfaces in a solvent medium occurs whenever there are strong solvent–solvent or solute–solvent interactions (e.g, strong ion–dipole interactions in water) and has some important consequences. First, it affects the local dielectric constant of the solvent since the solvent molecules no longer respond to an electric field as they would in the bulk. The restricted mobility of water molecules around small ions would suggest that the effective dielectric constant should be lower in the solvation zone than in the bulk liquid. However, since the dielectric constant of ice is actually higher than that of liquid water (see Fig. 8.1), the reverse may occur. At present these effects are not well understood, but it is clear that solvation zones cannot be treated entirely in terms of continuum theories, since they arise from highly specific solute–solvent and modified solvent–solvent interactions occurring at the molecular level.

Second, when the solvation zones of two solvated molecules or surfaces overlap, a short-range force arises that again cannot be treated in terms of continuum models. For example, if the local dielectric constant of water around strongly solvated ions differs from 80, the short-range Coulomb interaction would be modified. But this is only one aspect of the problem. There has been much recent theoretical and experimental activity aimed at unravelling all the subtle effects associated with these solvent-mediated interactions that are now usually referred to as *solvation* or *structural forces* and, when water is the solvent, *hydration* forces. The nature of solvation forces is investigated further in Chapters 7 and 13.

4.7   DIPOLE–DIPOLE INTERACTIONS

When two polar molecules are near each other there is a dipole–dipole interaction between them that is analogous to that between two magnets.

For two point dipoles of moments u_1 and u_2 at a distance r apart and oriented relative to each other as shown in Fig. 2.2, the interaction energy may be derived by a procedure similar to that used to obtain the energy for the charge–dipole interaction, and we find

$$w(r, \theta_1, \theta_2, \phi) = -\frac{u_1 u_2}{4\pi\varepsilon_0 \varepsilon r^3} [2 \cos\theta_1 \cos\theta_2 - \sin\theta_1 \sin\theta_2 \cos\phi]. \quad (4.7)$$

Equation (4.7) shows that maximum attraction occurs when two dipoles are lying in line, when the energy is given by

$$w(r, 0, 0, \phi) = -2u_1 u_2/4\pi\varepsilon_0 \varepsilon r^3, \quad (4.8)$$

while for two dipoles aligned parallel to each other the energy $w(r, 90°, 90°, 180°)$ is half of this value at the same interdipole separation, r. Equation (4.7) also shows that for two equal dipoles of moment 1 D, their interaction energy in vacuum will equal kT at $r = 0.36$ nm when the dipoles are in line and at $r = 0.29$ nm when parallel. Since these distances are of the order of molecular separations in solids and liquids, we see that at normal temperatures dipolar interactions (alone) are strong enough to bind only very polar molecules.

Figure 4.3 shows the variation of the pair interaction energy with distance for two point dipoles of moments 1 D approaching each other at different orientations; the solid curves are the exact solutions for finite-sized dipoles, here assumed to be of length $l = 0.1$ nm, while the dashed curves are based on Eq. (4.7) for two point dipoles. In general, we find that, even more than for the charge–dipole interaction, significant deviations from the ideal behaviour now occur for $r < 3l$, when Eq. (4.7) can no longer be used. At these smaller separations it is again necessary to analyse the interaction in terms of its individual charge–charge (Coulomb) contributions of which there will be four such terms for each pair of dipoles.

The above calculations, and Fig. 4.3, appear to indicate that two dipoles always prefer to mutually orient themselves in line, but this is true only at the same value of r. Most dipolar molecules are also anisotropic in shape—being longer along the direction of the dipole, so that in practice the centres of two such cigar-shaped molecules can come significantly closer together when they align in parallel, thereby making this interaction the more favourable one.

The dipole–dipole interaction is not as strong as the previous two electrostatic interactions we considered, and for dipole moments of order ~ 1 D, it is already weaker than kT at distances of about 0.35 nm in vacuum, while in a solvent medium this distance will be even smaller. This means

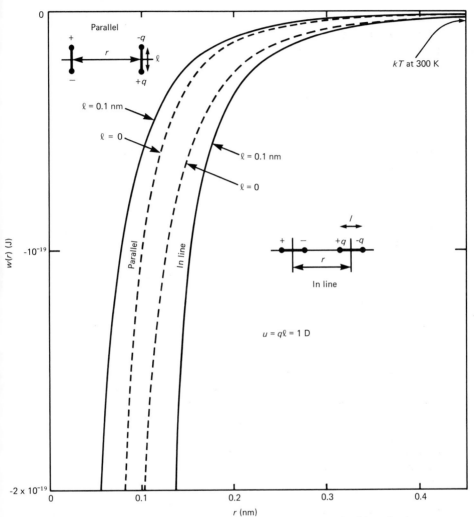

Fig. 4.3. Dipole–dipole interaction energy in vacuum between two dipoles each of moment 1 D. Note how much weaker this interaction is compared to the charge–dipole interaction (Fig. 4.2) and the large effect of finite dipole size.

that the dipole–dipole interaction, unlike the ion–dipole interaction, is usually not strong enough to lead to any strong mutual alignment of polar molecules in the liquid state. There are some exceptions, however, such as water whose small size and large dipole moment does lead to short range association in the liquid. A glance at Table 4.1 shows that the bond moments of O^-—H^+, N^-—H^+ and F^-—H^+ are unusually large. Since the electron-depleted H atom also has a particularly small size this means that other *electronegative* atoms such as —O, —N and —F can get quite close to these highly polar X^-—H^+ groups and thus experience a very strong field. This results in a strong attractive force that can align neighbouring molecules possessing such groups (e.g., H_2O, NH_3, HF, C=O, and many others) in both the liquid and crystalline state. Such liquids are called *associated liquids*, and the special type of interaction they experience is known as the *hydrogen-bonding* interaction.

The hydrogen bonding interaction is no more than a particularly strong type of directional dipole–dipole interaction. Because of the small size of the —H^+ group it is far stronger than that predicted by the point dipole approximation (compare the dashed and solid lines in Fig. 4.3). This interaction will be discussed further in Chapter 8. For the moment, let us investigate the opposite situation: when the orientation dependence of dipole–dipole and ion–dipole interaction energies are much *weaker* than the thermal energy kT, and therefore unable to align molecules fully.

4.8 ROTATING DIPOLES AND ANGLE-AVERAGED POTENTIALS

At large separations or in a medium of high ε, when the angle dependence of the interaction energy falls below the thermal energy kT, dipoles can now rotate more or less freely. However, even though the values of $\cos \theta$, $\sin \theta$, etc. when averaged over all of space are zero, the angle-averaged *potentials* are not zero since there is always a Boltzmann weighting factor that gives more weight to those orientations that have a lower (more negative) energy.

In general, the angle-averaged (Helmholtz) free energy $w(r)$ of an instantaneous orientation-dependent free energy $w(r, \Omega)$ is given by the *potential distribution theorem* (Landau and Lifshitz, 1980, Ch. III; Rushbrooke, 1940; Widom, 1963)

$$e^{-w(r)/kT} = \int e^{-w(r,\Omega)/kT} \, d\Omega \bigg/ \int d\Omega = \langle e^{-w(r,\Omega)/kT} \rangle, \qquad (4.9)$$

where $d\Omega = \sin \theta \, d\theta \, d\phi$ corresponds to the polar and azimuthal angles θ and ϕ (see Fig. 2.2) and the integration is over all of angular space. Accordingly, the denominator in Eq. (4.9) becomes

$$\int d\Omega = \int_0^{2\pi} d\phi \int_0^{\pi} \sin \theta \, d\theta = 4\pi,$$

so that in general we may write

$$e^{-w(r)/kT} = \langle e^{-w(r,\theta,\phi)/kT} \rangle = \frac{1}{4\pi} \int_0^{2\pi} d\phi \int_0^{\pi} e^{-w(r,\theta,\phi)/kT} \sin\theta \, d\theta. \quad (4.10)$$

We may note the spatially averaged values of some angles:

$$\langle \cos^2\theta \rangle = \frac{1}{4\pi} \int_0^{\pi} \cos^2\theta \sin\theta \, d\theta \int_0^{2\pi} d\phi = \tfrac{1}{3},$$

$$\langle \sin^2\theta \rangle = \tfrac{2}{3},$$

$$\langle \sin^2\phi \rangle = \langle \cos^2\phi \rangle = \tfrac{1}{2},$$

$$\langle \sin\theta \rangle = \langle \cos\theta \rangle = \langle \sin\theta\cos\theta \rangle = 0,$$

$$\langle \sin\phi \rangle = \langle \cos\phi \rangle = \langle \sin\phi\cos\phi \rangle = 0. \quad (4.11)$$

When $w(r, \Omega)$ is less than kT we can expand Eq. (4.9):

$$e^{-w(r)/kT} = 1 - \frac{w(r)}{kT} + \dots = \left\langle 1 - \frac{w(r,\Omega)}{kT} + \frac{1}{2}\left(\frac{w(r,\Omega)}{kT}\right)^2 - \dots \right\rangle,$$

thus

$$w(r) = \left\langle w(r,\Omega) - \frac{w(r,\Omega)^2}{2kT} + \dots \right\rangle. \quad (4.12)$$

The angle-averaged free energy for the *charge–dipole interaction* is therefore, using Eq. (4.5) for $w(r, \Omega)$,

$$w(r) = \left\langle -\frac{Qu\cos\theta}{4\pi\varepsilon_0\varepsilon r^2} - \left(\frac{Qu}{4\pi\varepsilon_0\varepsilon r^2}\right)^2 \frac{\cos^2\theta}{2kT} + \dots \right\rangle$$

$$\approx -\frac{Q^2 u^2}{6(4\pi\varepsilon_0\varepsilon)^2 kTr^4} \quad \text{for} \quad kT > \frac{Qu}{4\pi\varepsilon_0\varepsilon r^2}, \quad (4.13)$$

which is attractive and temperature dependent. Thus, for a monovalent ion interacting with the polar solvent molecules of a medium of dielectric constant ε, Eq. (4.13) will supersede Eq. (4.5) at distances larger than $r = \sqrt{Qu/4\pi\varepsilon_0\varepsilon kT}$, which for a monovalent ion in water, setting $Q = e$, $u = 1.85$ D, $\varepsilon = 80$, becomes roughly 0.2 nm (i.e., about 0.1 nm out from an ion of radius 0.1 nm). We can now see why only water molecules of the first shell around ions sometimes become strongly restricted in their motion, and we may anticipate that this should be the sort of range around an ion over which the properties of the solvent may be substantially different from the bulk values.

For the *dipole–dipole interaction*, a similar Boltzmann averaging of the interaction energy, Eq. (4.7), over all orientations $(\theta_1, \theta_2, \phi)$ leads to an angle-averaged interaction free energy of

$$w(r) = -\frac{u_1^2 u_2^2}{3(4\pi\varepsilon_0\varepsilon)^2 kTr^6} \quad \text{for} \quad kT > \frac{u_1 u_2}{4\pi\varepsilon_0\varepsilon r^3}. \quad (4.14)$$

The Boltzmann-averaged interaction between two permanent dipoles is usually referred to as the *orientation* or *Keesom* interaction. It is one of three important interactions, each varying with the inverse sixth power of the distance, that together contribute to the total *van der Waals* interaction between atoms and molecules. Van der Waals forces will be discussed collectively in Chapter 6.

Equations (4.13) and (4.14) show that beyond a certain distance the interaction energies fall faster than $1/r^3$. In view of the analysis of Section 1.3 this confirms that neither ion–dipole nor dipole–dipole forces can produce long-range alignment effects in liquids. Note that the expressions for these interactions become modified in two dimensions, e.g. for molecules interacting on a surface (see Problem 4.3), but the distance dependence does not change and so neither does the above conclusion.

All the expressions derived so far give the *free* energies of the interactions, strictly the Helmholtz free energies since the interactions are implicitly assumed to occur at constant volume. From basic thermodynamics the Helmholtz free energy A of any system or interaction is related to the *total* internal energy U by the well-known Gibbs–Helmholtz equation

$$U = A + TS = A - T(\partial A/\partial T)_v = -T^2\partial(A/T)/\partial T, \quad (4.15)$$

where S is the entropy of the system. Thus, for the angle-averaged dipole–dipole (Keesom) interaction in vacuum $(\varepsilon = 1)$, we find

$$U = -\frac{2u_1^2 u_2^2}{3(4\pi\varepsilon_0)^2 kTr^6}, \quad (4.16)$$

which is twice the *free* energy, Eq. (4.14), with which it is often confused in the literature. Note, too, that in condensed phases the Helmholtz and Gibbs free energies, A and G, are essentially the same since they are related by $G = A + PV$ where the PV term is usually small.

The distinction between A and U does not arise for temperature-independent pair interactions, since then $U = A - T(\partial A/\partial T)_v = A$.

4.9   ENTROPIC EFFECTS

The reason why the interaction free energy (the energy available for doing work or the energy that gives the force) is less than the internal energy of two interacting dipoles is because some of the energy is taken up in aligning the dipoles as they approach each other, which ultimately transforms into heat. This unavailable part of the energy is associated with the entropic contribution to the interaction. Let us complete this chapter by considering these entropic effects a bit further.

From Eq. (4.15) the free energy may be written as

$$A = U - TS = U + T(\partial A/\partial T), \tag{4.17}$$

so that there is an entropic contribution $T(\partial A/\partial T)$ that must be added to the total energy before we can know the free energy of an interaction. For example, for both the charge–dipole and dipole–dipole angle-averaged interactions in vacuum, we find that since $A \propto -1/T$, then

$$T(\partial A/\partial T) = -A, \tag{4.18}$$

so that

$$A = U - A = \tfrac{1}{2}U. \tag{4.19}$$

Thus, half the total energy is absorbed internally during the interaction. This energy is taken up in decreasing the rotational freedom of the dipoles as they become progressively more aligned on approach. Since A is negative, the entropic contribution $T(\partial A/\partial T)$ to the free energy is *positive*—i.e., unfavourable—and since $S = -T(\partial A/\partial T)$ is negative we would say that the interaction is associated with a *loss of entropy*.

For interactions in a solvent medium, the medium dielectric constant generally appears in the denominator as ε or ε^2. Thus, if ε is temperature dependent, as occurs for polar solvents such as water (but not for non-polar hydrocarbon solvents), the interaction energy now has an entropic component even if the interaction potential does not have an explicit temperature dependence. For example, for the Born energy $A = Q^2/8\pi\varepsilon_0\varepsilon a$, and the

Coulomb energy $A = Q_1 Q_2 / 4\pi\varepsilon_0 \varepsilon a$, we find that

$$TS = -T\left(\frac{\partial A}{\partial T}\right) = \frac{Q^2}{8\pi\varepsilon_0 \varepsilon a}\left(\frac{T}{\varepsilon}\frac{\partial \varepsilon}{\partial T}\right) = A\left(\frac{T}{\varepsilon}\frac{\partial \varepsilon}{\partial T}\right), \qquad (4.20)$$

and likewise for the Coulomb energy,

$$TS = -T\left(\frac{\partial A}{\partial T}\right) = \frac{Q_1 Q_2}{4\pi\varepsilon_0 \varepsilon r}\left(\frac{T}{\varepsilon}\frac{\partial \varepsilon}{\partial T}\right) = A\left(\frac{T}{\varepsilon}\frac{\partial \varepsilon}{\partial T}\right), \qquad (4.21)$$

while for the angle-averaged charge–dipole and dipole–dipole interactions in a medium we obtain

$$TS = -T\left(\frac{\partial A}{\partial T}\right) = A\left(1 + \frac{2T}{\varepsilon}\frac{\partial \varepsilon}{\partial T}\right), \qquad (4.22)$$

where the first contribution to the entropy is associated with the orientational motion of the interacting *solute* dipoles as before (cf. Eq. (4.18)) and the second with the *solvent* molecules.

All the above entropic contributions that depend on $\partial \varepsilon / \partial T$ arise from changes in the configurations of the solvent molecules associated with these interactions. For water,

$$\frac{T}{\varepsilon}\frac{\partial \varepsilon}{\partial T} = -1.36 \qquad \text{at } 25°\text{C}, \qquad (4.23)$$

which is negative. This allows us to make a number of predictions concerning the effects of interactions on the surrounding water molecules. Thus, for the Born interaction, Eq. (4.20) shows that S is negative. We may therefore conclude that the solvation of ions by water is accompanied by a *decrease* in entropy, which indicates once again that water molecules become restricted in their translational and rotational freedom when they solvate ions. On the other hand, Eq. (4.21) shows that when two ions of opposite sign approach each other in a medium the entropy of the solvent *increases*.

The above considerations bring out the remarkable feature of the dielectric constant of a solvent in that it contains information on the entropic changes of the solvent molecules involved in an interaction. We shall make use of this property again when we consider other types of interactions.

PROBLEMS AND DISCUSSION TOPICS

4.1 Look at Fig. 3.1. Assume glycine to be a molecule with a dipole of unit charges $\pm e$ at a distance l apart. Estimate l from the data of Fig. 3.1. Also, from the intercept of the line estimate the additional non-electrostatic free energy contribution associated with the transfer of a molecule of glycine into a polar medium. If the glycine molecule is assumed to be spherical with diameter σ equal to l (as obtained above), what would be the free energy per unit surface area γ of this non-electrostatic transfer process? Are the values you obtain for l, σ and γ reasonable?

4.2 (i) What is the energy of a macroscopic sphere of material A and radius r in a liquid medium B, where γ_i is the interfacial energy of the $A-B$ interface? For water and hydrocarbon liquids (e.g. alkanes, oils, etc.) the value of γ_i is about 50 mJ m^{-2}. Estimate the solubility of small hydrocarbon molecules such as methane in water at 20°C, in mole fraction units, assuming that these molecules behave as small macroscopic spheres of radius $r = 0.2$ nm.

(ii) An inert non-polar oil is in contact with an aqueous 1:1 electrolyte solution (i.e., a salt of monovalent ions such as NaCl) whose cations and anions have the same bare-ion radii of 0.10 nm. If the salt concentration in the water is 1 mM (10^{-3} mol l^{-1}), calculate the ionic concentration in the oil phase. Assume that the dielectric constants of oil and water are 2.0 and 80, respectively, and that the temperature is $T = 20$°C. What would be the concentration if the ions had a radius of 0.08 nm?

(iii) In practice, the concentration of ions in non-polar hydrocarbon liquids is found to be much higher than any value calculated based on bare-ion radii because ions can enter these liquids surrounded by a certain number of water molecules, i.e. as hydrated ions. By considering the equations used in parts (i) and (ii), estimate the optimal radius of these hydrated ions in oil, their hydration number, and then calculate a new (and more realistic) value for the concentration of ions in the oil. Does your answer also explain why the partitioning of monovalent ions in hydrocarbon liquids is roughly the same for ions of different size? Would you expect the experimental values of the solubilities or partitioning of ions to be higher or lower than your 'theoretical' result?

4.3 Derive (i) the angle-averaged interaction potential for two dipolar molecules in free space, Eq. (4.14); (ii) the angle-averaged potential for two dipolar molecules constrained to interact on a surface with their dipoles freely rotating but always lying in the plane of the surface—the two-dimensional Keesom equation. (*Hint*: first calculate the values of $\langle \cos^2 \theta \rangle$, $\langle \sin^2 \theta \rangle$, etc., that replace those of Eq. (4.11) in two dimensions); and (iii) the mean interaction energy per molecule for a surface monolayer consisting of

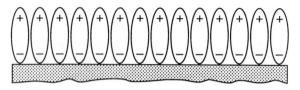

Fig. 4.4.

mobile molecules whose dipoles are aligned normal to the surface, as sketched in Fig. 4.4, where each molecule occupies a mean area A. Assume the dipoles have charges $\pm e$ a distance l apart (the dipole moment being $u = el$). What is the main difference between the last two interactions (ii) and (iii) concerning their contributions to two-dimensional phase transitions (*Hint*: consider the sign of their contribution to the two-dimensional van der Waals equation of state.)

4.4 The angle-averaged Helmholtz free energy of the dipole–dipole interaction $w(r)$ is related to the angle-dependent potential $w(r, \Omega)$ via Eq. (4.9): $e^{-w(r)/kT} = \langle e^{-w(r,\Omega)/kT} \rangle$. Show that for $w(r, \Omega) < kT$ the internal energy $U(r)$ is related to $w(r, \Omega)$ via

$$U(r) = \langle w(r, \Omega) e^{-w(r, \Omega)/kT} \rangle, \tag{4.24}$$

and use this to derive Eq. (4.16).

CHAPTER 5

INTERACTIONS INVOLVING THE POLARIZATION OF MOLECULES

5.1 THE POLARIZABILITY OF ATOMS AND MOLECULES

We now enter the last category of electrostatic interactions we shall be considering—those that involve molecular *polarization*, that is, the dipole moments induced in molecules by the electric fields emanating from nearby molecules. Actually, we have already been much involved with polarization effects: whenever the macroscopic dielectric constant of a medium entered into our consideration this was no more than a reflection of the way the molecules of the medium are polarized by the local electric field. Here we shall look at these effects in more detail, starting at the molecular level. We shall find that apart from the purely Coulombic interaction between two charges or dipoles in vacuum, all the other interactions we have so far considered are essentially polarization-type interactions.

All atoms and molecules are polarizable. Their (dipole) *polarizability* α is defined according to the strength of the *induced* dipole moment u_{ind} they acquire in a field E, viz.,

$$u_{ind} = \alpha E. \tag{5.1}$$

For a non-polar molecule, the polarizability arises from the displacement of its negatively charged electron cloud relative to the positively charged nucleus under the influence of an external electric field. For polar molecules, there are other contributions to the polarizability, discussed in the next section. For the moment, we shall concentrate on the polarizabilities of non-polar molecules, which we shall denote by α_0.

As a simple illustrative example of how polarizability arises, let us imagine a one-electron atom whose electron of charge $-e$ circles the nucleus at a distance R; this would also define the radius of the atom. If under the influence of an external field E the electron orbit is shifted by a distance l from the

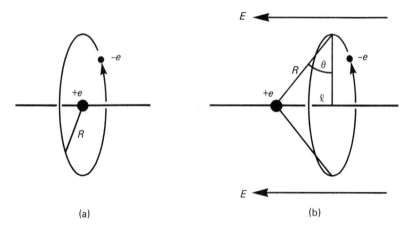

Fig. 5.1. Induced dipole in a one-electron atom. (a) No external fluid, $u_{ind} = 0$. (b) In an external field E the electron's orbit is shifted by a distance l from the positive nucleus so that the induced dipole moment is $u_{ind} = le = \alpha_0 E$, where the polarizability for this case is $\alpha_0 = 4\pi\varepsilon_0 R^3$.

nucleus (Fig. 5.1), then we have, for the induced dipole moment,

$$u_{ind} = \alpha_0 E = le. \tag{5.2}$$

Now the *external* force F_{ext} on the electron due to the field E is

$$F_{ext} = eE,$$

which must be balanced at equilibrium by the attractive force between the displaced electron orbit and the nucleus. This is none other than the Coulomb force $e^2/4\pi\varepsilon_0 R^2$ resolved along the direction of the field (Fig. 5.1). The *internal* (restoring) force is therefore

$$F_{int} = \frac{e^2}{4\pi\varepsilon_0 R^2} \sin\theta \approx \frac{e^2 l}{4\pi\varepsilon_0 R^3} \approx \frac{e}{4\pi\varepsilon_0 R^3} u_{ind}. \tag{5.3}$$

At equilibrium, $F_{ext} = F_{int}$, which leads to

$$u_{ind} = 4\pi\varepsilon_0 R^3 E = \alpha_0 E,$$

whence we obtain for the polarizability

$$\alpha_0 = 4\pi\varepsilon_0 R^3. \tag{5.4}$$

TABLE 5.1 Electronic polarizabilities α_0 of atoms, molecules, bonds, and molecular groups[a]

Atoms and molecules					
He	0.20	NH_3	2.3	$CH_2{=}CH_2$	4.3
H_2	0.81	CH_4	2.6	C_2H_6	4.5
H_2O	1.48	HCl	2.6	Cl_2	4.6
O_2	1.60	CO_2	2.6	$CHCl_3$	8.2
Ar	1.63	CH_3OH	3.2	C_6H_6	10.3
CO	1.95	Xe	4.0	CCl_4	10.5
Bond polarizabilities					
Aliphatic	C—C	0.48		C—O	0.60
Aromatic	C∴C	1.07		C=O	1.36
	C=C	1.65		N—H	0.74
Aliphatic	C—H	0.65		C—Cl	2.60
	O—H	0.73		C—Br	3.75
Molecular groups					
C—O—H	1.28		CH_2	1.84	
C—O—C	1.13		Si—O—Si	1.39	
C—NH_2	2.03		Si—OH	1.60	

[a] Polarizabilities α_0 are given in units of $(4\pi\varepsilon_0)\text{Å}^3 = (4\pi\varepsilon_0)10^{-30}\,\text{m}^3 = 1.11 \times 10^{-40}\,\text{C}^2\,\text{m}^2\,\text{J}^{-1}$. Note that when molecules are dissolved in a solvent medium their polarizability can change by up to 10%. Data compiled from Denbigh (1940), Hirschfelder et al. (1954) and Smyth (1955).

The unit of polarizability is therefore $4\pi\varepsilon_0 \times$ (volume) or $\text{C}^2\,\text{m}^2\,\text{J}^{-1}$. The polarizability of atoms and molecules that arises from such electron displacements is known as *electronic polarizability*. Its magnitude, apart from the $4\pi\varepsilon_0$ term, is usually less than but of the order of the (radius)^3 of the atom or molecule. For example, for water, $\alpha_0/4\pi\varepsilon_0 = 1.48 \times 10^{-30}\,\text{m}^3 = (0.114\,\text{nm})^3$, where 0.114 nm is about 15% less than the radius of a water molecule (0.135 nm).

Table 5.1 lists the electronic polarizabilities of some common atoms and molecules. Since the electronic polarizability is associated with displacements of electron clouds, it has long been recognized that the polarizability of a molecule can be obtained by simply summing the characteristic polarizabilities of its covalent bonds, since these are where the polarizable electrons are mostly localized. Table 5.1 also lists some *bond polarizabilities*. As can be seen, the polarizability of methane, CH_4, is simply four times that of the C—H bond (i.e., $\alpha_{CH_4}{=}4\alpha_{C-H}$). Likewise, the polarizability of ethylene, $CH_2{=}CH_2$, is given by $4\alpha_{C-H} + \alpha_{C=H}$. This additivity procedure is often accurate to within a few percent, but it can fail for molecules in which the bonds are not independent of each other (delocalized electrons, as in benzene) or when non-bonded lone-pair electrons that also contribute to the polarizability are present, as in H_2O and other hydrogen-bonding groups. Under such circumstances it has been found useful to assign polarizabilities to certain molecular groups. Some *group polarizabilities* are also included in Table 5.1. For example, the polarizability of CH_3OH is $3\alpha_{C-H} + \alpha_{C-O-H} = 4\pi\varepsilon_0(3 \times 0.65 + 1.28) \times 10^{-30} = 4\pi\varepsilon_0(3.23 \times 10^{-30} \text{ m}^3)$.

5.2 THE POLARIZABILITY OF POLAR MOLECULES

In Section 5.1 we considered the polarizability arising solely from the electronic displacements in atoms and molecules. A freely rotating dipolar molecule (whose time-averaged dipole moment is zero) also has an *orientational polarizability*, arising from the effect of an external field on the Boltzmann-averaged orientations of the rotating dipole. Thus, in the presence of an electric field E these orientations will no longer time-average to zero but will be weighted along the field. If at any instant the permanent dipole u is at an angle θ to the field E, its resolved dipole moment along the field is $u \cos \theta$, and its energy in the field from Eq. (4.6) is $-uE \cos \theta$, so that the angle-averaged induced dipole moment is given by (cf. Section 4.8)

$$u_{\text{ind}} = \langle u \cos \theta e^{uE \cos \theta / kT} \rangle$$

$$= \frac{u^2 E}{kT} \langle \cos^2 \theta \rangle = \frac{u^2}{3kT} E, \qquad uE \ll kT. \qquad (5.5)$$

Since u_{ind} is proportional to the field E, we see that the factor $u^2/3kT$ provides an additional contribution to the molecular polarizability. This is known as the *orientational* polarizability:

$$\alpha_{\text{orient}} = u^2/3\,kT. \qquad (5.6)$$

The total polarizability of a polar molecule is therefore

$$\alpha = \alpha_0 + u^2/3\,kT \qquad \text{(the Debye–Langevin equation)}, \qquad (5.7)$$

where u is its permanent dipole moment. Thus, for example, a polar molecule of moment $u = 1$ D $= 3.336 \times 10^{-30}$ C m at 300 K will have an orientational polarizability of

$$\alpha_{\text{orient}} = \frac{(3.336 \times 10^{-30})^2}{3(1.38 \times 10^{-23})300} = 9 \times 10^{-40} \qquad \text{C}^2 \text{ m}^2 \text{ J}^{-1}$$

$$= (4\pi\varepsilon_0)8 \times 10^{-30} \text{ m}^3,$$

a value that is comparable to the electronic polarizabilities α_0 of molecules (cf. Table 5.1).

In very high fields or at sufficiently low temperatures such that $uE \gg kT$, a dipolar molecule will become completely aligned along the field. When this happens (e.g., water near a small ion), the concept of a molecule's orientational polarizability breaks down, but the electronic polarizability still applies.

5.3 INTERACTIONS BETWEEN IONS AND UNCHARGED MOLECULES

When a molecule is at a distance r away from an ion, the electric field of the ion

$$E = ze/4\pi\varepsilon_0\varepsilon r^2 \tag{5.8}$$

will induce in the molecule a dipole moment of

$$u_{\text{ind}} = \alpha E = \alpha ze/4\pi\varepsilon_0\varepsilon r^2. \tag{5.9}$$

☐   ● **WORKED EXAMPLE** ●

Question: Estimate the distance by which the electron cloud of a methane (CH$_4$) molecule is shifted relative to the centre of the molecule due to the presence of a sodium ion whose centre is 0.4 nm from the centre of the molecule.

Answer:
● For a monovalent ion such as Na$^+$, the electric field at a distance of 0.4 nm from its centre is $E = e/4\pi\varepsilon_0 r^2 = (1.602 \times 10^{-19})/(4 \times 3.142 \times 8.854 \times 10^{-12})(0.4 \times 10^{-9})^2 = 9.0 \times 10^9 \text{ V m}^{-1}$. The induced dipole moment on a methane molecule is therefore, using Table 5.1,

$$u_{\text{ind}} = \alpha_0 E = 4\pi\varepsilon_0(2.6 \times 10^{-30})(9.0 \times 10^9) = 2.60 \times 10^{-30} \text{ C m}$$

$$= 2.60 \times 10^{-30}/3.336 \times 10^{-30} = 0.78 \text{ D}.$$

From Eq. (5.2) this corresponds to a unit charge separation in the molecule of $l = u_{\text{ind}}/e = 0.016$ nm, which is about 8% of the molecular radius of methane (of 0.2 nm). However, if we consider that four or more of the hydrogen electrons in CH$_4$ may be displaced simultaneously this distance falls to 0.004 nm, or about 2% of the molecular radius.   ☐

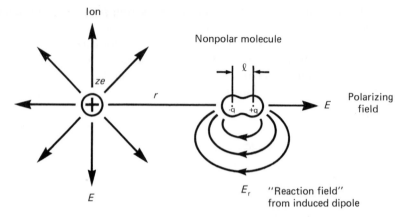

Fig. 5.2. A neutral (non-polar) molecule in a field E will acquire an induced dipole of moment $u_{ind} = ql = \alpha E$. The resulting force on the neutral molecule is therefore $F = q\Delta E$, where ΔE is the difference in E at either end of the dipole. Thus, $F = q(dE/dr)l = \alpha E(dE/dr)$, so that the interaction free energy is $w(r) = -\int F\,dr = -\frac{1}{2}\alpha E^2$. If the polarizing field E is due to an ionic charge, we obtain Eq. (5.12); if due to a dipole, we obtain Eq. (5.21).

From the above example it appears that induced dipole moments can be of order 1 Debye, which is quite large. We may therefore anticipate that the forces associated with induced dipole moments may likewise be quite large.

We shall now consider the interaction between an ion and an uncharged molecule (Fig. 5.2). The induced dipole will point away from the ion, if a cation, and towards the ion, if an anion. In either case this will lead to an attractive force between the ion and the polarized molecule. The reflected or 'reaction' field of the induced dipole acting back on the ion is, using Eq. (5.20),

$$E_r = -2u_{ind}/4\pi\varepsilon_0\varepsilon r^3 = -2\alpha E/4\pi\varepsilon_0\varepsilon r^3 = -2\alpha(ze)/(4\pi\varepsilon_0\varepsilon)^2 r^5, \quad (5.10)$$

so that the attractive force will be

$$F = -(ze)E_r = -2\alpha(ze)^2/(4\pi\varepsilon_0\varepsilon)^2 r^4 \quad (5.12)$$

$$= -\tfrac{1}{2}\alpha E^2, \quad (5.13)$$

where E is the polarizing field of the ion acting on the molecule (Eq. (5.8)).

It is important to note that this energy is half that expected for the interaction of an ion with a similarly aligned *permanent* dipole, which from Eq. (4.6) is

$$w(r) = -uE = -\alpha E^2. \quad (5.14)$$

The reason for this is that when a dipole moment is *induced* (rather than being *permanent* or *fixed*) some energy is taken up in polarizing the molecule. If we return to the example of Fig. 5.1, we see that this is the energy absorbed internally in displacing the positive and negative charges in the molecule and may be calculated by integrating the internal force F_{int} of Eq. (5.3) with respect to the charge separation from 0 to l, viz.,

$$w_{int}(r) = \int_0^l F_{int} \, dl = \int_0^l \frac{e^2 l \, dl}{4\pi\varepsilon_0 R^3} = \frac{(el)^2}{8\pi\varepsilon_0 R^3} = \frac{(\alpha E)^2}{2\alpha}$$

$$= \tfrac{1}{2}\alpha E^2. \tag{5.15}$$

Thus, a factor $+\tfrac{1}{2}\alpha E^2$ must be added to Eq. (5.14) to obtain the *free* energy for the ion-induced dipole interaction, i.e., Eq. (5.13). An alternative derivation of Eq. (5.13) is given in Fig. 5.2.

We may now insert the value for the total polarizability of polar molecules, $\alpha = \alpha_0 + u^2/3kT$, into Eq. (5.12) and obtain for the net ion-induced dipole interaction free energy

$$w(r) = -\frac{(ze)^2\alpha}{2(4\pi\varepsilon_0\varepsilon)^2 r^4} = -\frac{(ze)^2}{2(4\pi\varepsilon_0\varepsilon)^2 r^4}\left(\alpha_0 + \frac{u^2}{3kT}\right), \tag{5.16}$$

and we may note that the temperature-dependent term is identical to Eq. (4.13) derived earlier by a different method.

5.4 ION–SOLVENT MOLECULE INTERACTIONS AND THE BORN ENERGY

It is instructive to see how the ion-induced dipole interaction is related to the Born energy of an ion in a medium, which was previously discussed in Chapter 3. Let us first compute the total interaction energy of an ion in a medium with all the surrounding solvent molecules. For an ion of radius a, this can be calculated as before by integrating $w(r)$ of Eq. (5.12) or (5.16) over all of space:

$$\mu^i = \int_a^\infty w(r)\rho 4\pi r^2 \, dr = -\int_a^\infty \frac{\rho\alpha(ze)^2 4\pi r^2 \, dr}{2(4\pi\varepsilon_0\varepsilon)^2 r^4} = -\frac{\rho\alpha(ze)^2}{8\pi\varepsilon_0^2\varepsilon^2 a}, \tag{5.17}$$

where ρ is the number of solvent molecules per unit volume. To proceed further we have to make some connection between the molecular and continuum properties of the solvent. This requires us to find a relation between the molecular polarizability α and the dielectric constant ε of a medium. This is a very complex problem and not well understood. However, the value of $(\rho\alpha)/\varepsilon_0$ in the above equation may in a first approximation be associated with the *electric susceptibility* χ of a medium. This is the polarizability per unit volume of a medium and is related to the dielectric constant ε by

$$\chi = (\varepsilon - 1). \tag{5.18}$$

The change in free energy $\Delta\mu^i$ when an ion goes from a medium of dielectric constant ε_1 to one of ε_2 is therefore (since $d\chi = d\varepsilon$)

$$\Delta\mu^i = -\int_{\chi_1}^{\chi_2} \frac{(ze)^2}{8\pi\varepsilon_0\varepsilon^2 a}\, d\chi = -\int_{\varepsilon_1}^{\varepsilon_2} \frac{(ze)^2}{8\pi\varepsilon_0\varepsilon^2 a}\, d\varepsilon = -\frac{(ze)^2}{8\pi\varepsilon_0 a}\left[\frac{1}{\varepsilon_1} - \frac{1}{\varepsilon_2}\right],$$

$$\tag{5.19}$$

which is the change in Born energy, Eq. (3.8).

We have now seen how the Born energy—previously derived using a *continuum* analysis—can also be derived from a *molecular* approach. Furthermore, the molecular approach has the added advantage of providing insight into the limitations of the Born equation. Thus, for an ion in a polar solvent, we may expect the Born equation to be valid so long as the polarizabilities of the solvent molecules do not depend on their distance from the ion, i.e., so long as the polarizing field E and solvent dipole moment u are not so large that the Debye–Lengevin equation, Eq. (5.7), breaks down. We have already seen in Chapter 4 that for ions in a polar solvent, this condition is satisfied except for small or multivalent ions in very polar solvents such as water.

5.5   DIPOLE-INDUCED DIPOLE INTERACTIONS

The interaction between a polar molecule and a non-polar molecule is analogous to the ion-induced dipole interaction just discussed except that the polarizing field comes from a permanent dipole rather than a charge. For a fixed dipole u oriented at an angle θ to the line joining it to a polarizable molecule (Fig. 2.2), it may be shown that the magnitude of the electric field of the dipole acting on the molecule is

$$E = u(1 + 3\cos^2\theta)^{\frac{1}{2}}/4\pi\varepsilon_0\varepsilon r^3, \tag{5.20}$$

the interaction energy is therefore

$$w(r, \theta) = -\tfrac{1}{2}\alpha_0 E^2 = -u^2\alpha_0(1 + 3\cos^2\theta)/2(4\pi\varepsilon_0\varepsilon)^2 r^6. \tag{5.21}$$

For typical values of u and α_0, the strength of this interaction is not sufficient to mutually orient the molecules, as occurs in ion-dipole or strong dipole–dipole interactions. The effective interaction is therefore the angle-averaged energy. Since the angle average of $\cos^2\theta$ is $1/3$, Eq. (5.21) becomes

$$w(r) = -u^2\alpha_0/(4\pi\varepsilon_0\varepsilon)^2 r^6, \tag{5.22}$$

while more generally, for two different molecules each possessing a permanent dipole moment u_1 and u_2 and polarizabilities α_{01} and α_{02}, their net dipole-induced dipole energy is

$$w(r) = -\frac{[u_1^2\alpha_{02} + u_2^2\alpha_{01}]}{(4\pi\varepsilon_0\varepsilon)^2 r^6}. \tag{5.23}$$

This is often referred to as the *Debye interaction* or the *induction interaction*. It constitutes the second of three inverse sixth power contributions to the total *van der Waals* interaction energy between molecules. The first we have already encountered in the angle-averaged dipole–dipole or Keesom interaction, Eq. (4.14), which incidentally may also be obtained from Eq. (5.22) by replacing α_0 by $\alpha_{\text{orient}} = u^2/3kT$ so that for two dipoles u_1 and u_2, it gives the Keesom free energy:

$$w(r) = -\frac{u_1^2 u_2^2}{3(4\pi\varepsilon_0\varepsilon)^2 kTr^6}.$$

5.6 UNIFICATION OF POLARIZATION INTERACTIONS

Apart from the straight Coulomb interaction between two charges, all the other interactions we have considered have involved polarization effects, either explicitly for neutral molecules of polarizability α_0 or implicitly for rotating polar molecules that effectively behave as polarizable molecules of polarizability $\alpha = \alpha_0 + \alpha_{\text{orient}}$. For completeness, we may note that all these angle-averaged interactions may be expressed in one general equation; thus, for a charged polar molecule 1 interacting with a second polar molecule 2 we may write

$$w(r) = -\left(\frac{Q_1^2}{2r^4} + \frac{3kT\alpha_1}{r^6}\right)\frac{\alpha_2}{(4\pi\varepsilon_0\varepsilon)^2} \tag{5.24}$$

$$= -\left[\frac{Q_1^2}{2r^4} + \frac{3kT}{r^6}\left(\frac{u_1^2}{3kT} + \alpha_{01}\right)\right]\left(\frac{u_2^2}{3kT} + \alpha_{02}\right)\bigg/(4\pi\varepsilon_0\varepsilon)^2, \tag{5.25}$$

where Q_1 is the charge of the first molecule and $u_1, u_2, \alpha_{01}, \alpha_{02}$ are the dipole moments and electronic polarizabilities of the two molecules. Each of the six terms arising from the above equation may be identified (cf. Fig. 2.2) with a previous derivation (except for the small van der Waals $\alpha_{01}\alpha_{02}$ term discussed later). The unification of these various interactions is conceptually important, for it shows them all to be essentially polarization-type forces. If none of the molecules carries a net charge ($Q_1 = 0$), the above equation gives the Keesom-orientation and Debye-induction contributions to the total van der Waals force between two molecules. The third contributor to van der Waals forces—the *dispersion force*—will be introduced in the next chapter.

Equation (5.25) is also useful for rapidly estimating the relative strengths of charge, dipole, and electronic polarizability contributions in an interaction. Thus, for typical values of $Q_1 = e = 1.6 \times 10^{-19}$ C, $u_1 = 1D = 3.3 \times 10^{-30}$ C m, $\alpha_{01} = (4\pi\varepsilon_0)3 \times 10^{-30}$ m^3, and $r = 0.5$ nm, at $T = 300$ K the ratio of the three terms in the square brackets

$$(Q_1^2/2r^4):(u_1^2/r^6):(3kT\alpha_{01}/r^6)$$

is about 800:3:1. For water, however, because of its large dipole moment u and unusually small polarizability α_0, the ratio $u^2:3kT\alpha_0$ is about 20:1, i.e., the permanent-dipole-associated interactions of water always dominate over electronic polarization effects. Note, too, that the effectiveness of a dipolar interaction depends on $u^2/r^6 \sim (u/\sigma^3)^2$ rather than on the absolute value of the dipole moment. In other words, for a molecule to be considered as highly polar, it must have a high dipole moment *per molecular volume* and not simply a high u.

5.7 SOLVENT EFFECTS AND 'EXCESS POLARIZABILITIES'

The interaction between molecules or small particles in a solvent medium can be very different from that of isolated molecules in free space or in a gas. The presence of a suspending medium does more than simply reduce the interaction energy or force by a factor ε or ε^2, as might appear from the Eq. (5.25) and other equations derived earlier. First, the intrinsic dipole moment and polarizability of an isolated gas molecule may be different in the liquid state or when dissolved in a medium. This depends in a complicated way on its interactions with the surrounding solvent and can usually only be found by experiment. Second, as already discussed in Section 2.1, a dissolved molecule can move only by displacing an equal volume of solvent from its path; hence, the polarizability α in a medium must represent the *excess polarizability* of a molecule or particle over that of the solvent and must vanish when a dissolved particle has the same properties as the solvent. Qualitatively, we may say that if no electric field is reflected by a particle, it is 'invisible' in the solvent medium and consequently does not experience a force.

The problem of knowing the excess or effective polarizability can be approached by treating a dissolved molecule or a small particle as a separate dielectric medium of a given size and shape. This *continuum* approach has an obvious advantage since the dielectric constant of a medium is usually known. Accordingly, a molecule i may be modelled as a dielectric sphere of radius a_i and dielectric constant ε_i. Now in a medium of dielectric constant ε (Fig. 5.3a), such a dielectric sphere will be polarized by a field E and acquire an excess dipole moment given by (Landau and Lifshitz, 1963, p. 44)

$$u_{\text{ind}} = 4\pi\varepsilon_0\varepsilon\left(\frac{\varepsilon_i - \varepsilon}{\varepsilon_i + 2\varepsilon}\right)a_i^3 E$$

so that its effective or *excess polarizability* in the medium is

$$\alpha_i = 4\pi\varepsilon_0\varepsilon\left(\frac{\varepsilon_i - \varepsilon}{\varepsilon_i + 2\varepsilon}\right)a_i^3 = 3\varepsilon_0\varepsilon\left(\frac{\varepsilon_i - \varepsilon}{\varepsilon_i + 2\varepsilon}\right)v_i, \tag{5.26}$$

where $v_i = \frac{4}{3}\pi a_i^3$ is the volume of the molecule or sphere. This equation shows that for a dielectric sphere of high ε_i in free space (where $\varepsilon = 1$) its polarizability is roughly $\alpha_i \approx 4\pi\varepsilon_0 a^3$, as previously found for a simple one-electron atom (Eq. (5.4)). Further, if $\varepsilon > \varepsilon_i$, the polarizability is negative, implying that the direction of the induced dipole is opposite to that in free space.

If we substitute Eq. (5.26) into Eq. (5.24), we obtain the important result

$$w(r) = -\left[\frac{Q_1^2}{8\pi\varepsilon_0\varepsilon r^4} + \frac{3kT}{r^6}\left(\frac{\varepsilon_1 - \varepsilon}{\varepsilon_1 + 2\varepsilon}\right)a_1^3\right]\left(\frac{\varepsilon_2 - \varepsilon}{\varepsilon_2 + 2\varepsilon}\right)a_2^3, \tag{5.27}$$

which allows us to conclude the following:

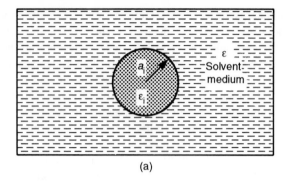

(a)

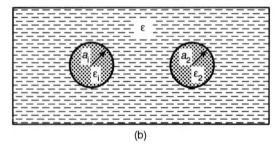

(b)

Fig. 5.3. A dissolved molecule or small solute particle can be modelled as a sphere of radius a_i and dielectric constant ε_i. Its total polarizability in a medium of dielectric constant ε is $\alpha_i = 4\pi\varepsilon_0\varepsilon a_i^3(\varepsilon_i - \varepsilon)/(\varepsilon_i + 2\varepsilon)$.

(i) The net force between dissolved molecules or small particles in a medium (Fig. 5.3b) can be zero, attractive, or repulsive, depending on the relative magnitudes of ε_1, ε_2 and ε.

(ii) Ions will be attracted to dissolved molecules of high dielectric constant (highly polar molecules where $\varepsilon_2 > \varepsilon$) but repelled from molecules of low dielectric constant (non-polar molecules where $\varepsilon_2 < \varepsilon$).

(iii) The interaction between any two identical uncharged molecules ($\varepsilon_1 = \varepsilon_2$) is always attractive regardless of the nature of the suspending medium. Interestingly, two microscopic air bubbles also attract each other in a liquid.

This approach will generally predict the right qualitative trends, but it is quantitatively somewhat model-dependent in that it treats solute molecules as if they were a uniform medium having bulk dielectric properties. It is essentially a continuum treatment where the molecular properties only appear in determining the molecular radius or volume. This may be valid for larger molecules, macromolecules, and small particles in solution but may fail for small molecules, especially when close together. Let us therefore end this chapter by looking into the reasonableness of treating a molecule as if it were a dielectric sphere. From Eq. (5.26) we see that for isolated molecules in the gas phase ($\varepsilon = 1$), their *total* polarizability should be given by

$$\frac{\alpha}{(4\pi\varepsilon_0)} = \left(\frac{\varepsilon - 1}{\varepsilon + 2}\right)\frac{3v}{4\pi}, \tag{5.28}$$

where ε is now the static dielectric constant of the molecules, assumed to be the same as that of the condensed state (e.g., the liquid state). Equation (5.28) is known as the Clausius–Mossotti equation.

If we are interested in the *electronic* polarizability α_0, then the value of ε in Eq. (5.28) is that measured in the visible range of frequencies and equals n^2, where n is the refractive index of the medium. Thus, for the *electronic* polarizability, we may write

$$\frac{\alpha_0}{(4\pi\varepsilon_0)} = \left(\frac{n^2 - 1}{n^2 + 2}\right)\frac{3v}{4\pi}, \tag{5.29}$$

which is known as the Lorenz–Lorentz equation.

Table 5.2 shows the experimental values for the electronic polarizabilities α_0 and dipole moments u of isolated molecules in the gas phase, from which their values of α may be obtained from Eq. (5.7):

$$\alpha = \alpha_0 + u^2/3kT.$$

Table 5.2 also shows the polarizabilities α_0 and α as calculated from Eqs (5.29) and (5.28) in terms of the purely bulk properties of the liquids: the refractive index n, the dielectric constant ε, the molecular weight M and mass density ρ, from which the volume occupied per molecule is given by

$$v = M/\rho N_0, \tag{5.30}$$

where N_0 is Avogadro's constant.

It is evident that for all the molecules listed in Table 5.2 their electronic polarizabilities α_0 are excellently described by Eq. (5.29). In contrast, the total polarizabilities α are well described by Eq. (5.28) only for weakly polar molecules such as $CHCl_3$ and C_6H_5OH. For small highly polar molecules such as H_2O and $(CH_3)_2CO$, the agreement is not good but improves for progressively larger molecules (cf. $CH_3OH \rightarrow C_2H_5OH \rightarrow C_6H_{13}OH$ and C_6H_5OH) as expected. It is not possible to ascertain whether the lack of agreement for highly polar molecules is due to the inapplicability of Eq.

TABLE 5.2 Molecular polarizabilities as determined from molecular or bulk properties[a,b]

Molecule	Polarizabilities deduced from gas (molecular) properties			Polarizabilities from condensed phase (continuum) properties					
	u_{gas} meas. (D)[c]	$\dfrac{\alpha_a}{(4\pi\varepsilon_0)}$ measured (10^{-30} m³)	$\dfrac{\alpha}{(4\pi\varepsilon_0)}$ calculated from $\alpha = \alpha_0 + \dfrac{u^2}{3kT}$ (10^{-30} m³)	M (10^3 kg mol⁻¹)	ρ (10^3 kg m⁻³)	n	ε	$\dfrac{\alpha_0}{(4\pi\varepsilon_0)}$ from $\left(\dfrac{n^2-1}{n^2+2}\right)\dfrac{3M}{4\pi N_0\rho}$ (10^{-30} m³)	$\dfrac{\alpha}{(4\pi\varepsilon_0)}$ from $\left(\dfrac{\varepsilon-1}{\varepsilon+2}\right)\dfrac{3M}{4\pi N_0\rho}$ (10^{-30} m³)
CCl₄ Carbon tetrachloride	0	10.5	10.5	153.8	1.59	1.460	2.2	10.5	11.2
C₆H₆ Benzene	0	10.3	10.3	78.1	0.88	1.601	2.3	10.4	10.5
CHCl₃ Chloroform	1.06	8.2	17.5 (20°C) 21.1 (−63°C)	119.4	1.48	1.446	4.8 (20°C) 6.8 (−63°C)	8.5	17.9 (20°C) 21.1 (−63°C)
H₂O Water	1.85	1.5	29.7	18.0	1.00	1.333	80	1.5	6.9
(CH₃)₂CO Acetone	2.85	6.4	73.4	58.1	0.79	1.359	21	6.4	25.3
CH₃OH Methanol	1.69	3.2	26.8	32.0	0.79	1.329	33	3.3	14.7
C₂H₅OH Ethanol	1.69	5.2	28.8	46.1	0.79	1.361	26	5.1	20.7
n-C₆H₁₃OH Hexanol	1.69	12.5	36.1	102.2	0.81	1.418	13	12.6	40.0
C₆H₅OH Phenol	1.45	11.2	26.4	94.1	1.07	1.551	10 (60°)	11.1	26.1

a All values at 20°C unless stated otherwise.

b α_0, electronic polarizability; α, total polarizability.

c 1 D = 3.336 × 10^{-30} C m.

(5.28) or to a changed dipole moment of the molecules in the liquid state. However, we may safely conclude that Eqs (5.28) and (5.29) serve as quantitatively reliable equations for determining the polarizabilities of all but very small highly polar molecules, and we may expect that Eq. (5.26) for the excess polarizability and Eq. (5.27) for the free energy should be likewise applicable.

But we are still not in a position to estimate the strength of the *total* interaction between neutral or polar molecules. There is one final contribution to the total force that must be considered before we can do that. This is the van der Waals dispersion force.

PROBLEMS AND DISCUSSION TOPICS

5.1 When an electric field is applied across a liquid containing large colloidal particles it is found that they align as shown in Fig. 5.4. Explain this phenomenon and suggest one condition when the particles will not align (ignore gravitational effects).

5.2 Derive Eq. (5.20).

5.3 What does Table 5.2 tell us about the ability of non-polar, polar and hydrogen bonding molecules to rotate freely in the gas phase compared to their mobility in the condensed phase?

5.4 The following room temperature properties of liquid chloroform ($CHCl_3$) are given in Table 5.2: electronic polarizability α_0, dipole moment u, refractive index n and density ρ. Assuming that none of these changes with temperature, calculate the dielectric constant of liquid chloroform at 20°C and at its freezing point of −63.5°C, and compare your answer with the measured values of $\varepsilon = 4.8$ and 6.8, respectively. What would you expect ε to be at a temperature just below the freezing point, i.e. for solid chloroform?

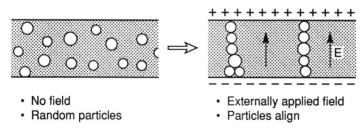

• No field   • Externally applied field
• Random particles   • Particles align

Fig. 5.4.

5.5 According to the Debye–Langevin and Clausius–Mossotti equations (Sections 5.2 and 5.7) a high dielectric constant arises for molecules with large dipole moments. This appears to be the case for water. However, when water freezes the dielectric constant of crystalline ice is actually higher than that of the liquid, even though one would expect a substantial decrease due to the restricted rotation of the now frozen molecules. Does this mean that another polarization mechanism must be operating with water (other than the electronic and rotating dipole mechanisms described in this chapter)? If so, propose a physically realistic theory for this phenomenon.

CHAPTER 6

VAN DER WAALS FORCES

6.1 ORIGIN OF THE VAN DER WAALS DISPERSION FORCE BETWEEN NEUTRAL MOLECULES: THE LONDON EQUATION

The various types of physical forces so far described are fairly easy to understand since they arise from straightforward electrostatic interactions involving charged or dipolar molecules. But there is a further type of force, which—like the gravitational force—acts between *all* atoms and molecules, even totally neutral ones such as helium, carbon dioxide and hydrocarbons. These forces have been variously known as dispersion forces, London forces, charge-fluctuation forces, electrodynamic forces, and induced-dipole–induced-dipole forces. We shall refer to them as *dispersion forces* since it is by this name that they are most widely known. The origin of this name has to do with their relation to the dispersion of light in the visible and UV regions of the spectrum, as we shall see. The literature on this subject is quite voluminous, and the reader is referred to books and reviews by London (1937), Hirschfelder *et al.* (1954), Moelwyn-Hughes (1961), Margenau and Kestner (1971), Israelachvili and Tabor (1973), Israelachvili (1974), and Mahanty and Ninham (1976).

Dispersion forces make up the third and perhaps most important contribution to the total van der Waals force between atoms and molecules, and because they are always present (in contrast to the other types of forces that may or may not be present depending on the properties of the molecules) they play a role in a host of important phenomena such as adhesion, surface tension, physical adsorption, wetting, the properties of gases, liquids and thin films, the strengths of solids, the flocculation of particles in liquids, and the structures of condensed macromolecules such as proteins and polymers. Their main features may be summarized as follows:

(1) They are long-range forces and, depending on the situation, can be effective from large distances (greater than 10 nm) down to interatomic spacings (about 0.2 nm).

(2) These forces may be repulsive or attractive, and in general the dispersion force between two molecules or large particles does not follow a simple power law.

(3) Dispersion forces not only bring molecules together but also tend to mutually align or orient them, though this orienting effect is usually weak.

(4) The dispersion interaction of two bodies is affected by the presence of other bodies nearby. This is known as the *non-additivity* of an interaction. ⌐

Dispersion forces are quantum mechanical in origin and amenable to a host of theoretical treatments of varying complexity, the most rigorous of which would take us into the world of quantum electrodynamics. Their origin may be understood intuitively as follows: for a non-polar atom such as helium, the time average of its dipole moment is zero, yet at any instant there exists a finite dipole moment given by the instantaneous positions of the electrons about the nuclear protons. This instantaneous dipole generates an electric field that polarizes any nearby neutral atom, inducing a dipole moment in it. The resulting interaction between the two dipoles gives rise to an instantaneous attractive force between the two atoms, and the time average of this force is finite. For a simple semiquantitative understanding of how these forces arise, let us consider the following model, based on the interaction between two Bohr atoms.

In the Bohr atom an electron is pictured as orbiting around a proton. The smallest distance between the electron and proton is known as the *first Bohr radius* a_0 and is the radius at which the Coulomb energy $e^2/4\pi\varepsilon_0 a_0$ is equal to $2h\nu$, that is,

$$a_0 = e^2/2(4\pi\varepsilon_0)h\nu = 0.053 \text{ nm}, \tag{6.1}$$

where h is the Planck constant and ν the orbiting frequency of the electron. For a Bohr atom, $\nu = 3.3 \times 10^{15} \text{ s}^{-1}$, so that $h\nu = 2.2 \times 10^{-18}$ J. This is the energy of an electron in the first Bohr radius and is equal to the energy needed to ionize the atom—the *first ionization potential, I*.

The Bohr atom has no permanent dipole moment. However, at any instant there exists an instantaneous dipole of moment

$$u = a_0 e$$

whose field will polarize a nearby neutral atom giving rise to an attractive interaction that is entirely analogous to the dipole-induced dipole interaction discussed in Chapter 5. The energy of this interaction in vacuum will therefore be given by Eq. (5.22) as

$$w(r) = -u^2\alpha_0/(4\pi\varepsilon_0)^2 r^6 = -(a_0 e)^2\alpha_0/(4\pi\varepsilon_0)^2 r^6,$$

where α_0 is the electronic polarizability of the second Bohr atom, which from Eq. (5.4) is approximately $4\pi\varepsilon_0 a_0^3$. Using this expression for α_0 and Eq. (6.1) for a_0, we immediately find that the above interaction energy can be written as

$$w(r) \approx -\alpha_0^2 h\nu/(4\pi\varepsilon_0)^2 r^6. \tag{6.2}$$

Except for a numerical factor, Eq. (6.2) is the same as that derived by London in 1930 using quantum mechanical perturbation theory. London's famous expression for the dispersion interaction energy between two identical atoms or molecules is (London, 1937)

$$w(r) = \frac{-C_{\text{disp}}}{r^6} = -\tfrac{3}{4}\,\alpha_0^2 h\nu/(4\pi\varepsilon_0)^2 r^6 = -\tfrac{3}{4}\,\alpha_0^2 I/(4\pi\varepsilon_0)^2 r^6, \tag{6.3}$$

and for two dissimilar atoms,

$$w(r) = -\tfrac{3}{2}\frac{\alpha_{01}\alpha_{02}}{(4\pi\varepsilon_0)^2 r^6}\frac{h\nu_1\nu_2}{(\nu_1 + \nu_2)} = -\tfrac{3}{2}\frac{\alpha_{01}\alpha_{02}}{(4\pi\varepsilon_0)^2 r^6}\frac{I_1 I_2}{(I_1 + I_2)}. \tag{6.4}$$

London's equation has since been superseded by more exact, though more complicated expressions (see Section 6.5), but it can be relied upon to give fairly accurate values, though these are usually lower than more rigorously determined ones.

From the above simple model we see that while dispersion forces are quantum mechanical (in determining the instantaneous, but fluctuating, dipole moments of neutral atoms), the ensuing interaction is still essentially electrostatic—a sort of quantum mechanical polarization force. And we may further note that the $1/r^6$ distance dependence is the same as that for the other two polarization interactions (the Keesom and Debye forces) that contribute to the net van der Waals force, discussed in Section 5.6. But before we consider these three interactions collectively let us first investigate the nature of dispersion forces.

6.2 STRENGTH OF DISPERSION FORCES: VAN DER WAALS SOLIDS AND LIQUIDS

To estimate the strength of the dispersion energy, we may consider two atoms or small molecules with $\alpha_0/4\pi\varepsilon_0 \approx 1.5 \times 10^{-30}\,\text{m}^3$ and $I = h\nu \approx$

2×10^{-18} J (a typical ionization potential in the UV). From Eq. (6.3) we find that for two atoms in contact at $r = \sigma \approx 0.3$ nm, $w(\sigma) = -4.6 \times 10^{-21}$ J $\approx 1kT$. This is very respectable energy, considering that the interaction appears at first sight to spring up from nowhere. But when we recall that the inducing (instantaneous) dipole moment of even a small hydrogen (Bohr) atom is of order $a_0 e \approx 2.4$ D, we can appreciate why the dispersion interaction is by no means negligible. Thus, [while very small non-polar atoms and molecules such as argon and methane are gaseous at room temperature and pressure, larger molecules such as hexane and higher molecular weight hydrocarbons are liquids or solids, held together solely by dispersion forces. The solids are referred to as *van der Waals solids*, and they are characterized by having weak undirected 'bonds', and therefore low melting points and low latent heats of melting.]

For spherically symmetrical inert molecules such as neon, argon and methane, the van der Waals solids they form at low temperatures are close-packed structures with 12 nearest neighbours per atom. Their lattice energy (12 shared 'bonds' or six full 'bonds' per molecule) is therefore approximately $6w(\sigma)$ per molecule, though if the attractions of more distant neighbours are also included the factor of six rises to 7.22. The expected *molar lattice energy* or *cohesive energy* of a van der Waals solid is therefore

$$U \approx 7.22 \ N_0 \left[\frac{3\alpha_0^2 \ hv}{4(4\pi\varepsilon_0)^2\sigma^6} \right], \tag{6.5}$$

where σ is now the equilibrium interatomic distance in the solid. Thus, for argon, since $\alpha_0/(4\pi\varepsilon_0) = 1.63 \times 10^{-30}$ m^3, $I = hv = 2.52 \times 10^{-18}$ J, and $\sigma = 0.376$ nm, we obtain $U \approx 7.7$ kJ mol^{-1}. This may be compared with the latent heat of melting plus vaporization for argon of $L_m + L_v = 7.7$ kJ mol^{-1}, which is approximately equal to the latent heat of sublimation or the cohesive energy U (ignoring the small PV term at this temperature).

Table 6.1 shows the calculated and experimental values for the cohesive energies of a number of van der Waals solids composed of small spherical atoms or molecules. The good agreement obtained is to some extent fortuitous since the computed values neglect

(i) the very short-range stabilizing repulsive forces (Chapter 7) that can *lower* the final binding energy at contact by up to 50%, and

(ii) other attractive forces, such as arise from other absorption frequencies and fluctuating quadrupole and higher multipole interactions, which can *raise* the final binding by up to 50% (see legend to Table 6.1).

These two opposing effects partially cancel each other so that the final results look more impressive than they deserve to.

TABLE 6.1 Strength of dispersion interaction between spherical non-polar molecules

Interacting molecules	Molecular diameter σ (nm) (From Fig. 7.1)	Polarizability $\alpha_0/4\pi\varepsilon_0$ (10^{-30} m³)	Ionization potential $I = h\nu_I$ (eV)[b]	London constant $C_{disp} = 3\alpha_0^2 h\nu_I/4(4\pi\varepsilon_0)^2$ (10^{-79} J m⁶)		Molar cohesive energy U (kJ mol⁻¹)		Boiling point T_B (K)	
				Theoretical Eq. (6.3)	Measured from gas law Eq. (6.14)[a]	Theoretical Eq. (6.5)	Measured $L_m + L_v$ (approx.)	Theoretical $\dfrac{3\alpha_0^2 h\nu_I}{4(4\pi\varepsilon_0)^2 \sigma^6 (1.5k)}$	Measured
Ne–Ne	0.308	0.39	21.6	3.9	3.8	2.0	2.1	22	27
Ar–Ar	0.376	1.63	15.8	50	45	7.7	7.7	85	87
CH_4–CH_4	0.400	2.60	12.6	102[c]	101[c]	10.9	9.8	121	112
Xe–Xe	0.432	4.01	12.1	233	225	15.6	14.9	173	165
CCl_4–CCl_4	0.550	10.5	11.5	1520	2960	23.9	32.6	265	350

[a] van der Waals constants a and b taken from the *Handbook of Chemistry and Physics*, CRC Press, 56th ed.

[b] 1 eV = 1.602×10^{-19} J.

[c] As an example of the reliability of the approximate equations for C_{disp}, Eqs (6.3) and (6.14), the most recent *ab initio* calculation for CH_4 (Fowler et al., 1989) gives 113×10^{-79} J m⁶, which is about 10% higher than the theoretical value given here. The most reliable experimental value, based on a number of different types of measurements (Thomas and Meath, 1977) is $C_{disp} = 124 \times 10^{-79}$ J m⁶, which is about 20% higher than the value given here.

☐   ● **WORKED EXAMPLE** ●

Question: Many atoms or small molecules have ionization potentials I close to 2×10^{-18} J. If their polarizability can be modelled in terms of the Bohr atom (Section 6.1) show that the strength of a typical van der Waals 'bond' is always approximately a few kT at room temperature, irrespective of the size or polarizability of the molecules.

Answer:

● The strength of a van der Waals bond is given by $w(r)$ at a separation $r = 2a$, where a is the molecular radius. From the Bohr atom model of polarizability $\alpha = 4\pi\varepsilon_0 a^3$, so that putting $r = 2a$ into the London equation, Eq. (6.3), we obtain $w(r = 2a) \approx \frac{3}{4} a^6 I/(2a)^6 \approx \frac{3}{4}(I/64) \approx \frac{3}{4}$ $(2 \times 10^{-18})/64 \approx 2 \times 10^{-20}$ J, which is a few kT and is independent of a or α.   ☐

For larger spherical molecules, with diameters greater than about 0.5 nm, the simple London equation can no longer be used. Clearly, as molecules grow in size their centre-to-centre distance ceases to have any significance as regards the strength of the cohesion energy. This is because the dispersion force no longer acts between the centres of the molecules, but between the centres of electronic polarization within each molecule, which, as we saw in Chapter 5, are located at the covalent bonds. Thus, for CCl_4, the calculated value for the cohesion energy is too small (Table 6.1) because the distance between the polarizable electrons of the two molecules is now significantly less than the equilibrium intermolecular separation of 0.55 nm. In Chapter 11 we shall investigate the interactions between spheres and particles whose radii are much larger than interatomic spacings.

Likewise, the simple London equation or Eq. (6.5) cannot be applied to asymmetric (non-spherical) molecules such as alkanes, polymers, and cyclic or planar molecules. To compute the binding energies within or between such complex molecules the molecular packing in the solid or liquid must be known (which, of course, transforms any theoretical endeavour from the predictive to the confirmatory), and the different contributions arising from different parts of the molecules must be considered separately. Under such conditions the exact dispersion interaction is difficult to compute, but some simplifying assumptions can often be made to arrive at reasonable working models. Let us consider one such model for normal alkanes, of general formula $CH_3-(CH_2)_n-CH_3$, where each molecule may be considered as a cylinder of diameter $\sigma = 0.40$ nm composed of CH_2 groups spaced linearly at intervals of $l = 0.127$ nm, corresponding to the normal CH_2-CH_2 distance along an alkane chain. If we now consider one such molecule surrounded by six

TABLE 6.2 Strength of dispersion interaction between non-spherical alkane molecules

Molecule	Number of CH$_2$ groups n	Molar cohesive energy (kJ mol^{-1})	
		Theoretical Eq. (6.7)	Measured $L_m + L_v$
CH$_4$	0	9.8 (measured)	9.8
C$_6$H$_{14}$	5	44.3	45.0
C$_{12}$H$_{26}$	11	85.7	86.1
C$_{18}$H$_{38}$	17	127.1	125.9

close-packed neighbouring cylinders, we may sum the dispersion energy of any one CH$_2$ group in the central molecule with all the CH$_2$ groups in the six surrounding molecules (similar to the lattice sum carried out in Section 3.4 for the ionic lattice energy). Thus, there will be six CH$_2$ groups at $r = \sigma$, 12 at $r = [\sigma^2 + l^2]^{1/2}$, 12 at $r = [\sigma^2 + (2l)^2]^{1/2}$, and so on. The molar cohesive energy per CH$_2$ group will therefore be given by the following rapidly converging series:

$$U = \frac{3\alpha_0^2 h\nu}{4(4\pi\varepsilon_0)^2}\left[\frac{6}{\sigma^6} + \frac{12}{[\sigma^2 + l^2]^3} + \frac{12}{[\sigma^2 + (2l)^2]^3} + \ldots\right]\frac{N_0}{2}. \quad (6.6)$$

Now for CH$_2$ groups, $\alpha_0/4\pi\varepsilon_0 = 1.84 \times 10^{-30}$ m^3, and $h\nu = 1.67 \times 10^{-18}$ J, so that we obtain

$$U \approx 6.9 \text{ kJ mol}^{-1} \text{ per CH}_2 \text{ group.} \quad (6.7)$$

Table 6.2 shows the experimental values for the latent heats of melting plus vaporization of alkanes together with the computed values based on Eq. (6.7). Experimentally, it is found that for straight chained alcohols, carboxylic acids, amides, esters, etc., the cohesive energy increases by between 6 and 7.5 kJ mol^{-1} per added CH$_2$ group. These good agreements between theory and experiment show two important aspects of dispersion forces: their non-directionality and near-additivity, though for interactions across a medium the dispersion force can be very non-additive (see Sections 6.6 and 6.8).

6.3   VAN DER WAALS EQUATION OF STATE

Let us now see how the van der Waals interaction potential between two molecules can be related to the constants a and b in the van der Waals equation of state

$$(P + a/V^2)(V - b) = RT. \tag{6.8}$$

We shall consider the molecules to be *hard spheres* of diameters σ, whose pair interaction energy is given by

$$w(r) = -\frac{3\alpha_0^2 h\nu}{4(4\pi\varepsilon_0)^2 r^6} = -\frac{C}{r^6}, \qquad r > \sigma,$$

$$= \infty, \qquad\qquad\qquad r < \sigma. \tag{6.9}$$

In Section 2.3 we found that the constant a is given by

$$a = 2\pi C/(n-3)\sigma^{n-3} = 2\pi C/3\sigma^3 \qquad \text{for } n = 6.$$

This value, however, was derived for the case where the van der Waals equation was expressed in terms of molecular parameters (Eq. (2.10)): $(P + a/v^2)(v - b) = kT$. Since $v = V/N_0$ we see that when V is the molar volume of the gas then

$$a = 2\pi N_0^2 C/3\sigma^3. \tag{6.10}$$

The London dispersion force coefficient C is therefore related to the van der Waals constant a by

$$C = 3\sigma^3 a/2\pi N_0^2. \tag{6.11}$$

The constant b is obtained from the volume unavailable for the molecules to move in, the 'excluded volume' per mole. Since one molecule cannot approach closer than σ to another, the excluded volume for the pair is $\frac{4}{3}\pi\sigma^3$, or $\frac{2}{3}\pi\sigma^3$ per molecule as previously derived in Section 2.3. Thus,

$$b = \frac{2}{3}\pi N_0 \sigma^3 \tag{6.12}$$

per mole, which is four times the molar volume of the molecules. The molecular diameter is therefore given by

$$\sigma = (3b/2\pi N_0)^{1/3}. \tag{6.13}$$

Combining the above with Eq. (6.11) we finally obtain for the interaction constant:

$$C = 9ab/4\pi^2 N_0^3 = 1.05 \times 10^{-76} \, ab \, \text{J m}^6, \tag{6.14}$$

where a is in $\text{dm}^6 \, \text{atm} \, \text{mol}^{-2}$ and b in $\text{dm}^3 \, \text{mol}^{-1}$. For example, for methane, CH_4, the experimental values are $a = 2.25 \, \text{dm}^6 \, \text{atm} \, \text{mol}^{-2}$, and $b = 0.0428 \, \text{dm}^3 \, \text{mol}^{-1}$, giving $C = 101 \times 10^{-79} \, \text{J m}^6$, which is in remarkably good agreement with the value for C of $3\alpha_0^2 h v/4(4\pi\varepsilon_0)^2 = 102 \times 10^{-79} \, \text{J m}^6$ calculated on the basis of the London equation. Table 6.1 also shows how good this agreement is for the other molecules listed, except for the largest molecule CCl_4, which, as already discussed, must have a stronger interaction than can be accounted for by applying the London equation between molecular centres.

For molecules constrained to interact on a surface, as occurs on adsorption and in surface monolayers, there is an analagous equation to the van der Waals equation of state. This two-dimensional analogue may be written as

$$(\Pi + a/A^2)(A - b) = RT, \tag{6.15}$$

where Π is the external surface pressure (in N m^{-1}), A the area, and a and b are constants as before. It is easy to verify that for an intermolecular pair potential of the form $w(r) = -C/r^6$, the constants a and b now become

$$a = \pi C N_0^2/4\sigma^4, \qquad b = \tfrac{1}{2}\pi N_0 \sigma^2. \tag{6.16}$$

6.4 GAS–LIQUID AND LIQUID–SOLID PHASE TRANSITIONS

Both the three-dimensional and two-dimensionl van der Waals equations of state can be applied to more complex systems, for example, to the interactions of small colloidal particles in a liquid and to the interactions of surfactant molecules on the surface of water or at an oil–water interface (Pallas and Pethica, 1985, 1987). Both equations predict the existence of a gas–liquid coexistence regime at some particular pressure so long as T is below the critical temperature. This first-order gas–liquid *phase transition* is primarily due to the attractive forces between the molecules or particles.

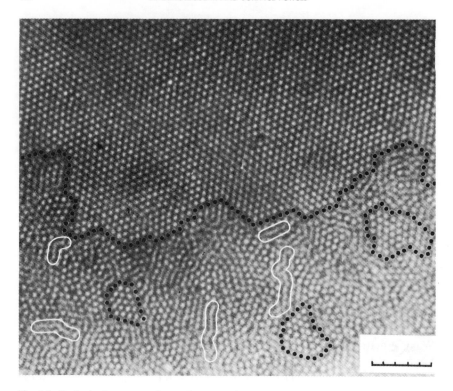

Fig. 6.1. Optical microscope views of 400 nm diameter colloidal particles interacting via a purely *repulsive* short-range force in aqueous solutions (scale bar = 5 μ). (a) Liquid–solid coexistence regime, i.e., a two-phase region, occurs at intermediate particle densities. In the ordered solid phase the particles are in rapid thermal (Brownian) motion but are constrained to remain within their lattice sites (i.e., the motion is highly localized). In the disordered liquid phase the particles can traverse over large distances and appear to do so in a snake-like fashion, known as 'reptation' (see the smeared out lines within the liquid domain arising from the finite exposure time). Note the diffuse nature of the solid–liquid 'interface' (dotted line), and bear in mind that it is continually fluctuating and will have a completely different shape a short time later. Molecular dynamics simulations indicate that the same concepts and behaviour apply at the molecular level (Glaser and Clark, 1990).

For molecules or particles that interact as hard spheres, or between which there is a purely *repulsive* force, the constant a in Eqs (6.8) and (6.15) becomes zero or negative. The resulting equations then predict a monotonic decrease of V with P, or A with Π, with no gas-to-liquid transition (i.e., no boiling temperature). However, at sufficiently high pressures where the density approaches the close-packing density there is always a liquid-to-solid transition from a disordered (liquid) state to an ordered crystalline (solid) state. This type of transition can arise even in the absence of any attractive

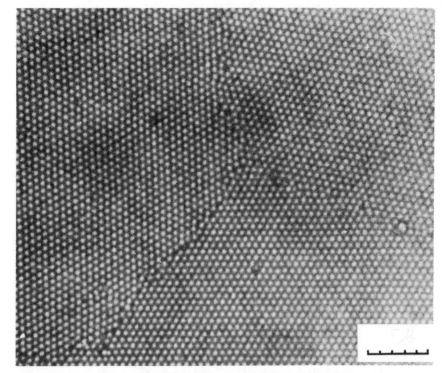

Fig. 6.1 (*continued*). (b) At a higher particle density there is only a single solid phase. However, in real systems there are grain boundaries between solid domains or grains, where defects and impurity molecules tend to collect and where the molecular motion is either more restricted (more solid-like) or, as is the case here, less restricted (more liquid-like). (Micrographs courtesy of Sei Hachisu; see also Okamoto and Hachisu, 1977.)

forces, when it is known as an *Alder transition* (Alder *et al.*, 1968). Alder transitions are intimately related to the excluded volume and geometry of molecules or particles, and arise because the only way these can get closer together is by going from a disordered or random liquid-like configuration to an ordered solid-like one (Fig. 6.1). The different roles played by attractive and repulsive forces, and of molecular shape, in determining the melting and boiling points of substances is further discussed in Section 7.4.

6.5 VAN DER WAALS BETWEEN POLAR MOLECULES

Three distinct types of force contribute to the total long-range interaction between polar molecules, collectively known as the van der Waals force:

these are the *induction* force, the *orientation* force and the *dispersion* force, each of which has an interaction free energy that varies with the inverse sixth power of the distance. Thus, for two dissimilar polar molecules, we have

$$w_{\text{VDW}}(r) = -C_{\text{VDW}}/r^6 = -[C_{\text{ind}} + C_{\text{orient}} + C_{\text{disp}}]/r^6$$

$$= -\left[(u_1^2\alpha_{02} + u_2^2\alpha_{01}) + \frac{u_1^2 u_2^2}{3kT} + \frac{3\alpha_{01}\alpha_{02}h\nu_1\nu_2}{2(\nu_1 + \nu_2)} \right] \bigg/ (4\pi\varepsilon_0)^2 r^6.$$

$$(6.17)$$

Table 6.3 shows how these three forces contribute to the net van der Waals energies of some molecules. The table reveals some interesting and important properties of van der Waals interactions.

(i) *Dominance of dispersion forces.* Dispersion forces generally exceed the dipole-dependent induction and orientation forces except for small highly polar molecules, such as water. The relative unimportance of dipolar forces is clearly seen in the hydrogen halides: as we go from HCl to HI the strength of the total van der Waals interaction increases even as the dipole moments diminish. Note, too, that in the interaction between two dissimilar molecules of which one is non-polar, the van der Waals energy is almost completely dominated by the dispersion contribution.

(ii) *Comparisons with experimental data.* The agreements between the computed (theoretical) values for C_{VDW} and those obtained from the gas law coefficients a and b are surprisingly good, even for NH_3 and H_2O. It is also possible to estimate the molar cohesive energies of polar molecules (London, 1937) though the agreement is not as good as for the non-polar spherical molecules listed in Table 6.1 partly because the effective diameters σ of polar (and therefore asymmetrical) molecules are not well defined. For example, for CH_3Cl of molecular diameter $\sigma \approx 0.43$ nm, using Eq. (6.5) we obtain $U \approx 29$ kJ mol^{-1} compared to the experimental value for $L_m + L_v$ of ~ 26 kJ mol^{-1}, while for water ($\sigma = 0.28$ nm) with only four nearest neighbours per molecule (see Chapter 8) we find $U \approx 70$ kJ mol^{-1} compared to the measured value of ~ 50 kJ mol^{-1}.

(iii) *Interactions of dissimilar molecules.* The van der Waals interaction energy between two dissimilar molecules A and B is usually intermediate between the values for $A-A$ and $B-B$. In fact, the coefficient C_{VDW} for $A-B$ is often close to the geometric mean of $A-A$ and $B-B$. Thus, for Ne–CH$_4$, the geometric mean (see Table 6.3) is $\sqrt{4 \times 102} = 20$, which may be compared with the computed value of 19, while for HCl–HI we obtain $\sqrt{123 \times 372} = 214$ compared to the computed value of 205. This procedure affords a convenient way of estimating the van der Waals interactions between

TABLE 6.3 Induction, orientation and dispersion free energy contributions to the total van der Waals energy in vacuum for various pairs of molecules at 293 K

Interacting molecules	Electronic polarizability $\dfrac{\alpha_0}{4\pi\varepsilon_0}$ (10^{-30} m³)	Permanent dipole moment u (D)[a]	Ionization potential $I = h\nu_I$ (eV)[b]	C_{ind} $\dfrac{2u^2\alpha_0}{(4\pi\varepsilon_0)^2}$	C_{orient} $\dfrac{u^4/3kT}{(4\pi\varepsilon_0)^2}$	C_{disp} $\dfrac{3\alpha_0^2 h\nu_I}{4(4\pi\varepsilon_0)^2}$	Total vdW energy C_{vdW} Theoretical Eq. (6.17)	Total vdW energy C_{vdW} From Gas law Eq. (6.14)	Dispersion energy contribution to total (theoretical) (%)
								van der Waals (vdW) energy coefficients C (10^{-79} J m⁶)	
Ne–Ne	0.39	0	21.6	0	0	4	4	4	100
CH₄–CH₄	2.60	0	12.6	0	0	102	102	101	100
HCl–HCl	2.63	1.08	12.7	6	11	106	123	157	86
HBr–HBr	3.61	0.78	11.6	4	3	182	189	207	96
HI–HI	5.44	0.38	10.4	2	0.2	370	372	350	99
CH₃Cl–CH₃Cl	4.56	1.87	11.3	32	101	282	415	509	68
NH₃–NH₃	2.26	1.47	10.2	10	38	63	111	162	57
H₂O–H₂O	1.48	1.85	12.6	10	96	33	139	175	24
Dissimilar molecules				$\dfrac{u_1^2\alpha_{02} + u_2^2\alpha_{01}}{(4\pi\varepsilon_0)^2}$	$\dfrac{u_1^2 u_2^2/3kT}{(4\pi\varepsilon_0)^2}$	$\dfrac{3\alpha_{01}\alpha_{02}h\nu_1\nu_2}{2(4\pi\varepsilon_0)^2(\nu_1 + \nu_2)}$			
Ne–CH₄				0	0	19	19[c]	—	100
HCl–HI				7	1	197	205	—	96
H₂O–Ne				1	0	11	12	—	92
H₂O–CH₄				9	0	58	67	—	87

[a] 1 D = 3.336 × 10⁻³⁰ C m.

[b] 1 eV = 1.602 × 10⁻¹⁹ J.

[c] This approximate value may be compared with the most recent ab initio calculation by Fowler et al. (1989) which gives 23 × 10⁻⁷⁹ J m⁶.

unlike molecules when direct experimental data are not available. However, for interactions involving highly polar molecules such as water it breaks down: for example, as can be seen from Table 6.3 for H_2O-CH_4 the net interaction is actually much less than that of H_2O-H_2O or CH_4-CH_4. Thus water and methane are attracted to themselves more strongly than they are attracted to each other. It is partly for this reason that non-polar molecules are not miscible (do not mix) with water but separate out into different phases in water. Such compounds (hydrocarbons, fluorocompounds, oils, fats) are known as *hydrophobic* (from the Greek 'water fearing'), and their low water solubility and their propensity to separate into clusters or phases in water is known as the *hydrophobic effect*. Table 6.3, however, reveals only part of the story: in liquid water the temperature-dependent orientation (dipole–dipole) interaction and the induction (dipole–neutral molecule) interaction are greatly modified, as will be discussed later.

6.6   GENERAL THEORY OF VAN DER WAALS FORCES BETWEEN MOLECULES

The London theory of dispersion forces has two serious shortcomings. It assumes that atoms and molecules have only a single ionization potential (one absorption frequency), and it cannot handle the interactions of molecules in a solvent. In 1963 McLachlan presented a generalized theory of van der Waals forces, which included in one equation the induction, orientation and dispersion force, and which could also be applied to interactions in a solvent medium. McLachlan's expression for the van der Waals free energy of two molecules or small particles 1 and 2 in a medium 3 is given by the series (McLachlan, 1963a, b)

$$w(r) = - \frac{6kT}{(4\pi\varepsilon_0)^2 r^6} \sum_{n=0,1,2,\ldots}^{\infty}{}' \frac{\alpha_1(iv_n)\alpha_2(iv_n)}{\varepsilon_3^2(iv_n)}, \qquad (6.18)$$

where $\alpha_1(iv_n)$ and $\alpha_2(iv_n)$ are the polarizabilities of molecules 1 and 2 and $\varepsilon_3(iv_n)$ the dielectric permittivity of medium 3 at *imaginary* frequencies iv_n, where[1]

$$v_n = (2\pi kT/h)n \approx 4 \times 10^{13} \, n \, \mathrm{s}^{-1} \qquad \text{at 300 K} \qquad (6.19)$$

and where the prime over the summation (Σ') indicates that the zero frequency $n = 0$ term is multiplied by $\frac{1}{2}$. McLachlan's equation looks complicated, but

---

[1] In the literature, frequencies are often denoted by ω where $\omega = 2\pi v$, Planck's constant by $\hbar$, where $\hbar = h/2\pi$.

it is actually quite straightforward to compute once we realize that $\alpha(iv_n)$ and $\varepsilon(iv_n)$ are real quantities, easily related to measurable properties, as we shall now see. For a molecule with one absorption frequency (the ionization frequency) v_I, its electronic polarizability at *real* frequencies v is given by the simple harmonic oscillator model (Von Hippel, 1958)

$$\alpha(v) = \alpha_0/[1 - (v/v_I)^2]. \tag{6.20}$$

Note that since most absorption frequencies are in the ultraviolet region (typically, $v_I \approx 3 \times 10^{15} \text{ s}^{-1}$) the value of α in the visible range of frequencies $v_{\text{vis}} \approx 5 \times 10^{14} \text{ s}^{-1}$ is essentially the same as α_0 because $(v_{\text{vis}}/v_I)^2 \ll 1$.

If the molecule also possesses a permanent dipole moment u there is an additional *orientational* or dipole relaxation polarizability contribution (Von Hippel, 1958) given by

$$\alpha_{\text{orient}}(v) = u^2/3kT(1 - iv/v_{\text{rot}}), \tag{6.21}$$

where v_{rot} is some average rotational relaxation frequency for the molecule, usually in the far infrared or microwave region of frequencies (typically, $v_{\text{rot}} \approx 10^{11} \text{ s}^{-1}$).

The total polarizability of a molecule in free space as a function of iv_n (replacing v by iv_n) may therefore be written as

$$\alpha(iv_n) = \frac{u^2}{3kT(1 + v_n/v_{\text{rot}})} + \frac{\alpha_0}{1 + (v_n/v_I)^2}, \tag{6.22}$$

which is a *real* function of v_n. At zero frequency ($v_n = 0$) this reduces to the Debye–Langevin equation, Eq. (5.7):

$$\alpha(0) = u^2/3kT + \alpha_0. \tag{6.23}$$

The variation of $\alpha(iv_n)$ with frequency v for a simple polar molecule is shown schematically in Fig. 6.2.

We may now return to McLachlan's expression, Eq. (6.18). The first term in the series is that for $n = 0$ (i.e., $v_n = 0$) so that from the above we immediately obtain the 'zero frequency contribution' to $w(r)$:

$$
\begin{aligned}
w(r)_{v=0} &= -\frac{3kT}{(4\pi\varepsilon_0)^2 r^6} \alpha_1(0)\alpha_2(0) \\
&= -\frac{3kT}{(4\pi\varepsilon_0)^2 r^6} \left[\frac{u_1^2}{3kT} + \alpha_{01}\right]\left[\frac{u_2^2}{3kT} + \alpha_{02}\right],
\end{aligned}
\tag{6.24}
$$

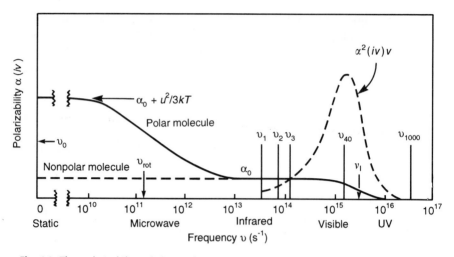

Fig. 6.2. The polarizability $\alpha(iv)$ as a function of frequency v of a simple polar molecule as given by Eq. (6.22). The frequencies $v_n = 4 \times 10^{13} \, n \, s^{-1}$ at which $\alpha(iv)$ contributes to the van der Waals energy (at 300 K) are shown as a series of vertical lines. The area under the curve of $\alpha^2(iv)v$ shows how different regions of the spectrum contribute to the net dispersion energy.

which is identical to Eq. (5.25) derived in Section 5.6 and includes both the orientation (Keesom) and induction (Debye) interaction energies discussed earlier.

Let us now turn to the non-zero frequency terms $(n = 1, 2, 3, \ldots)$ of Eq. (6.18). The summation entails calculating the values of $\alpha(iv_n)$ at the discrete frequencies given by Eq. (6.19). Now at normal temperatures the first frequency $v_{n=1} \approx 4 \times 10^{13} \, s^{-1}$ is already much greater than v_{rot}, so that $\alpha(iv_n)$ is effectively determined solely by the electronic polarizability contribution in Eq. (6.22). Further, since $v_{n=1}$ is still much smaller than a typical absorption frequency $v_1 \approx 3 \times 10^{15} \, s^{-1}$, it is clear that the frequencies v_n are very close together in the UV region. We may therefore replace the sum of discrete frequencies Σ by an integration over n, viz., $dn = (h/2\pi kT) \, dv$, so that

$$kT \sum_{n=1,2,\ldots}^{\infty} \rightarrow \frac{h}{2\pi} \int_{v_1}^{\infty} dv. \qquad (6.25)$$

Applying this to Eq. (6.18) we obtain for the 'finite frequency' free energy contribution to $w(r)$:

$$w(r)_{v>0} = -\frac{3h}{(4\pi\varepsilon_0)^2 \pi r^6} \int_{v_1}^{\infty} \alpha_1(iv)\alpha_2(iv) \, dv. \qquad (6.26)$$

Finally, by substituting the electronic polarizability as expressed by Eq. (6.22) into the above and integrating using the definite integral

$$\int_0^\infty \frac{dx}{(a^2 + x^2)(b^2 + x^2)} = \frac{\pi}{2ab(a + b)}, \tag{6.27}$$

we obtain

$$w(r)_{n>0} \approx -\frac{3\alpha_{01}\alpha_{02}}{2(4\pi\varepsilon_0)^2 r^6} \frac{h\nu_{11}\nu_{12}}{(\nu_{11} + \nu_{12})} \tag{6.28}$$

which is the London equation. The complete McLachlan formula is particularly suitable for computing the dispersion forces between molecules that have a number of different absorption frequencies or ionization potentials, for which the simple London expression breaks down.

6.7 VAN DER WAALS FORCES IN A MEDIUM

The theory of McLachlan is also naturally applicable to the interactions of molecules or small particles in a medium (McLachlan, 1965). In this case the polarizabilities $\alpha(i\nu)$ in Eq. (6.18) are the *excess polarizabilities* of the molecules as discussed in Section 5.7, and for a small spherical molecule 1 of radius a_1 in a medium 3 (Fig. 6.3a), the excess polarizability is given approximately by

$$\alpha_1(\nu) = 4\pi\varepsilon_0\varepsilon_3(\nu)\left(\frac{\varepsilon_1(\nu) - \varepsilon_3(\nu)}{\varepsilon_1(\nu) + 2\varepsilon_3(\nu)}\right)a_1^3. \tag{6.29}$$

The zero-frequency contribution to $w(r)$ in Eq. (6.18) is therefore

$$w(r)_{\nu=0} = -\frac{3kTa_1^3 a_2^3}{r^6}\left(\frac{\varepsilon_1(0) - \varepsilon_3(0)}{\varepsilon_1(0) + 2\varepsilon_3(0)}\right)\left(\frac{\varepsilon_2(0) - \varepsilon_3(0)}{\varepsilon_2(0) + 2\varepsilon_3(0)}\right), \tag{6.30}$$

where $\varepsilon_1(0)$, $\varepsilon_2(0)$ and $\varepsilon_3(0)$ are the static dielectric constants of the three media. Equation (6.30) is the same as that previously derived and discussed in Section 5.7 in connection with the orientation and induction forces in a medium. We shall return to it again after first considering the finite frequency dispersion contribution. Substituting Eq. (6.29) into Eq. (6.18), and replacing the summation by the integral of Eq. (6.25), the dispersion energy may be

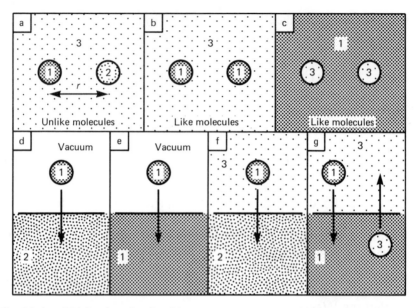

Fig. 6.3. (a–c) Molecules of radii a (diameter $\sigma = 2a$) interacting in a solvent medium. (d–g) Transfer of molecules from one medium to another.

written as

$$w(r)_{v > 0} = -\frac{3ha_1^3 a_2^3}{\pi r^6} \int_0^\infty \left(\frac{\varepsilon_1(iv) - \varepsilon_3(iv)}{\varepsilon_1(iv) + 2\varepsilon_3(iv)}\right)\left(\frac{\varepsilon_2(iv) - \varepsilon_3(iv)}{\varepsilon_2(iv) + 2\varepsilon_3(iv)}\right) dv.$$

$$(6.31)$$

Unfortunately, complete data on the absorption spectra of most materials are not available, and it is necessary to adopt a model that suitably represents the dielectric behaviour of the media. In the previous section we saw how it was possible to derive London's result for the dispersion force by treating molecules as simple harmonic oscillators. Similarly, for a dielectric medium that has one strong absorption peak at a frequency v_e (which is usually slightly different from that of the isolated molecules in the gas v_I) we may express its dielectric permittivity at frequencies $v > v_{n=1}$ as (Von Hippel, 1958)

$$\varepsilon(v) = 1 + (n^2 - 1)/[1 - (v/v_e)^2] \qquad (6.32)$$

so that

$$\varepsilon(iv) = 1 + (n^2 - 1)/[1 + (v/v_e)^2], \qquad (6.33)$$

where n is the refractive index, roughly equal to $\sqrt{\varepsilon(v_{vis})}$. Substituting the above into Eq. (6.31) and integrating as before we obtain

$$w(r)_{v>0} = -\frac{\sqrt{3}hv_e a_1^3 a_2^3}{2r^6}$$

$$\times \frac{(n_1^2 - n_3^2)(n_2^2 - n_3^2)}{(n_1^2 + 2n_3^2)^{1/2}(n_2^2 + 2n_3^2)^{1/2}[(n_1^2 + 2n_3^2)^{1/2} + (n_2^2 + 2n_3^2)^{1/2}]},$$

$$(6.34)$$

where for simplicity, it is assumed that all three media have the same absorption frequency v_e. The total van der Waals interaction free energy of two identical molecules 1 in medium 3 (Fig. 6.3b) is therefore

$$w(r) = w(r)_{v=0} + w(r)_{v>0}$$

$$\approx -\left[3kT\left(\frac{\varepsilon_1(0) - \varepsilon_3(0)}{\varepsilon_1(0) + 2\varepsilon_3(0)}\right)^2 + \frac{\sqrt{3}hv_e}{4}\frac{(n_1^2 - n_3^2)^2}{(n_1^2 + 2n_3^2)^{3/2}} \right]\frac{a_1^6}{r^6}, \quad (6.35)$$

which is strictly valid only for $r \gg a_1$.

The above equations highlight a number of important aspects of van der Waals forces in a solvent medium.

(i) Since $hv_e \gg kT$ we find, as for interactions in free space, that the $v > 0$ dispersion force contribution is usually greater than the $v = 0$ dipolar contribution.

(ii) The van der Waals force is much reduced in a solvent medium. For example, for two non-polar molecules of refractive index $n_1 = 1.5$, the strength of the dispersion force in a solvent of $n_3 = 1.4$ will be reduced from its value in free space (where $n_3 = 1$) by a factor of

$$\frac{(1.5^2 - 1)^2(1.5^2 + 2)^{-3/2}}{(1.5^2 - 1.4^2)^2(1.5^2 + 2 \times 1.4^2)^{-3/2}} \approx 32.$$

(iii) It is worth comparing the $v > 0$ dispersion contribution of Eq. (6.35) with London's equation for the free space interaction. If we put $n_3 = 1$, we obtain

$$w(r)_{v>0} = -\frac{\sqrt{3}hv_e a_1^6}{4r^6}\frac{(n_1^2 - 1)^2}{(n_1^2 + 2)^{3/2}} \quad (6.36)$$

whereas London's result is

$$w(r) = -\frac{3h\nu_1\alpha_0^2}{4(4\pi\varepsilon_0)^2 r^6} = -\frac{3h\nu_1 a_1^6}{4r^6}\frac{(n_1^2 - 1)^2}{(n_1^2 + 2)^2}.$$ (6.37)

The reason for the apparent discrepancy is that the absorption frequency of an isolated molecule ν_1 is different from that of the condensed phase ν_e. It is a simple matter to ascertain that if $\alpha(\nu)$ and $\varepsilon(\nu)$ are related by the Clausius–Mossotti–Lorenz–Lorentz equation, $\alpha(\nu) \propto [\varepsilon(\nu) - 1]/[\varepsilon(\nu) + 2]$, then the two absorption frequencies are formally related by

$$\nu_e = \nu_1\sqrt{3/(n_1^2 + 2)},$$ (6.38)

so that when this is put into Eq. (6.36) we obtain the result based on London's formula, Eq. (6.37).

(iv) The dispersion force between dissimilar molecules can be attractive or repulsive. The latter occurs when n_3 is intermediate between n_1 and n_2 in Eq. (6.34). Examples of repulsive van der Waals forces will be given in Chapter 11. However, the interaction between identical molecules is always attractive due to the symmetry of Eq. (6.35), and we may further note that if the solute and solvent molecules, 1 and 3, are interchanged (Fig. 6.3b, c), the expression for the van der Waals force between the two solute (ex-solvent) molecules remains practically the same.

The above equations provide a semiquantitative criterion for determining which liquids are likely to be miscible and which are not. As a rule of thumb it is known that 'like dissolves like'. This is already apparent from Eq. (6.35) where we see that the smaller the difference between n_1 and n_3, the smaller the attraction between two solute molecules (Fig. 6.3b, c) and the less will be their tendency to associate (i.e., separate out into different phases). Since Eq. (6.35) basically depends on the magnitude of $(n_1^2 - n_3^2)^2$ we may expect two liquids to become immiscible once $(n_1^2 - n_3^2)^2$ becomes too large. Now this can be written as $[\sqrt{(n_1^2 - 1)^2} - \sqrt{(n_3^2 - 1)^2}]^2$, which from the Lorenz–Lorentz equation, Eq. (5.29), is roughly proportional to $[\sqrt{(\alpha_{01}^2/a_1^6)^2} - \sqrt{(\alpha_{03}^2/a_3^6)}]^2$ or

$$w(r) \propto [\sqrt{U_1} - \sqrt{U_3}]^2$$ (6.39)

where U_1 and U_3 are the cohesive energies or latent heats of vaporization of the two liquids (cf. Eq. (6.5)). This semiquantitative derivation forms the basis of Hildebrand's 'solvent solubility parameter' δ, which is equal to the square root of the cohesive energy density of a liquid (Small, 1953). It is

found that if two non-polar liquids have similar values for δ, i.e., if $(\delta_1 - \delta_2)^2$ is small, they are miscible and the binary mixture is nearly ideal, e.g., showing little deviation from Raoult's law (for a review of solubility relationships see Kumar and Prausnitz, 1975).

(v) In some instances the dispersion force between two molecules in a solvent is very small, and then the zero-frequency contribution dominates the interaction. The best-known case concerns the lower molecular weight alkanes in water whose refractive indices are very close to that of water (compare $n_{CH_4} \approx 1.30$, $n_{C_4H_{10}} \approx 1.33$ and $n_{C_5H_{12}} \approx 1.36$ with $n_{H_2O} \approx 1.33$). When these molecules interact in water (or vice versa) the major contribution is now not the dispersion force but rather the first term in Eq. (6.35), and since $\varepsilon_{H_2O}(0) \approx 80$ while $\varepsilon_{alkane}(0) \approx 2$, this reduces to approximately

$$w(r)_{v=0} \approx \frac{-kTa_1^6}{r^6}, \tag{6.40}$$

which is purely entropic. Equation (6.40) shows that there is an increase in entropy as two alkane molecules approach each other in water, indicative of an increase in the motional freedom of the water molecules. It has long been known that the 'hydrophobic interaction' between small hydrocarbon molecules in water is mainly entropic, but the few measured values to date suggest a far stronger interaction than would be expected from Eq. (6.40). Thus, for two small molecules of radius a, we would expect a free energy of dimerization of order $kT(a/2a)^6 \approx kT/64$, or about 0.04 kJ mol^{-1} at 25°C (not enough to induce immiscibility), whereas the experimental values are at least 100 times larger, viz., ~ 10 kJ mol^{-1} for CH_4, C_6H_6 and C_6H_{12} (Tucker et al., 1981; Ben Naim et al., 1973). This lack of agreement is related to the breakdown in the simple model of the excess polarizability involving highly polar solvent or solute molecules, as discussed in Section 5.7 (and Table 5.2). The unique and unusual properties of water, both as a solvent and as a medium for solute–solute interactions, are further investigated in Chapters 8 and in Part II.

6.8   DISPERSION SELF-ENERGY OF A MOLECULE IN A MEDIUM

The concept of a dispersion self-energy of a molecule is analogous to the Born self-energy of an ion (Chapter 3) and, as for ions, it provides insight into the solubility and partitioning of molecules in different solvents. There are a number of different approaches to this problem (Mahanty and Ninham, 1976); we shall adopt the simplest, which nevertheless brings out the essential physics. Let us consider the transfer of a molecule 1 of diameter σ from free space into a medium 2 (Fig. 6.3d), where it becomes surrounded by 12 solvent molecules of similar diameter. The free energy change is therefore given by the (positive) energy needed to first create a cavity, which involves breaking six solvent–solvent 'bonds', plus the (negative) energy of placing molecule 1 in the cavity, which involves the formation of 12 new solute–solvent 'bonds'. Thus, for this process,

$$\Delta\mu_{disp}^i \approx \frac{3h\nu_1}{4(4\pi\varepsilon_0)^2\sigma^6}(6\alpha_{02}^2 - 12\alpha_{01}\alpha_{02}) \tag{6.41}$$

where ν_1 is assumed to be the same for the solute and solvent molecules. We may already note that if medium 2 = medium 1 (Fig. 6.3e), the above reduces to the result based on the London equation, Eq. (6.5), which gives the cohesive energy of the pure liquid or solid.

More generally, the free energy of transfer of a molecule 1 from medium 3 into medium 2 (Fig. 6.3f) is therefore given by

$$\Delta\mu_{disp}^i \approx \frac{3h\nu_1}{4(4\pi\varepsilon_0)^2\sigma^6}[-(6\alpha_{03}^2 - 12\alpha_{01}\alpha_{03}) + (6\alpha_{02}^2 - 12\alpha_{01}\alpha_{02})], \tag{6.42}$$

which, using Eq. (5.29), is roughly proportional to

$$\Delta\mu_{disp}^i \propto (\alpha_{03} - \alpha_{02})(2\alpha_{01} - \alpha_{02} - \alpha_{03})$$
$$\propto (n_3^2 - n_2^2)(2n_1^2 - n_2^2 - n_3^2), \tag{6.43}$$

which can be positive or negative. In particular, the above shows that solute transfer is always energetically favoured ($\Delta\mu^i$ negative) into the solvent whose refractive index is closer to that of the solute molecule—a further manifestation of the 'like dissolves like' rule. If medium 2 = medium 1 (Fig. 6.3g), the above reduces to

$$\Delta\mu_{disp}^i \approx -\frac{6(3h\nu_1)}{4(4\pi\varepsilon_0)^2\sigma^6}[\alpha_{01} - \alpha_{03}]^2 \tag{6.44}$$

$$\propto -(\sqrt{U_1} - \sqrt{U_3})^2 \propto -(n_1^2 - n_3^2)^2, \tag{6.45}$$

which is always negative and which applies equally to the two transfer processes shown in Fig. 6.3g. Thus, it is always *energetically* favourable for a solute molecule to move into its own environment. It is interesting to note that the above is essentially the same result that was obtained in the previous section, Eq. (6.39), and leads to the same semiempirical solubility relations as contained in the 'solubility parameter' criterion. In particular, the very low solubility of alkanes and hydrocarbons in water, where $n_1 \approx n_3$, is once again seen not to be governed by their dispersion interaction, but by other factors.

The above relationships and rule-of-thumb criteria, while useful, are limited to dispersion interactions and thus require that the solutes and solvents be non-polar and have no strong electrostatic or hydrogen-bonding interactions. In Chapter 9 we shall develop similar relationships, based on cohesive energies, which can be applied to other phenomena as well (such as wetting).

6.9   Further aspects of van der Waals forces: anisotropy, non-additivity and retardation effects

Anisotropy of dispersion forces

The polarizabilities of all but spherically symmetric molecules are anisotropic, having different values along different molecular directions. This arises because the electronic polarizabilities of bonds, which are a measure of the response of the electrons to an external field, are usually anisotropic (Hirschfelder et al., 1954). For example, the longitudinal and transverse polarizabilities of the C—H bond are $\alpha_\parallel/4\pi\varepsilon_0 = 0.79 \times 10^{-30}$ m^3 and $\alpha_\perp/4\pi\varepsilon_0 = 0.58 \times 10^{-30}$ m^3; the mean value (quoted in Table 5.1) is $\frac{1}{3}(\alpha_\parallel + 2\alpha_\perp)/4\pi\varepsilon_0 = 0.65 \times 10^{-30}$ m^3. One consequence of the anisotropy in α is that the dispersion force between molecules becomes dependent on their mutual orientation (see Hirschfelder et al., 1954; Israelachvili and Tabor, 1973; Israelachvili, 1974). In non-polar liquids the effect is not important, since the molecules are tumbling rapidly, and their *mean* polarizability is what matters. But in solids and liquid crystals the anisotropic attractive forces can sometimes be an important factor in driving molecules or molecular groups into favourable mutual orientations; for example, in determining the specific configurations of polymers and proteins, the ordering of molecules in liquid crystals, and the temperature-dependent phase transitions of lipid bilayers and monolayers (Zwanzig, 1963; Marcelja, 1973).

The orienting effects of the anisotropic dispersion forces are, however, rarely as important as those of dipole–dipole and hydrogen-bonding forces. It should also be remembered that for asymmetric molecules, the repulsive steric forces are also orientation-dependent (reflecting their asymmetric shape), and this is usually by far the most important factor in determining how molecules mutually align themselves in solids and liquid crystals (see Chapter 7).

Non-additivity of van der Waals forces and many-body effects

Unlike gravitational and Coulomb forces, van der Waals forces are not generally *pairwise additive*: the force between any two molecules is affected by the presence of other molecules nearby, so that one cannot simply add all the pair potentials of a molecule to obtain its net interaction energy with all the other molecules. This is because the field emanating from any one molecule reaches a second molecule both directly and by 'reflection' from other molecules since they, too, are polarized by the field. The net effect is that the van der Waals energy between two molecules in a medium is different from that calculated from a consideration of two-body (pair) interactions alone. The effect on the energy can be positive or negative and is usually small, less than 20% (see Problem 6.2), but can be important: for example, the total energy contributions, including many-body effects, to the lattice energies of the rare gas atoms, such as argon and xenon, determine that they pack as face-centred cubic (FCC) solids and not as hexagonal close-packed (HCP) solids. The non-additive property of van der Waals forces is particularly important in the interactions between large particles and surfaces in a medium (Chapter 11).

Retardation effects

When two atoms are an appreciable distance apart, the time taken for the electric field of the first atom to reach the second and return can become comparable with the period of the fluctuating dipole itself (cf. the distance travelled by light during one rotation of a Bohr atom electron is about $3 \times 10^8 \text{ m s}^{-1}/3 \times 10^{15} \text{ s}^{-1} \approx 10^{-7} \text{ m}$ or 100 nm). When this happens the field returns to find that the direction of the instantaneous dipole of the first atom is now different from the original and less favourably disposed to an attractive interaction. Thus, with increasing separation the dispersion energy between two atoms begins to decay even faster than $-1/r^6$, approaching a $-1/r^7$ dependence at $r > 100$ nm (see Problem 6.3). This is called the *retardation effect*, and the dispersion forces between molecules and particles at large separations are called *retarded forces*.

For two molecules in free space, retardation effects begin at separations above 5 nm and are therefore of little interest. However, in a medium, where the speed of light is slower, retardation effects come in at smaller distances, and they become particularly important when macroscopic bodies or surfaces interact in a liquid medium as we shall see in later chapters. Note that it is only the dispersion energy that suffers retardation; the zero-frequency orientation and induction energies remain non-retarded at all separations, so that as the separation increases these initially weak contributions ultimately dominate the interaction. Thus, as r increases to very large separations, the distance-dependence of the van der Waals energy between two molecules has the curious progression: $-1/r^6 \rightarrow -1/r^7 \rightarrow -1/r^6$.

PROBLEMS AND DISCUSSION TOPICS

6.1 Two non-polar solute molecules 1 and 2 interact in a slightly polar medium 3 at 298 K. The interaction is dominated entirely by van der Waals forces, and the molecules may be considered to behave as small spheres with the same dielectric properties as the bulk materials whose properties are: $n_1 = 1.40$, $n_2 = 1.50$, $n_3 = 1.45$, $\varepsilon_1 = n_1^2$, $\varepsilon_2 = n_2^2$, $\varepsilon_3 = 6.0$. If the absorption frequency v_e of all three media is the same and equal to $3 \times 10^{15}\,\mathrm{s}^{-1}$, will the equilibrium interaction potential be (i) attractive at all separations, including contact, (ii) repulsive at all separations, including contact, (iii) repulsive at small separations but attractive (adhesive) at some finite separation?

If your answer is (i) or (iii), is the adhesion energy strong enough to lead to aggregation or phase separation of the solute in the solvent above some critical concentration (the solubility)?

6.2 The simple treatment used to derive the London equation for the dispersion interaction energy between two isolated molecules A and B (Section 6.1) can be readily extended to include the effect of a third molecule C. This arises because an additional component of the electric field from A reaches B by 'reflection' from C (Chapter 5). Consider three identical molecules of polarizabilities α and ionization potentials $I = hv$ sitting at the corners of a triangle with sides of length r_1, r_2 and r_3. Ignoring numerical factors, derive a simple approximate expression (similar to Eq. (6.2)) for the additional three-body interaction energy. On the basis of this equation obtain a rough estimate for the magnitude of the three-body interaction energy relative to the two-body interaction for the case when the three molecules are in contact, i.e., when $r_1 = r_2 = r_3 = \sigma$.

In general, allowing that three molecules may be displaced relative to each other in a variety of ways, would you expect the three-body energy to be always negative (favourable), always positive (unfavourable), or one or other of these depending on the relative positions of the molecules? (For further reading on three-, four- and many-body effects, see Margenau and Kestner, 1971.)

6.3 Once two molecules (or particles) are sufficiently far apart the simple model, based on two Bohr atoms, used to derive the London equation (Section 6.1), breaks down. This is because the field reflected by the second atom returns to find that the direction of the original dipole has changed during the time it takes light to travel the distance $2r$. The extent of this change depends on the time it takes the Bohr electron to rotate about the nucleus, i.e. it depends on the frequency v. Thus in general, the *inducing* and *induced* dipoles become increasingly less correlated the farther two atoms or molecules are from each other, and this results in a weaker (retarded) dispersion attraction than given by the London formula. By considering the McLachlan theory (Section 6.6) it is clear that at any given separation r not all frequencies will contribute to the interaction so that Eq. (6.26) should not really be integrated to $v = \infty$. By considering how some appropriate cut-off frequency must replace the upper integration limit in the derivation of the London equation obtain an expression for the dispersion interaction potential between two identical molecules valid at all separations. Check that your equation predicts that the retarded dispersion energy is about half the non-retarded value at a separation of $r \approx 0.7c/v_1$.

Hint. In the limit of very large separations the rigorously derived retarded dispersion interaction is given by the *Casimir–Polder* equation:

$$w(r) = -23\alpha_0^2 hc/8\pi^2(4\pi\varepsilon_0)^2 r^7. \tag{6.46}$$

CHAPTER 7

REPULSIVE FORCES, TOTAL INTERMOLECULAR PAIR POTENTIALS AND LIQUID STRUCTURE

7.1 SIZES OF ATOMS, MOLECULES AND IONS

At very small interatomic distances the electron clouds of atoms overlap, and there arises a strong repulsive force that determines how close two atoms or molecules can ultimately approach each other. These repulsive forces are sometimes referred to as *exchange repulsion,* *hard core repulsion, steric repulsion,* or—for ions—the *Born repulsion,* and they are characterized by having very short range and increasing very sharply as two molecules come together. Strictly, they belong to the category of quantum mechanical or chemical forces discussed earlier, and unfortunately there is no general equation for describing their distance dependence. Instead, a number of empirical potential functions have been introduced over the years, all of which appear to be satisfactory so long as they have the property of a steeply rising repulsion at small separations. The three most common of such potentials are the *hard sphere potential*, the inverse *power-law potential* and the *exponential potential.* Let us begin by considering the first.

If atoms are considered as hard spheres, i.e., incompressible, the repulsive force suddenly becomes infinite at a certain interatomic separation. This simple model reflects the observation that when different atoms pack together in liquids and solids they often do behave as hard spheres, or 'billiard balls', of fixed radii characteristic for each atom. Defined in this way the radius of an atom (whether isolated or covalently bound) or a spherical molecule is then called its *hard sphere radius* or *van der Waals packing radius.* Results obtained from x-ray and neutron diffraction studies on solids (especially crystals) and from gas solubility, viscosity and self-diffusion data on liquids often yield values that agree to within a few percentage points. The van der Waals packing radii of most atoms and small molecules lie between 0.1 and 0.2 nm, some of which are illustrated in Fig. 7.1.

Similar concepts apply to ions in ionic crystals, where the characteristic

Monovalent cations	Li^+	Na^+	K^+	NH_4^+	Cs^+	$N(CH_3)_4^+$
	0.068	0.095	0.133	0.148	0.169	0.347

Divalent cations	Be^{2+}	Mg^{2+}	Fe^{2+}	Ca^{2+}	Ba^{2+}	
	0.031	0.065	0.076	0.099	0.135	

Trivalent cations	Al^{3+}	Fe^{3+}	La^{3+}			
	0.050	0.064	0.104			

Monovalent anions	F^-	OH^-	Cl^-	Br^-	I^-	NO_3^-
	0.136	0.176	0.181	0.195	0.216	0.264

Spherical molecules	Ne	Ar	Kr	Xe	CH_4	CCl_4
	0.154	0.188	0.201	0.216	0.20	0.275

Effective radii of nonspherical molecules and groups (approximate)	H_2O 0.14	O_2 0.18	NH_3 0.18	HCl 0.18	HBr 0.19
	CH_3OH 0.21	CH_3Cl 0.215	$CHCl_3$ 0.255	C_6H_6 0.265	C_6H_{12} 0.285
	$-CH_3$ group 0.20	$-CH_2-$ group 0.20	$-NH_2$ group 0.17	$-OH$ group 0.145	Aromatic ring thickness 0.37

Exposed radii of atoms covalently bonded in molecules	H 0.11	F 0.14	O 0.15	N 0.15	C 0.17
	Cl 0.18	S 0.18	Br 0.19	P 0.19	I 0.20

Covalent bond radii of atoms	$-H$ 0.03	$-C$ 0.077	$-N$ 0.070	$-O$ 0.066	$-F$ 0.064	$-S$ 0.104
		$=C$ 0.067	$=N$ 0.062	$=O$ 0.062	$-Cl$ 0.099	$-P$ 0.110
		$\equiv C$ 0.060	$\equiv N$ 0.055		$-Br$ 0.114	$-Si$ 0.117

Fig. 7.1. Effective packing radii of atoms, molecules and ions (nm).

packing radius is referred to as the *bare ion radius*. The bare ion radius is quite different from the hydrated ion radius of an ion in water, discussed in Section 4.5. Some bare ion radii are also shown in Fig. 7.1. Note that anions are generally bigger than cations; this is because they have gained rather than lost electrons, which also causes an additional electrostatic repulsion tending to further inflate the ion.

It is also recognized that the distance between the atomic centres of two atoms connected by a covalent bond can be expressed as the sum of their *covalent bond radii* (Fig. 7.1). Single-bond covalent radii are usually about 0.08 nm shorter than the non-bonded (van der Waals) radii.

By considering both the covalent and van der Waals radii of individual atoms in a molecule and the bond angles between them, one can establish the effective van der Waals radius of a molecule. This concept is strictly valid only for small, nearly spherical molecules such as CH_4, but Bondi (1968) has described procedures for calculating the effective van der Waals radii of molcules and molecular groups, some of which are included in Fig. 7.1. For very asymmetric molecules that cannot be considered even as quasi-spheres, one has to consider their *van der Waals dimensions* along different molecular axes. For example, in alkanes the effective radius of the cylinder-like paraffin chain is about 0.20 nm, while its total length is 0.127 nm per $CH_2—CH_2$ link along the chain plus ~ 0.20 nm for each hemispherical CH_3 group at each end.

There are many different methods for measuring the radii of atoms and molecules. These include PVT and spectroscopic data on gases, viscosity, solubility and diffusion data on both gases and liquids, and compressibility, electron, x-ray and neutron diffraction data on liquids and solids. The values for molecular sizes deduced from these different methods can differ by as much as 30%, as illustrated in Table 7.1. This is because each method measures a slightly different property. Thus, molecular radii $\sigma/2$ determined

TABLE 7.1 Radii of molecules deduced from different methods

Molecule	Minimum radius, from van der Waals coefficient b^a Eq. (7.1) (nm)	Mean radius (van der Waals packing radius, from Fig. 7.1) (nm)	Maximum radius, from volume occupied in liquid at 20°C, Eq. (7.2) (nm)
CH_3OH	0.19	0.21	0.23
$CHCl_3$	0.22	0.26	0.29
C_6H_6	0.23	0.27	0.30
CCl_4	0.24	0.28	0.305

[a] Constants b taken from the *Handbook of Chemistry and Physics*, CRC Press, 56th ed.

from the van der Waals equation of state coefficient b, Eq. (6.13),

$$\sigma/2 = 0.463b^{1/3} \text{ nm} \qquad (b \text{ in } \text{dm}^3 \text{ mol}^{-1}) \qquad (7.1)$$

gives the smallest value since it reflects the sizes of molecules during collisions when they approach each other closer than their equilibrium separation. Values determined from the viscosity and self-diffusion of molecules in liquids (Ertl and Dullien, 1973; Dymond, 1981; Evans et al., 1981) usually yield results similar to those obtained from crystal packing (i.e., the van der Waals radii), while radii calculated from the mean molecular volume occupied in the liquid state (even assuming close packing),

$$\tfrac{4}{3}\pi(\sigma/2)^3 = 0.7405(M/N_0\rho), \qquad (7.2)$$

yield the highest values because in the liquid the mobile molecules are on average 5–10% farther apart than in the close-packed crystalline solid.

7.2   REPULSIVE POTENTIALS

Returning now to our discussion of repulsive potentials, the *hard sphere potential* can be described by

$$w(r) = + (\sigma/r)^n, \qquad \text{where } n = \infty. \qquad (7.3)$$

Since for $r > \sigma$ the value of $w(r)$ is effectively zero while for $r < \sigma$ it is infinite, this expression nicely describes the hard sphere repulsion where σ is the hard sphere diameter of an atom or molecule, which may be associated with twice the van der Waals radius.

Two other repulsive potentials are worthy of note: the *power-law potential*

$$w(r) = + (\sigma/r)^n, \qquad (7.4)$$

where n is now an integer (usually taken between 9 and 16), and the *exponential potential*

$$w(r) = + ce^{-r/\sigma_0}, \qquad (7.5)$$

where c and σ_0 are adjustable constants, with σ_0 of the order of 0.02 nm. Both of these potentials are more realistic in that they allow for the finite compressibility or 'softness' of atoms. The power-law potential has little theoretical basis, while the exponential potential has some theoretical justification. It is generally recognized, however, that their common usage is due mainly to their mathematical convenience.

7.3 Total Intermolecular Pair Potentials

The total intermolecular pair potential is obtained by summing the attractive and repulsive potentials. Figure 7.2a illustrates the two main types of repulsive potentials discussed above and the shapes of the potential functions they lead to when a long-range inverse power attractive term is added (Fig. 7.2b). The best known of these is the *Lennard–Jones or '6–12' potential*:

$$w(r) = A/r^{12} - B/r^6 = 4\epsilon[(\sigma/r)^{12} - (\sigma/r)^6], \qquad (7.6)$$

which is widely used because of its simplicity and inverse sixth-power attractive van der Waals term. The particular mathematical form given for the Lennard–Jones potential above is commonly used though it is well to note that the parameter σ is different from the molecular diameter. Thus, for the Lennard–Jones potential, $w(r) = 0$ at $r = \sigma$, and the minimum energy occurs at $r = 2^{1/6}\sigma = 1.12\sigma$ (Fig. 7.2b). The minimum energy is $w(r) = -\epsilon$, where the attractive van der Waals contribution is -2ϵ while the repulsive energy contribution is $+\epsilon$. Thus the inverse 12th-power repulsive term decreases the strength of the binding energy at equilibrium by 50% (from -2ϵ to $-\epsilon$). This may be compared to the hard sphere repulsion potential where the binding energy at equilibrium would be just the van der Waals energy at contact (at $r = \sigma$). It is for this reason that the lattice energies of ionic crystals and van der Waals solids, calculated on the basis of a hard sphere repulsion, are sometimes too large (cf. Section 3.4). But quite often the hard sphere potential together with a London dispersion attraction gives excellent results when compared with the experimental molar lattice energies of non-polar solids (cf. Table 6.1). This is because the attractive forces are usually stronger than predicted by the simple London equation, which ignores contributions from other absorption frequencies, quadrupole interactions, etc. On the other hand, the repulsive potentials are believed to be steeper than $1/r^{12}$ and therefore weaker at the equilibrium separation than given by the Lennard–Jones potential. All these effects conspire to make the hard sphere potential a reasonable one for predicting contact energies. Figure 7.2c nicely illustrates this fortuitous cancellation of errors: here we see the experimentally determined pair potential for argon (Parsons *et al.*, 1972) together with the London dispersion energy, $w(r) = -50 \times 10^{-79}/r^6$ J,

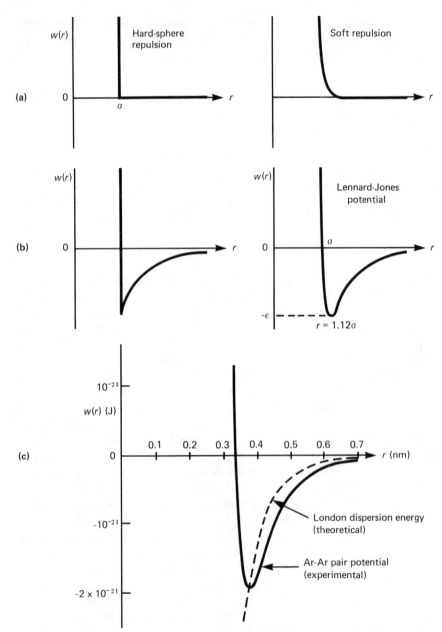

Fig. 7.2. (a) Repulsive potentials. (b) Full pair potentials obtained by adding an attractive long-range potential to the above repulsive potentials. The packing diameter of a molecule is given by the value of r at the potential energy minimum. (c) Experimental argon–argon potential compared with the London dispersion energy. ((c) From Parsons *et al.,* 1972.)

taken from Table 6.1. Note how the (theoretical) dispersion attraction passes through the experimental curve almost exactly at its minimum value.

7.4   ROLE OF REPULSIVE FORCES IN NON-COVALENTLY BONDED SOLIDS

Let us now turn our attention to the role of repulsive forces in solids. More often than not it is these that determine many of their properties. The reason for this is that for many types of attractive interactions, their orientation or angle dependence is weak (i.e., they are non-directional). This is particularly so for van der Waals forces. In contrast, the orientation dependence of the stabilizing repulsive force, which simply reflects the asymmetric shape of a molecule, often has a large effect. Thus, when molecules or ions come together in the condensed state, the way they can pack together—reflected in their relative sizes and shapes—now becomes a major consideration in determining their lattice structure, density, rigidity, etc. They also determine their melting points, but not their boiling points. To see why this is so, recall that latent heats of vaporization are closely associated with the cohesive energies in solids and liquids (Table 2.1). Thus it is the *attractive* forces that mainly determine latent heats of vaporization and, by Trouton's rule, also their boiling points. Melting points, however, are determined by the geometry of molecules. If their shape allows them to comfortably pack together into a lattice, they will tend to remain in this state and will therefore have a high melting point, If their shape does not allow for good packing, the melting point will be low. (Note too that latent heats of melting are usually fairly small, an indication that on melting few 'bonds' are broken and that the main effect is a rearrangement of molecular ordering.) Since the shape of a molecule is determined by its van der Waals dimensions, which is effectively a statement about its repulsive force, we can now see why *repulsive forces* are mainly responsible for melting while *attractive forces* are mainly responsible for boiling.

This somewhat broad generalization can be illustrated by the melting points and boiling points of some hydrocarbons, shown in Table 7.2. Here we clearly see that the T_B values are not much affected by branching or by replacing a single C—C bond by a double C=C bond, since the intermolecular attractive forces are not much changed. But the T_M values are lowered dramatically, since the regular crystalline packing of the all-*trans* chains is no longer possible when a 'kink' is present somewhere along the otherwise linear molecules. Note how the effect of a single kink persists even for the 16-carbon hexadecane molecule.

For some molecules, however, the orientation dependence of their *attractive*

TABLE 7.2 Effect of double bonds and chain branching on melting and boiling points of hydrocarbons

Hydrocarbon		T_M (°C)	T_B^a (°C)
C_6H_{14}	n-Hexane	−95	69
C_6H_{12}	1-Hexene	−140	63
C_6H_{12}	2-Hexene (cis)	−141	69
C_6H_{14}	2-Methyl pentane	−154	60
C_8H_{18}	n-Octane	−57	126
C_8H_{16}	1-Octene	−102	121
C_8H_{16}	2-Octane (cis)	−100	126
C_8H_{18}	2-Methyl heptane	−120	119
$C_{12}H_{26}$	n-Dodecane	−10	216
$C_{12}H_{24}$	1-Dodecene	−35	213
$C_{16}H_{34}$	n-Hexadecane	18	287
$C_{16}H_{32}$	1-Hexadecene	4	284
$C_{18}H_{32}$	n-Octadecane	28.2	316
$C_{18}H_{30}$	1-Octadecene	17.5	
$C_{18}H_{30}$	9-Octadecene	−30.5	

^a At 760 mmHg.

forces plays the dominant role in determining their solid- and liquid-state properties. This is, of course, especially true of covalent bonds, but also when strongly directional dipolar and hydrogen-bonding interactions are involved. A particularly notable representative of this phenomenon is the water molecule whose interactions are so remarkable (and important) that they are given special attention in the next chapter.

The sizes and shapes of large complex molecules such surfactants, polymers and biological macromolecules are particularly important in determining how they can pack, or 'self-assemble', into aggregated structures such as micelles and membranes in water, as will be discussed in Part III.

Packing mismatches remain important even for the interactions of macroscopic particles. For example, a concentrated dispersion of spherical colloidal particles of uniform diameter can order into a solid-like lattice, but when a few larger spheres are present these will tend to associate into clusters or collect at certain interfacial regions or grain boundaries (Fig. 6.1b) rather than be randomly dispersed. This effect also occurs at the atomic level where impurity atoms usually migrate to the grain boundaries of solids. Both these effects are further manifestations of the role of short-range repulsive forces, and illustrate how they naturally lead to geometric packing considerations at both the molecular and macroscopic levels.

7.5   ROLE OF REPULSIVE FORCES IN LIQUIDS: LIQUID STRUCTURE

When a solid melts, the ordered molecular structure that existed in the solid is not completely lost in the liquid. This phenomenon has led to such concepts as *liquid structure* and *molecular ordering* in liquids, with important consequences both for our understanding of the liquid state and for the way molecules and particles interact in liquids (Pryde, 1966; Kohler, 1972; Croxton, 1975; Kruus, 1977; Maitland *et al.*, 1981; Chandler, 1987; Allen and Tildesley, 1987; Ciccotti *et al.*, 1987). The occurrence of 'structure' in liquids arises first and foremost from the geometry of molecules, and as such reflects the repulsive forces between them. Let us see what it is all about by first considering the molecular events that take place as a solid is heated through its melting point.

Imagine a close-packed FCC lattice at 0 K where each molecule is surrounded by 12 nearest neighbours at a distance $r = \sigma$, six next-nearest neighbours at $r = \sqrt{2}\sigma$, 24 at $r = \sqrt{3}\sigma$, and so on. Figure 7.3a shows the number n of molecules to be found at a radial distance r away from any central or reference molecule. Ideally, the ordered crystalline structure extends indefinitely, i.e., there is *long-range order*.

At a higher temperature, but still below the melting point, the mean distance between the molecules increases slightly and the amplitude and frequency of molecular vibrations also increases; but the ordered lattice structure is maintained. The number density of molecules around the reference molecule now looks like Fig. 7.3b, and since it is now spread out in space rather than being a set of discrete lines one can no longer talk of the number n at r, but more of the probability of finding a molecule at r. In other words, the discrete values have become replaced by a *density distribution function*.

At the melting point the ordered lattice structure abruptly breaks down. This occurs once the magnitude of the molecular vibrations reaches a point where the molecules can move out of their confined lattice sites, i.e., they can now move past each other. Theoretically, for hard spheres, this ability should occur at a density of about 0.86 of the close-packed (van der Waals) solid at 0 K, corresponding to a volume expansion of about 16%. This is borne out by experiments where on melting, the increased volume is usually of this order, especially for spherical molecules (cf. 15–16% for Ne, Ar, Kr, Xe and CH_4, 17.5% for *n*-hexane, 20% for benzene). But it can be much larger (e.g., 31% for polyethylene) or much smaller, and even negative as for water (Bondi, 1968).

But even in the liquid state the molecules are still very much restricted in their motion and in the way they can position themselves with respect to each other. The 16% increase in the space available is not much—it corresponds to only a 5% increase in the mean intermolecular separation,

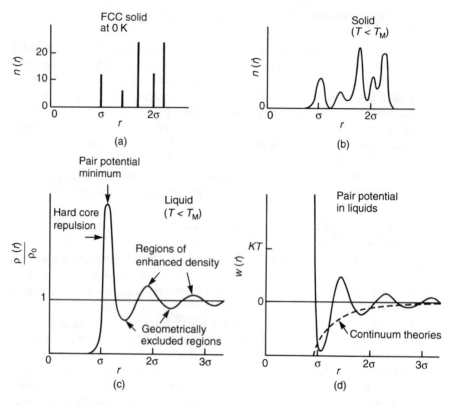

Fig. 7.3. Radial density distribution functions: (a) for a close-packed FCC solid at 0 K; (b) for a solid at a finite temperature below its melting point T_M; (c) for a liquid; such data are obtained from x-ray and neutron scattering experiments on liquids. (d) Pair potential for two molecules in a liquid whose density distribution function is shown on the left, computed from $\rho(r) = \rho_0 \exp[-w(r)/kT]$.

and the tendency to pack into an ordered lattice persists in the liquid. Thus, our spherical reference molecule will now have only slightly fewer nearest neighbours, about nine to 11 instead of 12, which will tend to group around it in a non-random way. The next-nearest neighbours will likewise order around the first group but with a smaller *degree of correlation*. Eventually, at larger distances, there will be no correlation at all with respect to our reference molecule. This *short-range order* extends over a few molecular diameters and characterizes the liquid state (recall that in the crystalline solid the order extends indefinitely).

Thus, radially away from any reference molecule the liquid density profile $\rho(r)$ looks like Fig. 7.3c. The density profile is usually plotted as $\rho(r)/\rho_0$ against r, where ρ_0 is the density of the bulk liquid. $\rho(r)/\rho_0$ is commonly referred to as the *radial distribution function* or the *pair correlation function*. At large distances it always approaches unity as $\rho(r)$ approaches ρ_0.

The magnitude and range of molecular ordering in liquids is enhanced by increasing the external pressure and lowering the temperature, as may be expected. Rigid hard-sphere molecules exhibit more short-range order or structure than easily deformable molecules, such as hexane, where internal bond flexibility allows for greater configurational freedom to pack in different ways. The net effect, however, is that the density $\rho(r)$ usually oscillates with distance away from any molecule and only reaches the bulk liquid value ρ_0 at some distance away. Thus, in general, a liquid medium near a molecule or surface will not have the properties of a structureless continuum, for which the density would be equal to the bulk value right up to contact.

7.6   EFFECT OF LIQUID STRUCTURE ON MOLECULAR FORCES

For two solute molecules or particles 1 dissolved in a liquid medium 2, the problem becomes exceedingly complicated. Three pair correlation functions may now be identified: $\rho_{11}(r)$ for the *solute–solute* density profile (i.e., the variation of $\rho_1(r)$ away from a solute molecule within the solvent medium), $\rho_{22}(r)$ for the *solvent–solvent* density profile, and $\rho_{12}(r)$ for the *solute–solvent* density profile. Each of these will exhibit oscillatory behaviours that are interdependent, depending on the sizes of the molecules and on the nature of the solute–solute, solute–solvent and solvent–solvent forces. For example, solute–solvent interactions will affect the ordering of solvent molecules around a solute molecule, which in turn affects the solute density profile around the solute molecule.

How does all this affect the interaction between two dissolved molecules? In Chapter 2 we saw that the effective pair potential (or potential of mean force) between two solute molecules at a distance r apart is related to their density $\rho(r)$ at r by the Boltzmann relation:

$$\rho_{11}(r)/\rho_0 = \exp[-w_{11}(r)/kT]. \tag{7.7}$$

Thus, $w_{11}(r)$ tends to zero at large r, where $\rho(r)/\rho_0 \to 1$, but oscillates with distance at smaller r. $w_{22}(r)$ behaves similarly. Such a pair potential is shown schematically in Fig. 7.3d. We see therefore how liquid structure can dramatically influence the interaction between dissolved molecules already at large distances. Unfortunately there is a lack of any direct experimental data concerning the effective pair potentials $w(r)$ of solute molecules and small particles interacting in a solvent medium and, in particular, how reliable continuum theories are in determining the first potential-energy minimum (see Fig. 7.3d and Problem 7.5). But this is not the case for interacting surfaces, and in Chapter 13 we shall see how solvent structure arises and affects the forces between macroscopic surfaces at distances below a few molecular diameters, for which both theoretical and experimental data are now available.

PROBLEMS AND DISCUSSION TOPICS

7.1 Derive Eq. (7.2).
(Note that for a FCC or HCP lattice of close-packed spheres the volume fraction occupied by the spheres is 0.7405, which is the maximum possible for an assembly of identical spheres. However, the maximum volume fraction of randomly packed spheres is only 0.64. This is known as *random close packing*, which is 86% of *close packing*.)

7.2 Plot temperature–composition phase diagrams for the four cases where the forces between the solute molecules in a solvent are as drawn in Fig. 7.4.

7.3 A colloidal dispersion consists of irregularly shaped particles in aqueous solution. Initially the forces between the particles are repulsive so that they remain dispersed. The pH of the solution is then changed so that the forces become attractive (see Chapter 12) and the particles aggregate into 'flocks'

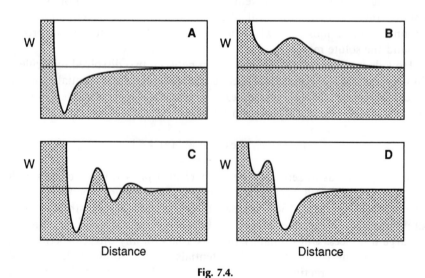

Fig. 7.4.

and fall out of the solution. It is commonly observed that when the precipitation occurs rapidly the density of the particles in the flocks is low, while if the precipitation is slow the particle density is higher and approaches that of random close packing. Explain this correlation between flocculation time and particle density.

7.4 Describe, with sketches, the different types of molecular ordering that occur in (i) smectic liquid crystals, (ii) nematic liquid crystals and (iii) a condensed liquid composed of spherical 0.5 nm diameter molecules confined between two solid surfaces where the gap width is 0.5 nm, 0.6 nm, 0.7 nm, 0.8 nm, 0.9 nm and 1.0 nm, respectively.

7.5 Pettitt and Rossky (1986) and Dang *et al.* (1990) used analytical and computer simulation techniques to calculate the pair potential between a Na^+ and a Cl^- ion in water. Compare their oscillatory potentials with that expected from the continuum Coulomb interaction in water using the ionic radii of Fig. 7.1. Comment on your findings.

CHAPTER 8

SPECIAL INTERACTIONS: HYDROGEN-BONDING, HYDROPHOBIC AND HYDROPHILIC INTERACTIONS

8.1 THE UNIQUE PROPERTIES OF WATER

Water is such an unusual substance that it has been accorded a special place in the annals of phenomena dealing with intermolecular forces, and two types of 'special interactions'—the *hydrogen bond* and the *hydrophobic effect*—are particularly relevant to the interactions of water. The literature on the subject is vast (Franks, 1972–82), not only because water is the most important liquid on earth, but also because it has so many interesting and anomalous properties.

For a liquid of such a low molecular weight, water has unexpectedly high melting and boiling points and latent heat of vaporization. There are, of course, many other substances of low molecular weight and high melting and boiling points, but these are invariably ionic crystals or metals whose atoms are held together by strong Coulombic or metallic bonds. These properties of water point to the existence of an intermolecular interaction that is stronger than that expected for ordinary, even highly polar, liquids.

The density maximum at 4°C exhibited by liquid water, and the unusual phenomenon that the solid (ice) is lighter than the liquid, indicates that in the ice lattice the molecules prefer to be farther apart than in the liquid. We may further conclude that the strong intermolecular bonds formed in ice persist into the liquid state and that they must be strongly orientation-dependent since water adopts a tetrahedral coordination (four nearest neighbours per molecule) rather than a higher packing density (cf. 12 nearest neighbours characteristic of close-packed van der Waals solids where the 'bonds' are non-directional). Water has other unusual properties, such as a very low compressibility and unusual solubility properties both as a solute and as a solvent (discussed later in this chapter).

If liquid water is strange, solid water (ice) is even stranger (Hobbs, 1974). The high molecular dipole moment and high dielectric constant of liquid

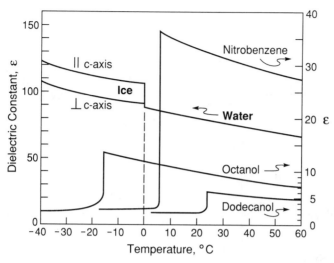

Fig. 8.1. Static dielectric constants as a function of temperature of some 'normal' polar liquids and water. For all the 'normal' liquids ε falls abruptly on freezing to a value that is close to the square of the refractive index, indicating that the thermal rotations of the dipolar molecules have stopped. However, for water ε rises on freezing, and continues to rise down to −70°C, after which it falls. The high polarizability of water is believed to arise from proton hopping along the H-bond network (see Fig. 8.2a) rather than molecular rotations. Data compiled from Landolt-Börnstein (1982), Hasted (1973) and Hobbs (1974). For theories of the dielectric constant of ice see Pauling (1935) and Hollins (1964).

water may at first appear to be related via the Debye–Langevin and Clausius–Mossotti equations (5.7) and (5.28). But unlike any typical polar liquid whose dielectric constant falls abruptly as it solidifies due to the freezing of the molecular (i.e., dipolar) rotations, when water freezes into ice the dielectric constant actually increases (Fig. 8.1), and it is still increasing at −70°C (Hasted, 1973). It is highly unlikely that this phenomenon can be explained in terms of the conventional picture of rotating dipolar molecules. The proton conductivity and mobility in ice is also higher than in the liquid. Both these phenomena suggest that the ice lattice affords some easy pathways for the movement of charges, particularly protons, perhaps by a 'proton hopping' mechanism along the hydrogen-bonding network. Such a mechanism could well persist into the liquid state. Thus, to understand the secrets of liquid water one may first have to unravel those of ice.

8.2   THE HYDROGEN BOND

The previous section indicates that some unusually strong and orientation-dependent bonds are involved in the interactions between water molecules.

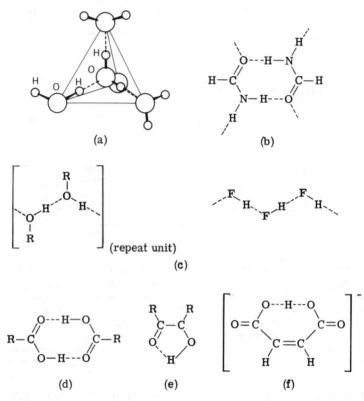

Fig. 8.2. Different types of hydrogen bonds and hydrogen-bonded structures. Linear hydrogen bonds have the lowest energy, but some H bonds with a —H ⋯ angle of 120° or less also occur. (a) Three-dimensional structures (e.g., ice). (b) Two-dimensional (layered) structures (e.g., formamide). (c) One-dimensional (chain and ring) structures (e.g., alcohols, HF). (d) Dimers (e.g., fatty acids). (e) Intramolecular H bond (not always linear). (f) Symmetric H bond (H atom shared).

It is a straightforward matter to ascertain which bond is responsible for this interaction by simply looking at the distances between various atomic centres in the ice lattice (Fig. 8.2a). The *intra*molecular O—H distance is about 0.10 nm, as expected for this covalent bond (see Fig. 7.1), but the *inter*molecular O ⋯ H distance is only 0.176 nm, much less than the 0.26 nm expected from summing the two van der Waals radii but still larger than the covalent distance of 0.10 nm. Thus, the intermolecular O ⋯ H bond is implicated, which at first sight appears to possess some covalent character. Such bonds are known as *hydrogen bonds*, and the reader is referred to Schuster *et al.* (1976) and Joesten and Schaad (1974) for the voluminous literature on the subject.

Hydrogen bonds are not unique to water; they exist to varying degrees between electronegative atoms (e.g., O, N, F and Cl) and H atoms covalently bound to similar electronegative atoms. These bonds are special in that they only involve hydrogen atoms, which, by virtue of their tendency to become positively polarized and their uniquely small size, can interact strongly with nearby electronegative atoms resulting in an effective H-mediated 'bond' between two electronegative atoms (see Section 4.7).

Originally, it was believed that the hydrogen bond was quasi-covalent and that it involved the sharing of an H atom or proton between two electronegative atoms. But it is now accepted (Coulson, 1961; Umeyama and Morokuma, 1977) that the hydrogen bond is predominantly an electrostatic interaction. With few exceptions, the H atom is not shared but remains closer to and covalently bound to its parent atom; accordingly, the hydrogen bond between two groups XH and Y is usually denoted by X—H $\cdots$ Y. Nevertheless, certain characteristics of hydrogen bonds do make them appear like weak covalent bonds. For example, they are not only fairly strong but also (fairly) directional. This endows them with the ability to form weak three-dimensional 'structures' in solids, whereas in liquids the short-range order can be of significantly longer range whenever hydrogen bonds are involved—hence the term *associated liquids*.

The strengths of most hydrogen bonds lie between 10 and 40 kJ mol^{-1} (Joesten and Schaad, 1974), which makes them stronger than a typical van der Waals 'bond' ($\sim$1 kJ mol^{-1}) but still much weaker than covalent or ionic bonds ($\sim$500 kJ mol^{-1}).

Hydrogen bonds can occur intermolecularly as well as intramolecularly and can happily exist in a non-polar environment. They are consequently particularly important in macromolecular and biological assemblies, such as in proteins, linking different segments together inside the molecules, and in nucleic acids, where they are responsible for the stability of the DNA molecule (holding the two helical strands together). Their involvement in setting up one-, two- and three-dimensional macromolecular structures is sometimes referred to as *hydrogen-bond polymerization* (illustrated in Fig. 8.2).

8.3   Models of water and associated liquids

Hydrogen bonds play a particularly prominent role in water since each oxygen atom with its two hydrogens can participate in four such linkages with other water molecules—two involving its own H atoms and two involving its unshared (lone-pair) electrons with other H atoms. To see exactly how this arises we require some picture of the charge distribution

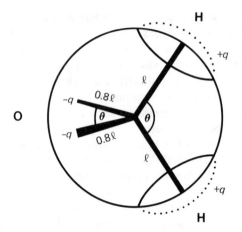

Fig. 8.3. ST2 model of water molecule; $q = 0.24e$, $l = 0.1\,nm$, $\theta = 109°$.

within the water molecule. Various models have been proposed of which the so-called ST2 model of water (named after a modified model by Stillinger and Rahman, 1974) will be described. Other models also exist (Allen and Tildesley, 1987; Ciccotti *et al.*, 1987), but it is recognized that as yet no single model has been able to account satisfactorily for the properties of water in all three phases (ice, liquid and vapour). However, the main features of ST2 water are simple and quite similar to some of the other models and so provide a good introduction to the various conceptual approaches that are being applied in attempts at understanding water.

In ST2 water (Fig. 8.3) the water molecule is modelled with charges of $+0.24e$ centred on each hydrogen atom and two compensating charges of $-0.24e$ on the opposite side of the oxygen atom, representing the two unshared electron pairs. The four charges are located along four tetrahedral arms radiating out from the centre of the O atom. The interaction between any two water molecules is assumed to involve an isotropic Lennard–Jones potential and 16 Coulombic terms representing the interaction between each of the four point charges on one molecule with the four on the other. The net Coulombic interaction obviously depends on the mutual orientation of the two molecules. When many molecules are involved their equilibrium configuration can only be solved on a computer, and when this is done the model can account for a number of the unusual properties of water, such as the highly open ice structure and the density maximum in the liquid state. Computer simulations show that this comes about because of the strong preference for the molecules to order into a lattice where each oxygen is tetrahedrally coordinated with four other oxygens, with a hydrogen atom lying in the line joining two oxygen atoms. It is this preferred linearity of the O—H $\cdots$ O bond in water that endows it with its strongly directional nature.

In liquid water the tendency to retain the ice-like tetrahedral network remains, but the structure is now disordered and labile. The average number of nearest neighbours per molecule *rises* to about five (hence the higher density of water on melting), but the mean number of H bonds per molecule *falls* to about 3.5 whose lifetimes are estimated to be about 10^{-11} s. Other strongly hydrogen-bonding molecules, such as formamide, ammonia and HF, also retain some of their ordered crystalline structure in the liquid state over short distances. Such liquids are known as *associated liquids*. It is also believed that the H-bond structure in such liquids is *cooperative* in the sense that the presence of H bonds between some molecules enhances their formation in nearby molecules, thereby tending to propagate the H-bonded network. If so, the interaction is non-pairwise additive, which presents serious problems in theoretical computations of aqueous and other systems involving cooperative associations.

It is instructive to note that the tetrahedral coordination of the water molecule is at the heart of the unusual properties of water, much more than the hydrogen bonds themselves. As a rule of thumb, molecules that can participate in only two H bonds can link up into a one-dimensional chain or ring (e.g., HF and alcohols, as shown in Fig. 8.2). Likewise, atoms of valence two such as selenium and tellurium can form long chains of covalently bonded atoms. Atoms that can participate in three bonds (e.g., arsenic, antimony and carbon in graphite), can form two-dimensional sheets or layered structures held together by weaker van der Waals forces. But only a tetrahedral, or higher, coordination allows for a *three*-dimensional network to form. For example, it is the tetrahedral coordination characteristic of carbon and silicon that results in their almost infinite variety of atomic associations whether in chain molecules (e.g., polymers, surfactants, polypeptides), cyclic compounds or two- and three-dimensional crystals (e.g., diamond, silica, sheet silicates).

8.4 RELATIVE STRENGTHS OF DIFFERENT TYPES OF INTERACTIONS

In earlier chapters we saw how for molecules where only one type of attractive force is present (e.g., pure Coulombic or pure dispersion) their molar cohesive energies and other properties can be computed reasonably accurately. Usually, however, two or more interactions occur simultaneously and it becomes difficult to apply simple potential functions, especially when orientation-dependent dipolar and H-bonding interactions are involved. In

TABLE 8.1 Relative strengths of different types of interactions as reflected in the boiling points of compounds[a]

Molecule		Molecular weight	Dipole moment (D)	Boiling point (°C)
Ethane	CH_3CH_3	30	0	−89
Formaldehyde	HCHO	30	2.3	−21
Methanol	CH_3OH	32	1.7	64
n-Butane	$CH_3CH_2CH_2CH_3$	58	0	−0.5
Acetone	CH_3COCH_3	58	3.0	56.5
Acetic acid	CH_3COOH	60	1.5	118
n-Hexane	$CH_3(CH_2)_4CH_3$	86	0	69
Ethyl propyl ether	$C_5H_{12}O$	88	1.2	64
1-Pentanol	$C_5H_{11}OH$	88	1.7	137

[a] In order to make comparisons meaningful, molecules have been put into three groups of similar molecular weights and size. Within each group the first molecule is non-polar and interacts purely via dispersion forces, the second is polar and the third also interacts via H-bonds.

spite of this complexity, some general patterns do emerge when we compare the boiling points of different compounds, which is a measure of the cohesive forces holding molecules together in condensed phases (Table 2.1). Such a comparison is made in Table 8.1, where we see that the weakest interactions are the dispersion and dipolar interactions, followed by H-bonding interactions. Then there is a large jump to the much stronger covalent and ionic interactions, not shown in Table 8.1. Note (i) the dominance of H-bonding forces even in very polar molecules such as acetone, and (ii) the increasing importance of dispersion forces for larger molecules. However, though dispersion forces are the ones mainly responsible for bringing molecules together, they lack the discrimination, specificity and directionality of dipolar and H-bonding interactions, and it is these that often determine the fine and subtle details of molecular and macromolecular structures, such as those of molecular crystals, polypeptides (e.g., proteins) polynucleotides (e.g., DNA and RNA), micelles and biological membranes.

8.5   THE HYDROPHOBIC EFFECT

So far in this chapter we have considered the interactions of water molecules with other water molecules. In the rest of this chapter we shall investigate the equally interesting interactions of water with other compounds, i.e., when water acts as a solvent or as a solute.

The strong inclination of water molecules to form H bonds with each other influences their interactions with non-polar molecules that are incapable of forming H bonds (e.g., alkanes, hydrocarbons, fluorocarbons, inert atoms). When water molecules come in contact with such a molecule they are faced with an apparent dilemma: whichever way the water molecules face, it would appear that one or more of the four charges per molecule will have to point towards the inert solute molecule and thus be lost to H-bond formation. Clearly the best configuration would have the least number of tetrahedral charges pointing towards the unaccommodating species so that the other charges can point towards the water phase and so be able to participate in H-bond attachments much as before. There are many options to salvaging lost H bonds. If the non-polar solute molecule is not too large, it is possible for water molecules to pack around it without giving up *any* of their hydrogen-bonding sites. Examples of such arrangements are shown in Fig. 8.4. Since we have already established (Chapter 6) that the dispersion interaction between water and hydrocarbons is not very different from that between water or, for that matter, between hydrocarbon and hydrocarbon, we see that the main effect of bringing water molecules and non-polar molecules together is the *reorientation* of the water molecules so that they can participate in H-bond formation more or less as in bulk water (i.e., without necessitating any breakage of H bonds).

Thus, thanks to the uncanny ability of tetrahedrally coordinated molecules to link themselves together around almost any inert molecule, whatever its size or shape, the apparent dilemma mentioned earlier is often easily solved. Indeed, since water molecules in the liquid state participate on average in ~ 3.0–3.5 H bonds, it would appear from Fig. 8.4 that around an inert solute molecule the water molecules actually have a higher coordination (of four) and thus be even more ordered than in the bulk liquid. It is also clear that the sizes and shapes of non-polar solute molecules are fairly critical in determining the water structure adopted around them. This is often referred to as *hydrophobic solvation* or *hydrophobic hydration*. At present there is no simple theory of such solute–solvent interactions. However, both theoretical and experimental studies do indicate that the reorientation, or restructuring, of water around non-polar solutes or surfaces is *entropically* very unfavourable, since it disrupts the existing water structure and imposes a new and *more ordered* structure on the surrounding water molecules.

It is for this reason that the hydrocarbons are so sparingly soluble in water, characterized by a highly unfavourable free energy of solubilization that is mainly entropic and, as we saw in Chapter 6, cannot be accounted for by continuum theories of van der Waals forces. For example, the free energy of transfer of methane and *n*-butane molecules from bulk liquid into water at 25°C is about 14.5 and 24.5 kJ mol^{-1}, respectively. For *n*-butane, this is split

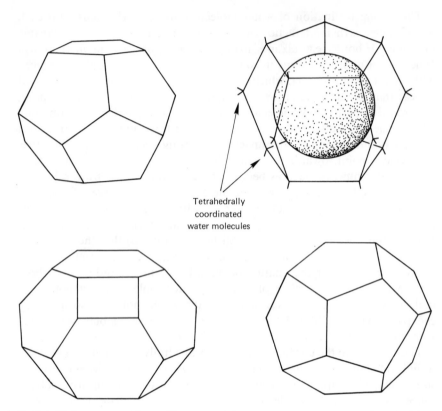

Tetrahedrally
coordinated
water molecules

Fig. 8.4. Clathrate 'cages' formed by water molecules around a dissolved non-polar solute molecule. Such structures are not rigid but labile, and their H bonds are not stronger than in pure water, but the water molecules forming these cages are more ordered than in the bulk liquid.

up as follows:

$$\Delta G = \Delta H - T\Delta S = -4.3 + 28.7 = +24.5 \text{ kJ mol}^{-1}.$$

Thus, the decrease in entropy contributes 85% to this interaction, and for many other hydrocarbons (e.g., benzene) the entropic contribution to ΔG is even higher.

It is also observed that for different hydrocarbons, the free energy of transfer is roughly proportional to the surface areas of the molecules, an indication that the number of reoriented water molecules is more or less determined by the non-H-bonding areas exposed to them. Thus, for methane, whose van der Waals radius is $a = \sigma/2 = 0.2$ nm, the surface area per molecule is about $4\pi a^2 \approx 0.50 \, \text{nm}^2$. The free energy of transfer, $\Delta G = 14.5 \, \text{kJ mol}^{-1}$, when calculated per unit surface area turns out to be about $48 \, \text{mJ m}^{-2}$. Similarly for the cylindrically-shaped n-butane molecule whose surface area is given by (see Section 7.1) $4\pi a^2 + 2\pi a(3 \times 0.1275) \approx 1.0 \, \text{nm}^2$. Using the above value of $\Delta G = 24.5 \, \text{kJ mol}^{-1}$, we deduce a corresponding surface free energy of $41 \, \text{mJ m}^{-2}$. These values are very close to the interfacial free energies γ_i of bulk hydrocarbon–water interfaces which generally lie between 40 and $50 \, \text{mJ m}^{-2}$. The high surface tension of water, about $72 \, \text{mJ m}^{-2}$, may also be taken as an example of this mainly entropic effect since air behaves as an inert, non-H-bonding 'medium' in this sense.

Conversely, water molecules are highly insoluble in inert solvents, but here the effective surface energy of transfer is significantly larger than $50 \, \text{mJ m}^{-2}$ (see Problem 8.2). Clearly, when water molecules in the bulk liquid have to rearrange their coordination so as to accommodate a foreign solute molecule the price is high; but when a single water molecule is completely extracted out from its water network the price is even higher. We shall return to considerations of surface energies and their relation to hydrophobic and other surface phenomena in later chapters.

The immiscibility of inert substances with water, and the mainly entropic nature of this incompatibility is known as the *hydrophobic effect* (Kauzmann, 1959; Tanford, 1980), and such substances, e.g., hydrocarbons and fluorocarbons, are known as *hydrophobic substances*.[1] Similarly, *hydrophobic surfaces* are not 'wetted' by water; when water comes into contact with such surfaces it rolls up into small lenses and and subtends a large contact angle on them (see Chapter 15).

8.6 THE HYDROPHOBIC INTERACTION

Closely related to the hydrophobic *effect* is the hydrophobic *interaction*, which describes the unusually strong attraction between hydrophobic molecules and surfaces in water—often stronger than their attraction in free space. For example, the van der Waals interaction energy between two contacting methane molecules in free space is $-2.5 \times 10^{-21} \, \text{J}$ (Table 6.1), while in water it is $-14 \times 10^{-21} \, \text{J}$. Similarly, the surface tensions of most saturated hydrocarbons lie in the range 15–$30 \, \text{mJ m}^{-2}$, while their interfacial tensions with water are in the range 40–$50 \, \text{mJ m}^{-2}$. In Section 6.7 we saw that this strong interaction in water cannot be accounted for by continuum theories

---

[1] Hydrophobic means 'water-fearing'. But note that the interaction between a hydrophobic molecule and water is actually attractive, due to the dispersion force. However, the interaction of water with itself is much more attractive. Water simply loves itself too much to let some substances get in its way.

of van der Waals forces, which actually predict a *reduced* interaction in water. Because of its strength it was originally believed that some sort of 'hydrophobic bond' was responsible for this interaction. But it should be clear from what has been described above that there is no bond associated with this mainly entropic phenomenon, which arises primarily from the rearrangements of H-bond configurations in the overlapping solvation zones as two hydrophobic species come together, and is therefore of much longer range than any typical bond.

To date there have been very few direct measurements of the hydrophobic interaction between dissolved non-polar molecules, mainly because they are so insoluble. Tucker *et al.* (1981) reported values of -8.4 and -11.3 kJ mol^{-1} for the free energies of dimerization of benzene–benzene and cyclohexane–cyclohexanol, respectively, and Ben Naim *et al.* (1973) deduced a value of about -8.5 kJ mol^{-1} for two methane molecules. On the theoretical side the problem is horrendously difficult because the hydrophobic interaction between two molecules is much more complex, involves many other molecules and is of longer range than that arising from any simple additive pair potential. The complex connectivities of the H-bonding network must also be included in any theory or computer simulation. Thus, there is as yet no satisfactory theory of the hydrophobic interaction, though a number of promising theoretical approaches have been proposed (Dashevsky and Sarkisov, 1974; Pratt and Chandler, 1977; Marcelja *et al.*, 1977; Pangali *et al.*, 1979; Luzar *et al.*, 1987; reviewed by Nicholson and Parsonage, 1982).

Israelachvili and Pashley (1982b) measured the hydrophobic force law between two macroscopic curved hydrophobic surfaces in water and found that in the range 0–10 nm the force decayed exponentially with distance with a decay length of about 1 nm (an exponential distance dependence for the interaction had previously been proposed by Marcelja *et al.*, 1977). Based on these findings Israelachvili and Pashley proposed that for small solute molecules, the hydrophobic free energy of dimerization increases in proportion with their diameter σ according to

$$\Delta G \text{ (hydrophobic pair potential)} \approx -20\sigma \text{ kJ mol}^{-1}, \qquad (8.1)$$

where σ is in nanometres. For example, for cyclohexane ($\sigma = 0.57$ nm), this gives $\Delta G \approx -11.4$ kJ mol^{-1} in agreement with the measured value of -11.3 kJ mol^{-1}.

The hydrophobic interaction plays a central role in many surface phenomena, in molecular self-assembly, in micelle formation, in biological membrane structure, and in determining the conformations of proteins (Kauzmann, 1959; Tanford, 1980; Israelachvili *et al.*, 1980a; Israelachvili, 1987a), and we shall encounter this interaction again in later chapters.

□   ● **WORKED EXAMPLE** ●

Question: The hydrophobic attraction between molecules such as surfactants, proteins and polymers in water results in their spontaneous self-assembly into large well-ordered structures such as micelles and biological organelles (see Fig. 13.12). This association appears to increase the order and thus lower the entropy of the universe, in contradiction with the second law of thermodynamics. Resolve this paradox.

Answer:
- This is more of a discussion topic and involves a number of issues.
- There is usually more than one contribution to the total entropy change during any process. While it is true that the coming together of the hydrophobic (solute) molecules is associated with a decrease in their entropy of mixing, this is more than offset by the increase in the configurational entropy of the water (solvent) molecules, as discussed at the end of Sections 4.9 and 6.7.
- The above statement, while true, nevertheless does not go to the heart of this problem which is actually much more fundamental, since it can be posed for any attractive pair potential that leads to association, whether physical or chemical, with or without a solvent. Association necessarily implies the formation of a two-phase system, with some molecules remaining behind in the dilute phase, and *their* entropy—which must also be taken into account—increases. When the entropy changes of all the molecules involved in this spontaneous aggregation process are added up, including any heat of reaction, the net result will always be an increase.

□

8.7   Hydrophilicity

While there is no phenomenon actually known as the hydrophilic effect or the hydrophilic interaction, such effects can be recognized in the propensity of certain molecules and groups to be water soluble and to *repel* each other strongly in water, in contrast to the strong attraction exhibited by hydrophobic groups. Hydrophilic (i.e., water-loving) groups prefer to be in contact with water rather than with each other, and they are often *hygroscopic* (taking up water from vapour). As might be expected, strongly hydrated ions and zwitterions (discussed in Chapter 4) are hydrophilic; but some uncharged and even non-polar molecules can be hydrophilic if they have the right geometry and if they contain electronegative atoms capable of associating

TABLE 8.2 Hydrophilic groups and surfaces

Molecules and ions
Alcohols (CH_3OH, C_2H_5OH, glycerol)
Sugars (glucose, sucrose), urea
Polyethylene oxide ($—CH_2CH_2O—)_n$

Molecular groups[a]		
Anionic		
Carboxylate	$—COO^-$	
Sulphonate	$—SO_3^-$	
Sulphate	$—SO_4^-$	
Phosphate ester	$—OPO_2\,O—$	
Cationic		
Trimethyl ammonium	$—N^+(CH_3)_3$	
Dimethyl ammonium	$>N^+(CH_3)_2$	

Soluble proteins
DNA
Li^+, Mg^{2+}, Ca^{2+}, La^{3+}

Zwitterionic
$—OPO_2^-\,OCH_2CH_2N^+(CH_3)_3$

Phosphatidylcholine (lecithin)

Polar (non-ionic)
$—NH_2$
$—NO(CH_3)_2$
$—SOCH_3$
$—PO(CH_3)_2$

Amine
Amine oxide
Sulphoxide
Phosphine oxide

Polar groups that are not hydrophilic when attached to a long hydrocarbon (hydrophobic) chain[a]

Alcohol	$—OH$	Amide	$—CONH_2$
Ether	$—OCH_3$	Nitroalkanes	$—NO$, $—NO_2$
Mercaptan	$—SH$	Aldehyde	$—CHO$
Amines	$—NH(CH_3)$, $—N(CH_3)_2$	Ketone	$—COCH_3$

Solid surfaces
Silica below 600°C (surface characterized by Si—OH groups)
Swelling clays (montmorillonite)
Chromium, gold (when clean)

[a] Compiled from a longer list given by Laughlin (1978, 1981).

with the H-bond network in water, for example, the O atoms in alcohols and polyethylene oxide and the N atoms in amines. Table 8.2 lists some common hydrophilic molecules and molecular groups, as well as some hydrophilic surfaces that are wetted by water or that repel each other in aqueous solutions. From this table we see that a polar group is not necessarily hydrophilic and that a non-polar group is not always hydrophobic!

While hydrophobic molecules tend to increase the ordering of water molecules around them, hydrophilic molecules are believed to have a disordering effect. Certain highly polar molecules are so effective in altering or disrupting the local water structure that when dissolved in water they can have a drastic effect on other solute molecules. For example, when urea, $(NH_2)_2C{=}O$, is dissolved in water it can cause proteins to unfold. Such non-ionic but highly potent compounds are commonly referred to as *chaotropic agents*, a term that was coined to convey the idea that their disruption of the local water structure leads to chaos (not the least of which being produced in the minds of those trying to understand this phenomenon).

The degree of hydrophilicity of molecular groups is important for understanding the interactions and associations of *amphiphilic molecules* such as surfactants, lipids and certain copolymers (Part III). In such molecules one end contains a hydrophilic group while the rest of the molecule is hydrophobic, usually a long hydrocarbon chain. The hydrophilicity of some hydrophilic groups, for instance the OH group, can be completely neutralized when they are attached to a long *n*-alkyl chain such as $-(CH_2)_{11}CH_3$ (see Table 8.2).

It appears therefore that the hydrophilic and hydrophobic interactions, unlike electrostatic and dispersion interactions, are interdependent and therefore not additive. Indeed, one would not expect them to be independent of each other, since both ultimately rely on the structure of the water H-bonds adopted around dissolved groups. In Section 16.6 we shall see that the hydrophobic energy per CH_2 group of an alkane chain is much reduced when a hydrophilic head-group is attached to the end of the chain. We shall also see that both the hydrophobic and the hydrophilic (hydration) interactions as measured between macroscopic surfaces in water are of relatively long range, extending well beyond a few molecular diameters, and that when both hydrophilic and hydrophobic groups are present on a surface the net interaction between two such surfaces is not necessarily the sum of the separate components.

PROBLEMS AND DISCUSSION TOPICS

8.1 Are there any compounds other than water that expand on freezing?

In what ways are their bonding and physical properties unusual and/or similar to those of water?

8.2   The solubilities of cyclohexane (C_6H_{12}) and benzene (C_6H_6) in water are, respectively, 55 and 1800 parts per million by weight at 20°C. Calculate the solubilities in mole fraction units. Assuming that these organic molecules can be considered as macroscopic spheres with radii as given in Fig. 7.1, calculate their interfacial energies per unit surface area exposed to water and compare your results with the known values for γ_{12} given in Table 15.1.

The measured solubility of water in cyclohexane and benzene is, respectively, 59 and 620 parts per million by weight at 20°C. By using the same argument as above and a value of 0.14 nm for the radius of the water molecule, calculate the values for γ_{12} obtained from this reciprocal set of solubility data. Comment on possible reasons for the total lack of agreement in this case.

8.3   The strength of the attractive hydrophobic interaction increases with temperature. This often leads to molecular association above some critical temperature, rather than the more typical case where increasing the temperature leads to dissociation. Draw and describe the full temperature-composition phase diagrams for (i) a 'typical' two-component liquid–liquid system and (ii) a hydrophobic solute in water.

8.4   Polymethylene oxide, $[-CH_2-O-]_n$, is hydrophobic, but polyethylene oxide, $[-CH_2-CH_2-O-]_n$, which has one more hydrophobic CH_2 group per segment is hydrophilic and miscible with water. Give possible reasons for this.

PART TWO

THE FORCES BETWEEN
PARTICLES AND SURFACES

CHAPTER 9

SOME UNIFYING CONCEPTS IN INTERMOLECULAR AND INTERPARTICLE FORCES

9.1 FACTORS FAVOURING THE ASSOCIATION OF LIKE MOLECULES OR PARTICLES IN A MEDIUM

In this second part we shall be looking at the physical forces between particles and surfaces. While the fundamental forces involved are the same as those already described (i.e., electrostatic, van der Waals, solvation forces), they can manifest themselves in quite different ways and lead to qualitatively new features when acting between large particles or extended surfaces. These differences will be discussed in the next chapter. In this chapter we shall look at the similarities, and see how the ideas developed in Part I also apply to the interactions of macroscopic particles and surfaces. We shall find that, independently of the type of interaction force involved, certain semiquantitative relations describing molecular forces—known as *combining relations*—are applicable quite generally to all systems (i.e., to the interactions of molecules, particles, surfaces, and even complex multicomponent systems).

Let us start by noting that for any type of interaction between two molecules A and B, the interaction energy at any given separation is always (to a good approximation) proportional to the product of some molecular property of A times some molecular property of B. Let us denote these properties by A and B. Referring to Fig. 2.2 we find, for example, that for the charge–non-polar-molecule interaction we may write $A \propto Q_A^2$ and $B \propto \alpha_B$; for the dipole–dipole interaction, $A \propto u_A$ or u_A^2 and $B \propto u_B$ or u_B^2; while for the dispersion interaction, we have $A \propto \alpha_A$ and $B \propto \alpha_B$. Note that even for the gravitational interaction (Eq. (1.1)), we may put $A \propto$ (mass of body A) and $B \propto$ (mass of body B).

Thus, for many different types of interactions, we may express the binding energies of molecules A and B in contact as

$$W_{AA} = -A^2, \qquad W_{BB} = -B^2 \qquad \text{(for like molecules)} \qquad (9.1)$$

and

$$W_{AB} = -AB \qquad \text{(for unlike molecules)}, \tag{9.2}$$

where only for the purely Coulombic charge–charge interaction are the negative signs reversed (cf. Fig. 2.2). Let us now consider a liquid consisting

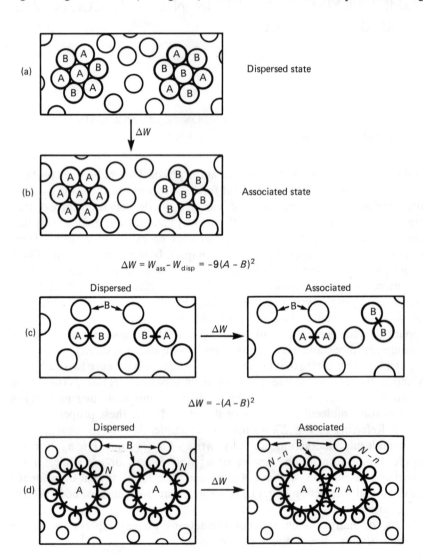

(a)   Dispersed state

$\downarrow \Delta W$

(b)   Associated state

$$\Delta W = W_{\text{ass}} - W_{\text{disp}} = -9(A - B)^2$$

Dispersed   Associated

(c) ΔW

$$\Delta W = -(A - B)^2$$

Dispersed   Associated

(d) ΔW

$$\Delta W = -n(A - B)^2$$

of a mixture of molecules A and B in equal amounts. If the molecules are randomly *dispersed* (Fig. 9.1a), then on average an A molecule will have both A and B molecules as nearest neighbours, and similarly for molecule B. However, if the molecules are *associated*, then the molecular organization of nearest neighbours around molecules A and B will be as in Fig. 9.1b. The difference in energy between the associated and dispersed clusters will therefore be $\Delta W = -9(A - B)^2$, where we note that in this two-dimensional case nine A—A 'bonds' and nine B—B 'bonds' have replaced 18 A—B 'bonds'. In three dimensions with 12 nearest neighbours around each central molecule, and starting with six A and six B molecules around each A and B molecule, we would find (Problem 9.1) that $\Delta W = -22(A - B)^2$ and that 22 A—A and 22 B—B 'bonds' have been formed on association. For the simplest case of two associating A molecules (Fig. 9.1c), we have $\Delta W = -(A - B)^2$.

Thus in general we may write

$$\Delta W = W_{ass} - W_{disp} = -n(A - B)^2 \tag{9.3}$$

where n is always equal to the number of like 'bonds' that have been formed in the process of association, irrespective of how many molecules are involved or their relative sizes (Fig. 9.1d). Since $(A - B)^2$ must always be positive we see that in general $\Delta W < 0$ (i.e., $W_{ass} < W_{disp}$). We may therefore conclude that the associated state of like molecules is energetically preferred to the dispersed state. In other words, *there is always an effective attraction between like molecules or particles in a binary mixture.*

Equation (9.3) can be developed further to provide more general insights into the interactions of like solute molecules and particles in a medium. First,

---

Fig. 9.1. (a) Two central molecules A and B surrounded by an equal number of A and B molecules in a solvent. Since there are three A—A 'bonds', three B—B 'bonds', and 18 A—B 'bonds' we may write $W_{disp} = -(3A^2 + 3B^2 + 18AB)$. (b) Seven A molecules and seven B molecules have associated in two small clusters. There are now 12 A—A 'bonds' and 12 B—B 'bonds' so that $W_{ass} = -12(A^2 + B^2)$. The net change in energy on going from the dispersed to the associated state is therefore $\Delta W = W_{ass} - W_{disp} = -9(A - B)^2$. Note that there is no change in the number of A and B molecules exposed to the solvent. Thus ΔW does not involve any term due to the interaction with the surrounding medium if it is unchanged during the redistribution of the A and B molecules. (c) Two A molecules associating in a medium of B molecules. Here $W_{disp} = -2AB$, and $W_{ass} = -(A^2 + B^2)$, so that $\Delta W = -(A - B)^2$. (d) Two large particles A associating in a medium of small solvent molecules B. This involves the replacement of $2n$ A—B 'bonds' by n A—A 'bonds' and n B—B 'bonds'. $W_{disp} = -2NAB$, $W_{ass} = -2(N - n)AB - nA^2 - nB^2$, so that $\Delta W = -n(A - B)^2$. Note that n is proportional to the area of adhesion or effective 'contact area' of the two particles which is proportional to their radii (see Worked Example in Section 9.1).

Eq. (9.3) may be expressed in a number of different forms:

$$\Delta W = -n(A - B)^2 = -n(\sqrt{-W_{AA}} - \sqrt{-W_{BB}})^2 \qquad (9.4)$$

$$= -n(A^2 + B^2 - 2AB) = n(W_{AA} + W_{BB} - 2W_{AB}), \qquad (9.5)$$

where ΔW may be readily seen to be the same as the interaction pair potential $w(\sigma)$ in the medium (at contact).

Second, since $-nW_{AA}$ and $-nW_{BB}$ are roughly proportional to the respective molar cohesion energies U_A and U_B we find that

$$\Delta W \propto -(\sqrt{U_A} - \sqrt{U_B})^2 \qquad (9.6)$$

which is essentially the same in Eq. (6.39), previously derived for the specific case of dispersion forces. If W_{AA} and W_{BB} are sufficiently different (i.e., if the molecules are very different; for example, A polar, B non-polar), then ΔW will be large enough to overcome the entropic drive to disorder resulting in a low solubility (immiscibility) or phase separation. The immiscibility of water and hydrocarbons and the 'like-dissolves-like' rule, previously discussed in Sections 6.7 and 6.8, are examples of this phenomenon. Furthermore, since $\Delta W \propto n$, larger particles or polymers of higher molecular weight are more likely to phase separate than smaller particles or molecules, and indeed the vast majority of polymers are immiscible with each other.

Third, Eq. (9.4) shows that the value of $\Delta W / n$ for molecules of type A coming into contact in medium B is the same as for the inverse case of molecules of type B associating in medium A. This reciprocity property was previously noted for the specific case of van der Waals forces (Section 6.7). Finally, Eq. (9.4) clearly shows that the interaction of two solute molecules in a solvent medium is intimately coupled to the strength of the solvent–solvent interaction. Thus, the attraction between two particles in water or between two protein molecules in a lipid bilayer is not independent of the surrounding water–water or lipid–lipid interactions.

While the above semiquantitative criteria have broad applicability, there are two very important exceptions: first, for the Coulomb interaction between charged atoms or ions, since the sign of ΔW is reversed, the dispersed state (Fig. 9.1a) is the favoured one. Thus, in ionic crystals (e.g., Na^+Cl^-), the cations and anions are nearest neighbours in the lattice. Second, hydrogen-bonding molecules do not readily fall into this simple behaviour pattern because the strength of the H bond between two different molecules cannot be expressed simply in terms of $W_{AB} = -AB$. For example, while an acetone molecule cannot form H bonds with another acetone molecule, it can do so

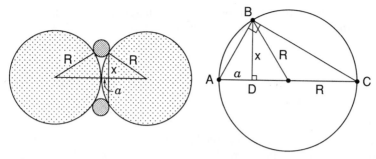

Fig. 9.2.

with water via its $C{=}O$ group, and for this reason, acetone is miscible with water. Such strongly hydrophilic groups or molecules (see Table 8.2) will repel each other in water due to their strong binding to water, and their specific solvation interactions in water, discussed in Chapters 8 and 13, cannot be described in terms of the simple equations presented here.

□   ● WORKED EXAMPLE ●

Question: Two macroscopic spherical particles of radius R are in adhesive contact as in Fig. 9.1d. What is their effective contact area?

Answer: This is a subtle problem. The answer depends on whether the forces are long-range or short-range compared to the sizes of the particles. However, since we are here considering adhesion, we may assume the forces to be of short range, effectively acting over a distance of the size of the solvent or particle molecules. Thus, referring to Fig. 9.2 (left), we need to determine the area that excludes solvent molecules, of radius a, between the surfaces. That is, we need to determine πx^2 in terms of R and a.

From the geometric construction of Fig. 9.2 (right) we apply Pythagoras' theorem: $AC^2 = AB^2 + BC^2 = AD^2 + BD^2 + BD^2 + DC^2$. Thus: $4R^2 = a^2 + 2x^2 + (2R - a)^2$, which simplifies to

$$x^2 = (2R - a)a \approx 2Ra \qquad \text{for } R \gg a. \qquad (9.7)$$

This relation, known as the 'chord theorem', is important for deriving many of the equations in later chapters. Thus, for two large spheres in contact their effective 'area of contact' or effective 'interaction area' is given by $\pi x^2 \approx 2\pi Ra$, where a is a measure of the range of the forces (usually of the order of molecular dimensions). Note that the effective area is proportional to R.

It is a simple matter to show that for two spheres of different radii, R_1 and R_2, their effective interaction area is $4\pi R_1 R_2 a/(R_1 + R_2)$.   □

9.2   TWO LIKE SURFACES COMING TOGETHER IN A MEDIUM: SURFACE AND INTERFACIAL ENERGY

The above approach may be readily applied to the interaction of two macroscopic surfaces in a liquid. Let us start with two flat surfaces of A, each of unit area, in a liquid B. We may equate ΔW of Eq. (9.5) with the (negative) free energy change of bringing these two surfaces into adhesive contact in the medium. This energy, or work, is defined as twice the *interfacial energy* γ_{AB} of the A–B interface, which is positive by convention. Thus,

$$\Delta W = -2\gamma_{AB}. \tag{9.8}$$

The factor of two arises because by bringing these two surfaces into contact the two initially separate media of A have merged into one, so that we have effectively eliminated *two* unit areas of the A–B interface. Now, let there be n bonds per unit area. In Eq. (9.5) nW_{AB} is therefore the energy change of bringing unit area of A into contact with unit area of B in vacuum. This is known as the *adhesion energy* or *work of adhesion* per unit area of the A–B interface. Likewise, nW_{AA} is the (negative) energy change of bringing unit area of A into contact with unit area of A in vacuum, known as the *cohesion energy* or *work of cohesion*. Note that two unit areas of A are eliminated in this process. By convention, the cohesion energy is related to the (positive) *surface energy* γ_A by $nW_{AA} = -2\gamma_A$. Similarly, $nW_{BB} = -2\gamma_B$. These different surface energies will be discussed in more detail in Chapter 15; for the moment, we simply note that for two surfaces Eq. (9.5) may be expressed in the form

$$\gamma_{AB} = \gamma_A + \gamma_B - W_{AB} \qquad \text{per unit area.} \tag{9.9}$$

This important thermodynamic relation is valid for both solid and liquid interfaces. It gives the free energy (always negative) of bringing unit areas of surfaces A into contact in liquid B, and vice versa, since $\gamma_{AB} = \gamma_{BA}$.

All the above equations belong to an important class of expressions known as *combining relations* or *combining laws*. They are extremely useful for deriving relationships between various energy terms in a complex system and are often used for obtaining approximate values for parameters that cannot be easily measured. For example, if we return to Eq. (9.4), we may also write it as

$$\Delta W = n(W_{AA} + W_{BB} - 2\sqrt{W_{AA}W_{BB}}) \qquad (9.10)$$

so that Eq. (9.9) becomes

$$\gamma_{AB} = \gamma_A + \gamma_B - 2\sqrt{\gamma_A\gamma_B} = (\sqrt{\gamma_A} - \sqrt{\gamma_B})^2, \qquad (9.11)$$

a useful expression that is often used to estimate the interfacial energy γ_{AB} solely from the surface energies or surface tensions of the pure liquids, γ_A and γ_B, in the absence of any data on the energy of adhesion W_{AB}. We shall encounter these and other combining relations again later.

9.3 FACTORS FAVOURING THE ASSOCIATION OF UNLIKE MOLECULES, PARTICLES OR SURFACES IN A THIRD MEDIUM

Let us now proceed from two-component to three-component systems, starting with a consideration of the mixture shown in Fig. 9.3. For two unlike molecules or particles A and B coming together in the solvent medium composed of molecules of type C (Fig. 9.3a, b), we find

$$\Delta W = W_{ass} - W_{disp} \propto -AB - C^2 + AC + BC \propto -(A - C)(B - C).$$

$$(9.12)$$

This is a very interesting result because it shows that the energy of association can now be positive or negative. If positive, the particles effectively repel each other and therefore remain dispersed in medium C.

Let us see how an effective repulsion has arisen from interaction potentials that are all attractive to begin with. The phenomenon may be thought of as *Archimedes' principle* being applied to intermolecular forces. In gravitational forces we all know that iron sinks while wood rises in water. Thus, wood is effectively experiencing a repulsion from the earth when in water. This is because it is lighter than water, and if it were to descend, it would have to displace an equal volume of water and drive it upwards to replace the space previously occupied by the wood. Since water is denser and so more attracted to the earth than wood, the whole process would be energetically unfavourable, viz., the energy lost in water going up is not recovered by the energy gained in wood coming down. This displacement principle applies to all interactions, including intermolecular interactions.

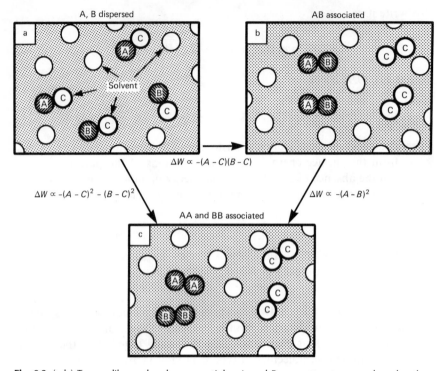

Fig. 9.3. (a,b) Two unlike molecules or particles A and B may attract or repel each other in a third medium C. Repulsion will occur (ΔW positive) if the properties of C are intermediate between those of A and B; for example, for gravitational forces, if the density of C (e.g., water) is between that of A (e.g., iron) and B (e.g., wood). In such cases the dispersed state (a) is energetically favoured. (c) The associated state of like molecules has a lower energy than either (a) or (b).

Equation (9.12) tells us that if C is intermediate between A and B, two particles (or surfaces) will repel each other, an effect that was previously noted for van der Waals forces (Section 6.7) where the operative properties of media A, B and C are their dielectric constants and refractive indices. However, whatever the relation between A, B and C, the most favoured final state will be that of particles A associating with particles A, B with B and C with C (Fig. 9.3c).

This procedure may be extended to mixtures with more species. We may therefore generalize our earlier conclusion, viz., *there is always an effective attraction between like molecules or particles in a multicomponent mixture* (again with the proviso that the interactions are not dominated by Coulombic or H-bonding forces). And in addition, *unlike particles may attract or repel each other in a solvent.*

9.4 PARTICLE–SURFACE INTERACTIONS

Let us now extend the above analysis to the case of a particle C near an interface dividing two immiscible liquid media A and B (Fig. 9.4). Four situations may arise:

(i) *Desorption*: the particle is repelled from the interface on either side of it (also known as 'negative adsorption');

(ii) *Adsorption*: the particle is attracted to the interface from either side; or

(iii) and (iv) *Engulfing*: the particle is attracted from one side (left or right) but repelled from the other (right or left).

Now using Eq. (9.12) we may write for the energy change for the particle coming up to the interface:

$$\text{from the left:} \qquad \overrightarrow{\Delta W} \propto -(C - A)(B - A), \qquad (9.13)$$

$$\text{from the right:} \qquad \overleftarrow{\Delta W} \propto -(C - B)(A - B). \qquad (9.14)$$

Thus the net energy change on taking the particle *across* the interface from right to left (Fig. 9.4b) is

$$\overleftarrow{\Delta W}_{\text{tot}} \propto \overleftarrow{\Delta W} - \overrightarrow{\Delta W} \propto (A - C)^2 - (B - C)^2 \propto \gamma_{\text{BC}} - \gamma_{\text{AC}}.$$

$$(9.15)$$

The above equation predicts the following: first, if $A > C > B$ or $A < C < B$, the particle will be attracted to the interface from either side, leading to adsorption. As can be seen, this will occur if the particle's properties C are intermediate between those of the two solvent media. The adsorption of amphiphilic molecules at hydrocarbon–water interfaces is an example of such a phenomenon. Amphiphilic molecules such as detergents and surfactants (from the words 'surface-active') are partly hydrophilic and partly hydrophobic and so have properties intermediate between the two liquids.

Second, if $A > B > C$ or $A < B < C$ (B intermediate), the particle will be attracted to the interface from the left but repelled from the right (engulfing), while if $B > A > C$ or $B < A < C$ (A intermediate), it will be attracted from the right but repelled from the left (reverse engulfing or ejection). Now since these six combinations exhaust all the possibilities, we see that repulsion from both sides of an interface (i.e., negative adsorption from both sides)

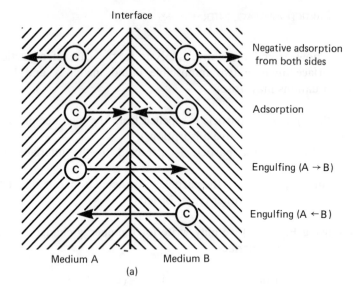

Interface

Negative adsorption
from both sides

Adsorption

Engulfing (A → B)

Engulfing (A ← B)

Medium A   Medium B
(a)

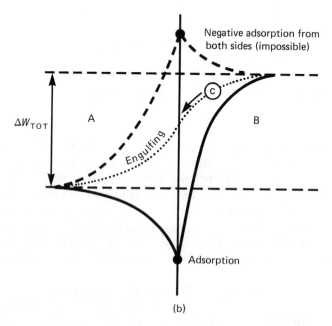

Negative adsorption from
both sides (impossible)

ΔW_{TOT}

A

Engulfing

B

Adsorption

(b)

Fig. 9.4. (a) The three possible modes of interaction of a particle C with a liquid–liquid interface. (b) Corresponding schematic energy versus distance profiles (assumed monotonic) for $\Delta W_{TOT} < 0$. Note that if medium A is solid, then particle C will adsorb on it from B since it cannot be engulfed by A.

cannot occur and that either adsorption or engulfing will be the rule. It is for this reason that surfaces are so prone to adsorbing molecules or particles from vapour or solution. Indeed, if a medium B is a vapour ($B \approx 0$), then we must have $A > C > B$ or $C > A > B$, both of which result in a net attraction of gaseous molecules C towards the surface of A, i.e., vapour molecules will always be attracted to a surface.

9.5   ADSORBED SURFACE FILMS: WETTING AND NON-WETTING

The above examples and Fig. 9.4 apply only to isolated molecules of C, in other words, they apply only to concentrations of C below its solubility in media A and B. At higher concentrations the molecules or particles of C may associate into a separate phase on one or the other side of the A–B interface, or they may build up at the interface. Which of these happens depends on the relative magnitudes of A, B and C. In this final section we shall consider the factors favouring the formation of thick adsorbed films on a solid surface.

In Fig. 9.5a we have, initially, a solid surface of A in contact with a binary liquid mixture of molecules C and B. Again there are three possibilities:

(i) For C intermediate between A and B, molecules C will be attracted to A while molecules B will be repelled from it, and an adsorbed monolayer or film of C will be energetically favourable (Fig. 9.5b). In this case C is said to *wet* the surface.

(ii) For B intermediate between A and C, the roles of B and C are reversed, favouring adsorption of B and negative adsorption of C (Fig. 9.5c). In this case C *unwets* the surface (while B will wet the surface if C is the solvent).

(iii) Finally, when A is intermediate, both molecules B and C are attracted to the solid surface. Under such circumstances no uniformly adsorbed film of B or C will form, but different regions of the interface will collect macroscopic droplets of the B or C phase (Fig. 9.5d). This is known as *partial wetting* or *non-wetting*.

It is left as an exercise for the interested reader to ascertain that when the total surface energies of the whole system is minimized, the *contact angle* θ formed by these droplets (Fig. 9.5d) is given by

$$\cos \theta = (B + C - 2A)/(B - C), \qquad (9.16)$$

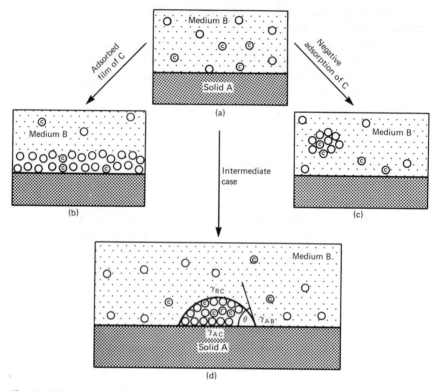

Fig. 9.5. (a) Low concentration of solute molecules C in medium B (i.e., below saturation). (b) *Wetting*: an adsorbed film of C develops and grows in thickness as the concentration of C in B approaches saturation. This corresponds to $\cos \theta > 1$ in Eq. (9.16). (c) *Unwetting*: resulting from repulsion between C and A in medium B above saturation. This corresponds to $\cos \theta < -1$ in Eq. (9.16). (d) *Partial wetting*: intermediate case between the two above, corresponds to $1 > \cos \theta > -1$.

which leads to values of $\cos \theta$ between 1 and -1 (θ between $0°$ and $180°$) only when A is intermediate between B and C. The above equation may also be written in the forms

$$\gamma_{AC} + \gamma_{BC} \cos \theta = \gamma_{AB} \qquad \text{(Young equation)}, \qquad (9.17)$$

$$\gamma_{BC}(1 + \cos \theta) = \Delta W_{ABC} \qquad \text{(Young–Dupré equation)}, \qquad (9.18)$$

where ΔW_{ABC} is the adhesion energy per unit areas of surfaces A and C adhering in medium B. These important fundamental equations will be derived and discussed further in Chapter 15.

The purpose of the phenomenological discussion of this chapter is to illustrate how a few basic notions concerning two-particle interaction energies can be applied to progressively more complex situations, and vice versa (i.e., how fairly complex situations can arise from, and be understood in terms of, the simplest possible pair potential, Eq. (9.2)). However, this type of non-specific approach, while conceptually useful, has its limits in that it does not take into account the way interaction energies vary with distance. Two particles or surfaces may have an adhesive energy minimum at contact, but if the force law is not monotonic—it may be repulsive before it becomes attractive closer in—the particles will remain separated (i.e., they effectively repel each other). Only a quantitative analysis in terms of the magnitudes of the operative forces and their distance dependence can provide a full understanding of interparticle and interfacial phenomena.

PROBLEMS AND DISCUSSION TOPICS

9.1   In the legend to Fig. 9.1(a) and (b) it is shown that in two dimensions the associated state (b) is energetically more favourable than the dispersed state (a) by a factor of $-9(A - B)^2$. Calculate this energy in the case of three dimensions, i.e. for a spherical cluster of 12 molecules surrounding the central molecule.   (*Answer*: $-22(A - B)^2$.)

9.2   Derive Eq. (9.16). (*Hint*: One way of solving this problem is to consider Fig. 9.5(d) and first show that the curved and flat areas, A_c and A_f, of a truncated sphere of constant volume are related to each other according to $dA_c/dA_f = \cos \theta$.)

9.3   Two immiscible liquids B and C are in contact with a solid surface A where there is a finite contact angle θ between the B–C interface and A (as in Fig. 9.5d). Would you expect θ to change as one approaches the critical point of the B–C system, and if so how?

9.4   Describe what happens to the distribution of B and C molecules near the solid surface of A in Fig. 9.5 if B and C are miscible?

CHAPTER 10

CONTRASTS BETWEEN INTERMOLECULAR, INTERPARTICLE AND INTERSURFACE FORCES

10.1 SHORT-RANGE AND LONG-RANGE EFFECTS OF A FORCE

We are told that when an apple fell on Newton's head it set in motion a thought process that eventually led Newton to formulate the law of gravity. The conceptual breakthrough in this discovery was the recognition that the force that causes apples to fall is the same force that holds the moon in a stable orbit around the earth.

On earth, gravity manifests itself in many different ways: in determining the height of the atmosphere; the capillary rise of liquids, and the behaviour of waves and ripples. In biology it decrees that animals that live in the sea (where the effect of gravity is almost negligible) can be larger than the largest possible land animal; that heavy land animals such as elephants must have short thick legs while a man or a spider can have proportionately long thin legs; that larger birds must have progressively larger wings (e.g., eagles and storks) while smaller birds, flies and bees can have relatively small light wings; that only small animals can carry many times their own weight (e.g., ants), and much else (Thompson, 1968). But beyond the immediate vicinity of the earth's surface and out to the outer reaches of space this same force now governs the orbits of planets, the shapes of nebulae and galaxies, the rate of expansion of the universe, and, ultimately, its age. The first group of phenomena—those occurring locally on the earth's surface—may be thought of as the *short-range* effects of the gravitational force, while the second and very different types of phenomena occurring on a cosmological scale may be thought of as the *long-range* effects of gravity.

Intermolecular forces are no less versatile in the way the same force can have very different effects at short and long range, though here 'short range' usually means at or very close to molecular contact (<1 nm) while 'long-range' forces are rarely important beyond 100 nm (0.1 μm). In Part I we saw that the properties of gases and the cohesive strengths of condensed

152

phases are determined mainly by the interaction energies of molecules in contact $\omega(\sigma)$, i.e., molecules interacting with their immediate neighbours. For example, the van der Waals pair energy of two neighbouring molecules is at least 64 times stronger than that between next nearest neighbours ($1/\sigma^6$ compared to $1/(2\sigma)^6$). Only the Coulomb interaction is effectively long ranged in that the energy decays slowly, as $1/r$, and remains strong at large distances. However, in a medium of high dielectric constant such as water the strength of the Coulomb interaction is much reduced as is its range due to *electroneutrality* and *ionic screening* effects (Chapter 12).

We may therefore conclude that apart from ionic crystals (Chapter 3) the properties of solids and liquids are determined mainly by the molecular binding forces, i.e. by the strength of the interactions at or very near molecular contact, the long-range nature of the interaction (e.g., the exact distance dependence of the force law) playing only a minor role.

A very different situation arises when we consider the interactions of macroscopic particles or surfaces, for now when all the pair potentials between the molecules in each body is summed we shall find (i) that the net interaction energy is proportional to the size (e.g. radius) of the particles, so that the energy can be very much larger than kT even at separations of 100 nm or more, and (ii) that the energy and force decay much more slowly with the separation. All these characteristics make the interactions between macroscopic bodies effectively of much longer range than those between molecules even though the same basic type of force may be operating in each case.

Furthermore, if the force law is not monotonic (not purely attractive or repulsive) then all manner of behaviour may arise depending on the specific form of the long-range distance dependence of the interaction. This is illustrated in Fig. 10.1. For example, consider the purely attractive interaction of Fig. 10.1a. If both molecules and particles experience the same type of interaction, both will be attracted to each other, and the thermodynamic properties of an assembly of molecules in the gas or condensed phase will be determined by the depth of the potential well at contact, as will the adhesion energy of two particles. However, for the energy law in Fig. 10.1b two molecules will still attract each other since the energy barrier is negligibly small compared to kT, but two macroscopic particles will effectively repel each other since the energy barrier is now too high to surmount. Under such circumstances particles dissolved in a medium will remain dispersed even though the ultimate *thermodynamic equilibrium state* is the aggregated state.

We thus encounter another important difference between molecular and particle interactions, namely, that particles can be (and often are) trapped in some *kinetic* or *metastable state* if there is a sufficiently high energy barrier that prevents them from accessing all parts of their interaction potential over

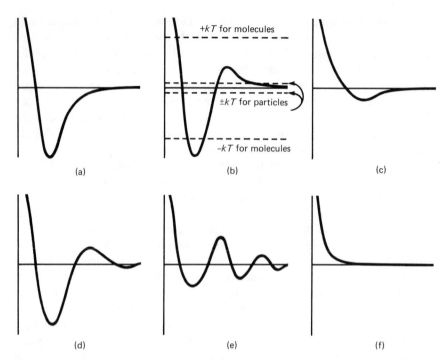

Fig. 10.1. Typical interaction potentials encountered between molecules and macroscopic particles in a medium. (a) This potential is typical of vacuum interactions but is also common in liquids. Both molecules and particles attract each other. (b) Molecules attract each other; particles effectively repel each other. (c) Weak minimum. Molecules repel, particles attract. (d) Molecules attract strongly, particles attract weakly. (e) Molecules attract weakly, particles attract strongly. (f) Molecules repel, particles repel.

some reasonable time period. Figure 10.1 illustrates some other types of commonly occurring intermolecular and intersurface potential functions and the different effects they can have on molecule–molecule and particle–particle interactions. In the following chapters we shall see how such interaction potentials actually arise in different systems.

☐ ● **WORKED EXAMPLE** ●

Question: Consider a 10% by weight dispersion of glass particles of radius $a = 100$ nm in water at room temperature interacting via a potential as shown in Fig. 10.1b. Estimate what the energy barrier would have to be (in units of kT) for the particles to remain dispersed over a period of one day.

Answer: The mean Brownian velocity, v, per particle will be given by $\frac{1}{2}mv^2 \approx kT$, where $m = 1.3 \times 10^{-17}$ kg is the particle mass (assuming a density of 3 × 10^3 kg m^{-3}). Thus we obtain $v = 0.025$ m s^{-1}. The number of particles per unit volume is 8 × 10^{18} m^{-3}, so that their mean separation is about 5 × 10^{-7} m (500 nm). Thus the time between collisions will be about 5 × $10^{-7}/0.025 = 2 × 10^{-5}$ s, so that the number of collisions per day will be 60 × 60 × $24/2 × 10^{-5} ≈ 4 × 10^9$. We therefore require the condition that the probability of two colliding particles overcoming their energy barrier ΔW should be less than $1/(4 × 10^9) = 2.5 × 10^{-10}$. Putting $2.5 × 10^{-10} = e^{-\Delta W/kT}$ we obtain $\Delta W/kT = 22$. Thus the energy barrier should be in excess of about $22kT$ to ensure kinetic stability, i.e. that most of the particles remain dispersed, over a 24 h period. (Note: a more rigorous calculation shows that the collision rate would be significantly lower due to viscous drag effects, and that an energy barrier of only 16 kT is needed to keep the system stable.)   □

The interactions of 'soft' particles, e.g. those composed of flexible surfactant or polymer molecules, are particularly subtle due to the interplay of the short-range and long-range forces. The sizes and shapes of such 'self-assembling' molecular aggregates are regulated by the short-range interactions between the molecules, which are sensitive to electrolyte concentration, pH, temperature, etc. On the other hand, the long-range interactions between these aggregates—those that determine whether they will attract or repel each other—are also sensitve to these variables. Thus, different parts of the intermolecular interaction potential govern very different properties in these systems. The interdependence of short-range and long-range forces and, consequently, of *intraparticle* and *interparticle* forces is discussed in Part III.

10.2   INTERACTION POTENTIALS BETWEEN MACROSCOPIC BODIES

In this section we shall relate the pair interaction between two molecules to the interaction between a molecule and a flat surface, between two spherical particles (or two curved surfaces), between a spherical particle and a flat surface, and between two flat surfaces.

Molecule–surface interaction

Let us once again assume that the pair potential between two atoms or small molecules is purely attractive and of the form $w(r) = -C/r^n$. Then, with the further assumption of *additivity*, the net interaction energy of a molecule and the planar surface of a solid made up of like molecules (Fig. 10.2a) will be the sum of its interactions with all the molecules in the body. For molecules in a circular ring of cross-sectional area dx dz and radius x, the ring volume

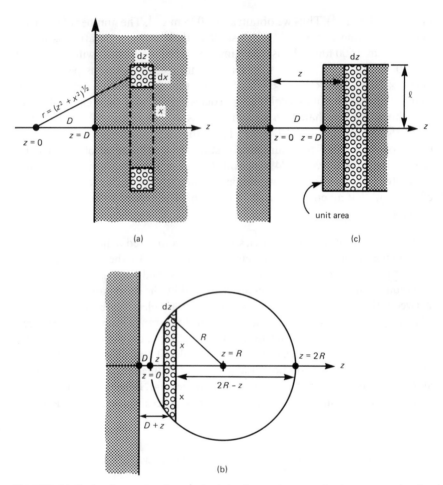

Fig. 10.2. Methods of summing (integrating) the interaction energies between molecules in condensed phases to obtain the interaction energies between macroscopic bodies. (a) Molecule near a flat surface or 'wall'. (b) Spherical particle near a wall ($R \gg D$). (c) Two planar surfaces ($l \gg D$).

is $2\pi x \, dx \, dz$, and the number of molecules in the ring will be $2\pi\rho x \, dx \, dz$, where ρ is the number density of molecules in the solid. The net interaction energy for a molecule at a distance D away from the surface will therefore be

$$w(D) = -2\pi C\rho \int_{z=D}^{z=\infty} dz \int_{x=0}^{x=\infty} \frac{x \, dx}{(z^2 + x^2)^{n/2}} = \frac{2\pi C\rho}{(n-2)} \int_D^\infty \frac{dz}{z^{n-2}}$$

$$= -2\pi C\rho/(n-2)(n-3)D^{n-3} \qquad \text{for } n > 3, \qquad (10.1)$$

which for $n = 6$ (van der Waals forces) becomes

$$w(D) = -\pi C\rho/6D^3. \tag{10.2}$$

The corresponding force, $F = \partial w(D)/\partial D = -\pi C\rho/2D^4$, could of course have been derived in a similar way by summing (integrating) all the pair forces resolved along the z axis.

Sphere–surface and sphere–sphere interaction

We can now calculate the interaction energy of a large sphere of radius R and a flat surface (Fig. 10.2b). First, from the chord theorem, Eq. (9.7), we know that for the circle: $x^2 = (2R - z)z$. The volume of a thin circular section of area πx^2 and thickness dz is therefore $\pi x^2\,dz = \pi(2R - z)z\,dz$, so that the number of molecules contained within this section is $\pi\rho(2R - z)z\,dz$, where ρ is the number density of molecules in the sphere. Since all these molecules are at a distance $(D + z)$ from the planar surface, the net interaction energy is, using Eq. (10.1),

$$W(D) = -\frac{2\pi^2 C\rho^2}{(n-2)(n-3)}\int_{z=0}^{z=2R}\frac{(2R-z)z\,dz}{(D+z)^{n-3}}. \tag{10.3}$$

For $D \ll R$, only small values of z ($z \approx D$) contribute to the integral, and we obtain[1]

$$W(D) = -\frac{2\pi^2 C\rho^2}{(n-2)(n-3)}\int_0^\infty \frac{2Rz\,dz}{(D+z)^{n-3}}$$

$$= -\frac{4\pi^2 C\rho^2 R}{(n-2)(n-3)(n-4)(n-5)D^{n-5}}, \tag{10.4}$$

which for $n = 6$ (van der Waals forces) becomes

$$W(D) = -\pi^2 C\rho^2 R/6D. \tag{10.5}$$

Note that the interaction energy is proportional to the radius of the sphere and that it decays as $1/D$, very much slower than the $1/r^6$ dependence of the intermolecular pair interaction.

[1] To avoid confusion we shall use W and D to denote the interaction free energies of macroscopic bodies whose surfaces are at a distance D apart, reserving w and r for the interactions of atoms and molecules.

For $D \gg R$, we may replace $(D + z)$ in the denominator of Eq. (10.3) by D, and we then obtain

$$W(D) = -\frac{2\pi^2 C\rho^2}{(n-2)(n-3)} \int_0^{2R} \frac{(2R-z)z \, dz}{D^{n-3}} = -\frac{2\pi C\rho(4\pi R^3\rho/3)}{(n-2)(n-3)D^{n-3}}.$$

(10.6)

Since $4\pi R^3\rho/3$ is simply the number of molecules in the sphere, the above is essentially the same as Eq. (10.1) for the interaction of a molecule (or small sphere) with a surface. It is left as an exercise for the interested reader to show that for two spheres of equal radii R whose surfaces are at a small distance D apart $(R \gg D)$, their interaction energy is one half that given by Eq. (10.4) or (10.5), while for two spheres far apart $(D \gg R)$ the energy varies as $-1/D^n$ as for two molecules. At intermediate separations $(R \approx D)$ the expression for the interaction potential is more complicated but remains analytic (Hamaker, 1937).

Surface–surface interactions

Let us now calculate the interaction energy of two planar surfaces a distance D apart. For two infinite surfaces, the result will be infinity, and so we have to consider the energy per unit surface area. Let us start with a thin sheet of molecules of unit area and thickness dz at a distance z away from an extended surface of *larger* area (Fig. 10.2c). From Eq. (10.1) the interaction energy of this sheet with the surface is $-2\pi C\rho(\rho \, dz)/(n-2)(n-3)z^{n-3}$. Thus, for the two surfaces, we have

$$W(D) = -\frac{2\pi C\rho^2}{(n-2)(n-3)} \int_D^\infty \frac{dz}{z^{n-3}} = -\frac{2\pi C\rho^2}{(n-2)(n-3)(n-4)D^{n-4}},$$

(10.7)

which for $n = 6$ becomes

$$W(D) = -\pi C\rho^2/12D^2 \qquad \text{per unit area.} \qquad (10.8)$$

It is important to note that Eqs (10.7) and (10.8) are for unit area of one surface interacting with an infinite area of another surface. In practice this usually amounts to two unit areas of both surfaces, but it is strictly applicable only when D is small compared to the lateral dimensions of the surfaces.

10.3   EFFECTIVE INTERACTION AREA OF TWO SPHERES: THE LANGBEIN
       APPROXIMATION

When two large spheres or a sphere and a flat surface are close together, one sometimes wants to know what is their 'effective area' of interaction. First, we may note that no matter how large a sphere becomes it *never* approaches the behaviour of a flat surface. Its interaction energy will increase linearly with radius R (Eq. (10.4)), but the distance dependence of the interaction will not change to that corresponding to two planar surfaces. However, the concept of an effective 'interaction zone' and effective 'interaction area', A_{eff}, are useful. If we compare Eq. (10.4) for a sphere near a surface with Eq. (10.7) for two surfaces, we find that the interaction of a sphere and a surface is the same as that of two planar surfaces *at the same surface separation D* if their area is

$$A_{eff} = 2\pi R D/(n-5)$$

$$= 2\pi R D \qquad \text{for } n = 6. \qquad (10.9)$$

From Fig. 10.2b and Eq. (9.7) this area is simply equal to πx^2 when $z = D$ (so long as $R \gg D$). In other words, the effective area of interaction of a sphere with a surface is the circular zone centred at a distance $-D$ from the surface (inside the sphere). This is known as the *Langbein approximation*.

The effective area of interaction increases linearly with both R and D as may be expected. For example, for a sphere of radius $R = 1\ \mu m$ at a distance $D = 1$ nm from a wall, the effective interaction area is about $2\pi R D \approx 10^{-14}\ m^2$ or 10 000 nm^2, which corresponds to an area of radius ~ 45 nm. At contact, when $D \approx 0.3$ nm, this radius falls to 25 nm, though in practice elastic flattening generally significantly increases the contact area (see Section 15.5).

10.4   INTERACTIONS OF LARGE BODIES COMPARED TO THOSE BETWEEN
       MOLECULES

The geometries analysed above are the most commonly encountered. In the next chapter we shall look specifically at the van der Waals interactions of these and of other geometries as well (e.g., cylinders). Meanwhile, let us consider some of the implications of the interaction potentials obtained so far.

First we may note that for two macroscopic bodies, the interaction energy generally decays much more slowly with distance than it does for two molecules. For example, whereas the van der Waals energy between atoms and molecules is of short range, having an inverse sixth-power distance dependence, the van der Waals energy between large condensed bodies decays more slowly with distance (cf. $1/D$ for spheres, $1/D^2$ for planar surfaces) and is effectively of much longer range. This is yet another manifestation of the long-range nature of interparticle forces.

Second, the van der Waals interaction energy of a small molecule of diameter σ with a wall is given by Eq. (10.2) as $w(D) = -\pi C\rho/6D^3$. At contact we may put $D \approx \sigma$ and $\rho \approx \sqrt{2}/\sigma^3$ (corresponding to a close packed solid) and obtain

$$w(\sigma) \approx -\sqrt{2}\pi C/6\sigma^6 \approx -0.74C/\sigma^6 \tag{10.10}$$

which is of the same order as $-C/\sigma^6$ for two small molecules in contact. Likewise, for a sphere of atomic dimensions ($R = \sigma/2$) in contact with a wall ($D = \sigma$), we find from Eq. (10.5) that

$$W(\sigma) \approx -2\pi^2 C/12\sigma^6 \approx -1.6C/\sigma^6, \tag{10.11}$$

while for two spheres

$$W(\sigma) \approx -0.8C/\sigma^6, \tag{10.12}$$

which again are very close to the molecule–molecule pair potential at contact. However, once the size of a sphere increases above atomic dimensions (i.e., once $R > \sigma$), then at contact ($D = \sigma$) Eq. (10.5) becomes

$$W(\sigma) \approx -2\pi^2 CR/6\sigma^7 \approx -1.6(2R/\sigma)C/\sigma^6,$$

which reduces to Eq. (10.11) only for small radii, of order $R \approx \sigma/2 \approx 0.1$–0.2 nm, but which increases linearly with R for larger spheres. Likewise for the interaction of two spheres. This is an important result. It shows that for diameters $2R$ beyond about 0.5 nm, a molecule must already be considered as a (small) particle or else the strength of its interaction will be underestimated. The higher strength of the measured interaction between large molecules over that predicted on the assumption that $D = 2R$ was seen in Table 6.1 for CCl_4 but not for the smaller molecules. In conclusion, while for two atoms or small molecules the contact interaction energy has no explicit size dependence, that between larger particles increases linearly with their radius.

As regards the stabilizing repulsive forces between large bodies, these too manifest themselves in quite different ways when acting between macroscopic bodies. While the short-range steric or Born repulsion between individual molecules may remain the same (see Problem 10.5), large spheres will deform elastically or even plastically when in contact under a large adhesive force. The stabilizing repulsive forces of macroscopic particles therefore also involve the elastic properties of the materials (discussed further in Chapters 15 and 18).

10.5   INTERACTION ENERGIES AND INTERACTION FORCES: THE DERJAGUIN APPROXIMATION

So far we have been dealing mainly with interaction *energies* rather than the *forces* experienced by molecules and particles. This is because most experimental data on molecular interactions are of a thermodynamic nature and therefore more readily understood in terms of interaction energies, as we saw in Part I. However, between macroscopic bodies it is the forces between them that are often easier to measure, and of greater interest, than their interaction energies.

It is therefore desirable to be able to relate the force law $F(D)$ between two curved surfaces to the interaction free energy $W(D)$ between two planar surfaces. Luckily, a simple relation exists for the two geometries most commonly encountered, viz. two flat surfaces and two spheres (a sphere near a flat surface being a special case of two spheres with one sphere very much larger than the other). A glance at Eq. (10.4) shows that for the additive intermolecular pair potential $w(r) = -C/r^n$, the value of $F(D)$ for a sphere near a flat surface is

$$F(D) = -\frac{\partial W(D)}{\partial D} = -\frac{4\pi^2 C \rho^2 R}{(n-2)(n-3)(n-4)D^{n-4}}.$$ (10.13)

This force law can be seen to be simply related to $W(D)$ per unit area of two planar surfaces, Eq. (10.7), by

$$F(D)_{\text{sphere}} = 2\pi R W(D)_{\text{planes}}.$$ (10.14)

This is a very useful relationship, and while it was derived for the special case of an additive inverse power potential, it is in fact valid for any type of force law, as will now be shown.

Assume that we have two large spheres of radii R_1 and R_2 a small distance D apart (Fig. 10.3). If $R_1 \gg D$ and $R_2 \gg D$, then the force between the two spheres can be obtained by integrating the force between small circular regions of area $2\pi x \, dx$ on one surface and the opposite surface, which is assumed to be locally flat and at a distance $Z = D + z_1 + z_2$ away. The net

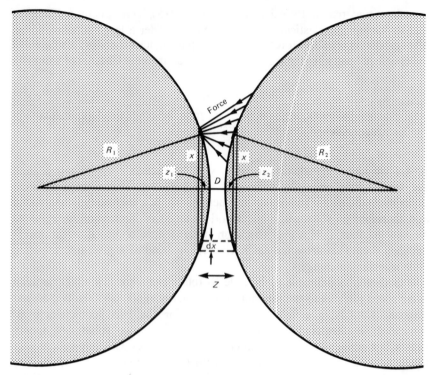

Fig. 10.3. The Derjaguin approximation (Derjaguin, 1934), which relates the force law $F(D)$ between two spheres to the energy per unit area $W(D)$ of two flat surfaces by $F(D) = 2\pi[R_1R_2/(R_1 + R_2)]W(D)$.

force between the two spheres (in the z direction) is therefore

$$F(D) = \int_{Z=D}^{Z=\infty} 2\pi x \, dx f(Z), \qquad (10.15)$$

where $f(Z)$ is the normal force per unit area between two flat surfaces. Since from the Chord Theorem $x^2 \approx 2R_1z_1 = 2R_2z_2$, we have

$$Z = D + z_1 + z_2 = D + \frac{x^2}{2}\left(\frac{1}{R_1} + \frac{1}{R_2}\right) \qquad (10.16)$$

and

$$dZ = \left(\frac{1}{R_1} + \frac{1}{R_2}\right)x \, dx, \qquad (10.17)$$

so that Eq. (10.15) becomes

$$F(D) \approx \int_D^\infty 2\pi \left(\frac{R_1 R_2}{R_1 + R_2} \right) f(Z) \, \mathrm{d}Z = 2\pi \left(\frac{R_1 R_2}{R_1 + R_2} \right) W(D), \qquad (10.18)$$

which gives the force between two spheres in terms of the energy per unit area of two flat surfaces at the same separation D. Equation (10.18) is known as the *Derjaguin approximation* (Derjaguin, 1934). It is applicable to any type of force law, whether attractive, repulsive or oscillatory, so long as the range of the interaction and the separation D is much less than the radii of the spheres. It is a useful theoretical tool, since it is usually easiest to derive the interaction energy for two planar surfaces (rather than for curved surfaces). It is also useful for interpreting experimental data and it has been well verified experimentally, as discussed in Chapters 11–15.

From the Derjaguin approximation, Eq. (10.18), we may deduce the following:

(i) If one sphere is very large so that $R_2 \gg R_1$, we obtain $F(D) = 2\pi R_1 W(D)$, which is the same as Eq. (10.14) and corresponds to the limiting case of a sphere near a flat surface.

(ii) For two equal spheres of radii $R = R_1 = R_2$, we obtain $F(D) = \pi R W(D)$, which is half the value for a sphere near a flat surface.

(iii) For two spheres in contact $(D = \sigma)$, the value of $W(\sigma)$ can be associated with 2γ, where γ is the conventional surface energy per unit area of a surface. Eq. (10.18) then becomes

$$F(\sigma) = F_{\mathrm{ad}} = \frac{4\pi\gamma R_1 R_2}{(R_1 + R_2)} \qquad (10.19)$$

which gives the adhesion force F_{ad} between two spheres in terms of their surface energy. Adhesion forces are discussed in Chapter 15.

(iv) Perhaps the most intriguing aspect of the Derjaguin approximation is that it tells us that the distance dependence of the force between two curved surfaces can be quite different from that between two flat surfaces even though the same type of force is operating in both. This is illustrated in Fig. 10.4, where we see that a purely repulsive force between two curved surfaces can be attractive between two planar surfaces (over a certain distance regime), with equilibrium at some finite separation (Fig. 10.4b). Conversely, a purely attractive force between curved surfaces can become repulsive between two planar surfaces (Fig. 10.4c).

(v) Finally, it may be readily shown that for two cylinders of radii R_1 and

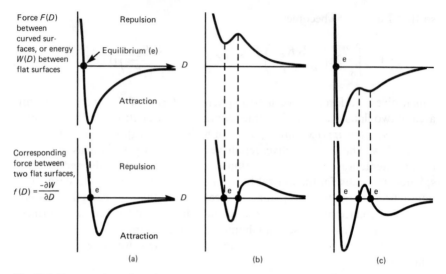

Fig. 10.4. Top row: force laws between two curved surfaces (e.g., two spherical particles). Bottom row: corresponding force laws between two flat surfaces. Note that stable equilibrium occurs only at points marked e where the force is zero ($f = 0$) *and* the force curve has negative slope; the other points where $f = 0$ are unstable.

R_2 crossed at an angle θ to each other, the Derjaguin approximation becomes

$$F(D) = 2\pi\sqrt{R_1 R_2}\, W(D)/\sin\theta, \qquad \text{for } D \ll R_1, R_2. \qquad (10.20)$$

Note that for two cylinders of equal radii $R = R_1 = R_2$, crossed at right angles to each other ($\theta = 90°$, $\sin\theta = 1$) the above reduces to the same result as for a sphere of radius R near a flat surface. In other words the interaction of two orthogonal cylinders is the same as that of a sphere and a wall if all three radii are the same.

□   ● WORKED EXAMPLE ●

Question: Two orthogonal cylinders of equal radius R are separated by a small distance D ($D \ll R$). Using purely geometrical arguments, show that the surface geometry around the contact region is the same, to first order, as for a sphere of the same radius R at the same distance D from a flat surface.

Answer: Referring to Fig. 10.5a, and using Eq. (9.7), we have $x^2 = 2RD_1$ and $y^2 = 2RD_2$. Referring to Fig. 10.5b, this implies $2R(D_1 + D_2) = x^2 + y^2 = r^2$. Thus $D_1 + D_2$ is constant if r is constant, i.e. if P describes a circle. Since the above equation is indistinguishable from that for a sphere of the same radius

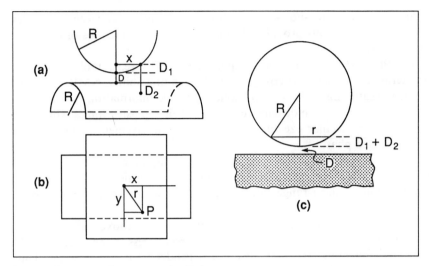

Fig. 10.5. Geometry of two crossed cylinders.

near a flat surface, Fig. 10.5c, we have proved that as far as interactions (and many other properties) are concerned the two geometries are locally equivalent. □

10.6   Experimental measurements of intermolecular and surface forces

When we come to consider experimental measurements of intermolecular forces we are confronted with a bewildering variety of data to draw upon. This is because almost any measurement in the below-10 eV category, whether in physics, chemistry or biology, is in some respect a measurement of intermolecular forces. For example, we have already seen how such a common property as the boiling point of a substance provides information on the strength of intermolecular binding energies. It is therefore difficult to make a list of experimental measurements of intermolecular forces; it is far better to draw upon whatever relevant data exist as the situation arises. This was done in Part I, and we shall continue with this practice in Parts II and III. However, different types of measurements do provide different insights and information, and in the rest of this chapter we shall categorize experiments according to the type of information they provide on intermolecular and intersurface interactions. First, let us recapitulate some of those already mentioned.

(i) Thermodynamic data on gases, liquids and solids (e.g., PVT data, boiling points, latent heats of vaporization, lattice energies) provide information on the short-range attractive potentials between molecules. Adsorption isotherms provide information on the interactions of molecules with surfaces.

(ii) Physical data on gases, liquids and solids (e.g., molecular beam scattering experiments, viscosity, diffusion, compressibility, NMR, x-ray and neutron scattering of liquids and solids) provide information on the short-range interactions of molecules, especially their repulsive forces which give insight into molecular size and shape, and their involvement in the structure of condensed phases.

(iii) Thermodynamic data on liquids and liquid mixtures (e.g., phase diagrams, solubility, partitioning, miscibility, osmotic pressure) provide information on short-range solute–solvent and solute–solute interactions.

Such experimental data as listed in (i)–(iii) above often provide thermodynamic information only, so that direct access to the intermolecular potential functions (i.e., their distance dependence) is not possible. Thus, experimental PVT data may be compared with the van der Waals equation of state (Chapter 6), which contains terms to account for both the attractive and repulsive forces, but it does not give any information on the nature and range of the force laws themselves. To gain this information, some more direct measurement of forces is required. Of the various methods that have been devised for measuring molecular forces, the most direct employ macroscopic bodies or extended surfaces, where distances can be measured to within 0.1 nm, where the forces are large and measurable, and where entropic (thermal) effects are negligible. It is from such experiments (illustrated in Fig. 10.6) that many hard data on intermolecular and surface interactions have emerged.

(iv) Particle detachment and peeling experiments (Fig. 10.6a, b) provide information on particle adhesion forces and the adhesion energies of solid surfaces in contact (i.e., attractive short-range forces). Such experiments are important in powder technology, xerography, ceramic processing, the making of adhesive films and in understanding how cracks propagate in solids.

(v) Measuring the force between two macroscopic surfaces as a function of surface separation can provide the full force law of an interaction (Fig. 10.6c). Such direct force measurements are described in the next section.

(vi) Various surface studies such as surface tension and contact angle measurements give information on liquid–liquid and solid–liquid adhesion energies (Fig. 10.6d). When contact angles are measured under different atmospheric environments or as a function of time, these relatively simple experiments can provide surprisingly valuable insights into the states of surfaces and adsorbed films, and of molecular reorientation times at interfaces.

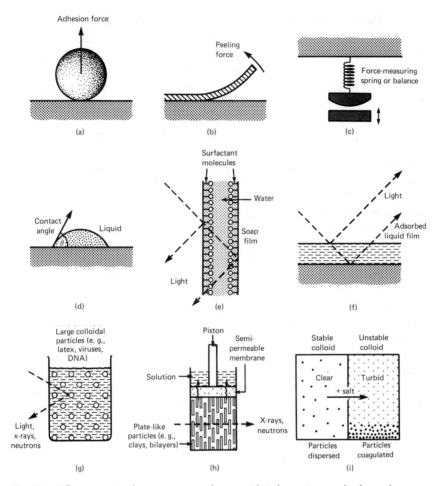

Fig. 10.6. Different types of measurements that provide information on the forces between particles and surfaces. (a) Adhesion measurements (practical applications: xerography, particle adhesion, powder technology, ceramic processing). (b) Peeling measurements (practical applications: adhesive tapes, material fracture and crack propagation). (c) Direct measurements of force as a function of surface separation (practical applications: testing theories of intermolecular forces). (d) Contact angle measurements (practical applications: testing wettability and stability of surface films, detergency). (e) Equilibrium thickness of thin free films (practical applications: soap films, foams). (f) Equilibrium thickness of thin adsorbed films (practical applications: wetting of hydrophilic surfaces by water, adsorption of molecules from vapour, protective surface coatings and lubricant layers, photographic films). (g) Interparticle spacing in liquids (practical applications: colloidal suspensions, paints, pharmaceutical dispersions). (h) Sheet-like particle spacings in liquids (practical applications: clay and soil swelling behaviour, microstructure of soaps and biological membranes). (i) Coagulation studies (practical application: basic experimental technique for testing the stability of colloidal preparations).

(vii) The thicknesses of free soap films and liquid films adsorbed on surfaces (Fig. 10.6e, f) can be measured as a function of salt concentration or vapour pressure. Such experiments provide information on the long-range repulsive forces stabilizing thick wetting films. Various optical techniques (e.g., reflected intensity, total internal reflection spectroscopy, or ellipsometry) have been used to measure film thickness to within 0.1 nm.

(viii) Dynamic interparticle separations and motions in liquids can be measured using NMR, light scattering, x-ray scattering and neutron scattering (Fig. 10.5g, h). In such experiments the particles can be globular or spherical (e.g., micelles, vesicles, colloidal particles, latex particles, viruses), sheet-like, (e.g., clays, lipid bilayers), or rod-like (e.g., DNA). The interparticle forces can be varied by changing the solution conditions, and their mean separation can be varied by changing the quantity of solvent, for example, by changing the hydrostatic or osmotic pressure via a semipermeable membrane. Notable among these techniques is the *Compression Cell* or *Osmotic Pressure Technique* (Homola and Robertson, 1976; LeNeveu *et al.*, 1976) from which the interaction force between particles can be obtained from the deviations from ideality in the PVT data. Such techniques are usually limited to measuring only the repulsive parts of a force law.

(ix) In coagulation studies on colloidal dispersions (Fig. 10.5i) the salt concentration, pH, or temperature of the suspending liquid medium (usually water) is changed until the dispersion becomes unstable and the particles coalesce (*coagulate* or *flocculate*). Coagulation rates can be very fast or very slow (see Worked Example in Section 10.1). Such studies provide information on the interplay of repulsive and attractive forces between particles in pure liquids as well as in surfactant and polymer solutions.

10.7   DIRECT MEASUREMENTS OF SURFACE AND INTERMOLECULAR FORCES

Most of the methods shown in Fig. 10.6 do not give the force law (the force as a function of distance) but rather the adhesion force or minimum energy at some particular state, e.g., the equilibrium state, of the system. Other methods, such as osmotic pressure measurements, involve the collective interactions of many molecules or particles so that the data gained tends to be of a thermodynamic nature and not directly translatable into a force law. The most unambiguous way to measure a force-law is to position two bodies close together and directly measure the force between them, e.g., from the deflection of a spring—very much as one would measure the force between two magnets. While the principle of direct force measurements is usually very straightforward, the challenge comes in measuring very weak forces at very small intermolecular or surface separations which must be controlled and measured to within 0.1 nm.

The first direct measurements of intermolecular forces were those of Derjaguin and coworkers (Derjaguin and Abrikossova, 1954; Derjaguin *et al.*, 1956) who measured the attractive van der Waals forces between a convex lens and a flat glass surface in vacuum. An electrobalance was used to measure the forces and an optical technique to measure the distance between two glass surfaces. Measurements were made in the distance regime 100–1000 nm, and the results fell within 50% of the predictions of the Lifshitz theory of van der Waals forces (Chapter 11).

These experiments opened the way for the slow but steady progress that has led to the highly sophisticated and versatile techniques that are used today for measuring the interactions between surfaces in vapours or liquids at the ångstrom resolution level. Both static (i.e. equilibrium) and dynamic (e.g. viscous) forces can now be studied with unprecedented precision providing information not only on the fundamental interactions in liquids but also into the structure of liquids adjacent to surfaces and other interfacial phenomena. Three techniques which can directly measure the force laws between two bodies of *macroscopic, colloidal* and *atomic* dimensions, respectively, will now be briefly described.

Measuring surface forces: the Surfaces Forces Apparatus (SFA)

During the last 20 years various direct force-measuring techniques have been developed which allow for the full force laws to be measured between two surfaces at the ångstrom resolution level (Israelachvili, 1989). Tabor and Winterton (1969) and Israelachvili and Tabor (1972, 1973) developed apparatuses for measuring the van der Waals forces between molecularly smooth mica surfaces in air or vacuum. The results using these new techniques confirmed the predictions of the Lifshitz theory of van der Waals forces (Chapter 11) down to surface separations as small as 1.5 nm. These techniques were then further developed for making measurements in liquids, which opened up a whole world of new phenomena of relevance to a much wider spectrum of surface science. We shall now describe one such apparatus which has become a standard research tool in many laboratories.

Figure 10.7 shows an SFA with which the force between two surfaces in controlled vapours or immersed in liquids can be directly measured (Israelachvili and Adams, 1978; Israelachvili, 1987b). The distance resolution is about 0.1 nm and the force sensitivity is about 10^{-8} N (10^{-6} g). Modified versions have been developed by Klein (1980), Parker *et al.* (1989a), Israelachvili and McGuiggan (1990), and Tonck *et al.* (1988) have extended the SFA method to opaque materials, replacing the optical technique for measuring distances (see below) by a capacitance method—the overall accuracy remaining about the same.

The SFA contains two curved molecularly smooth surfaces of mica

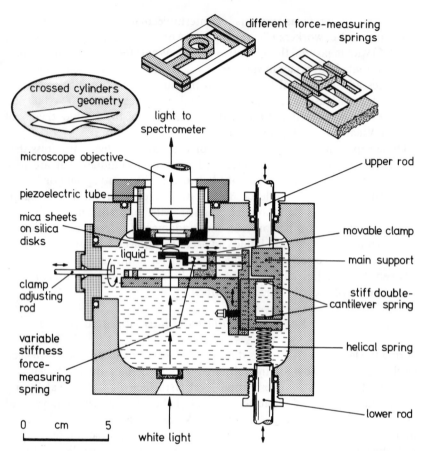

Fig. 10.7. Surface Forces Apparatus (SFA) for directly measuring the force laws between surfaces in liquids or vapours at the ångstrom resolution level. With the SFA technique two atomically smooth surfaces immersed in a liquid can be brought towards each other in a highly controlled way (the surface separation being controlled to 1 Å). As the surfaces approach each other they trap a very thin film of liquid between them and the forces between the two surfaces (across the liquid film) can be measured. In addition, the surfaces can be moved laterally past each other and the shear forces also measured during sliding. The results on many different liquids have revealed ultrathin film properties that are profoundly different from those of the bulk liquids, for example, that liquids can support both normal loads and shear stresses, and that molecular relaxations can take 10^{10} times longer in a 10 Å film that in the bulk liquid. Only molecular theories, rather than continuum theories, can explain such phenomena. However, most long-range interactions are adequately explained by continuum theories.

(of radius $R \approx 1$ cm) between which the interaction forces are measured using a variety of (interchangeable) force-measuring springs. The two surfaces are in a crossed cylinder configuration which is locally equivalent to a sphere near a flat surface or to two spheres close together (see Section 10.5).

The separation between the two surfaces from microns down to molecular contact can be measured by use of an optical technique using multiple beam interference fringes called Fringes of Equal Chromatic Order (FECO). Here the two transparent mica sheets (each about 2 μm thick) are first coated with a semireflecting 50–60 nm layer of pure silver before they are glued onto the curved silica discs (silvered sides down). Once in position in the apparatus, as shown in Fig. 10.7, white light is passed vertically up through the two surfaces and the emerging beam is then focused onto the slit of a grating spectrometer. From the positions and shapes of the coloured FECO fringes seen in the spectrogram the distance between the two surfaces can be measured, usually to better than 0.1 nm, as can the exact shapes of the two surfaces and the refractive index of the liquid (or material) between them; the latter allows for reasonably accurate determinations of the quantity of material (e.g., lipid or polymer) deposited or adsorbed on the surfaces.

The distance between the two surfaces is controlled by use of a three-stage mechanism of increasing sensitivity: the coarse control (upper rod) allows positioning to within about 1 μm, the medium control (lower rod, which depresses the helical spring and which in turn bends the much stiffer double-cantilever spring by 1/1000 of this amount) allows positioning to about 1 nm. Finally, a piezoelectric crystal tube—which expands or contracts vertically by about 1 nm per volt applied axially across its cylindrical wall—is used for positioning to 0.1 nm.

Given the facility for moving the surfaces towards or away from each other and, independently, for measuring their separation (each with a sensitivity or resolution of about 0.1 nm), the force measurements themselves now become straightforward. The force is measured by expanding or contracting the piezoelectric crystal by a known amount and then measuring optically how much the two surfaces have actually moved; any difference in the two values when multiplied by the stiffness of the force-measuring spring gives the force difference between the initial and final positions. In this way, both repulsive and attractive forces can be measured and a full force law can be obtained over any distance regime. The force-measuring spring can be either a single-cantilever or a double-cantilever fixed-stiffness spring, or—as shown in Fig. 10.7—the spring stiffness can be varied during an experiment (by a factor of 1000) by shifting the position of the dove-tailed clamp using the adjusting rod. Other spring attachments, two of which are shown at the top of the figure, can also be used. Each of these springs are interchangeable and can be attached to the main support allowing for greater versatility in measuring strong or weak, attractive or repulsive forces.

Once the force F as a function of distance D is known for the two surfaces (of radius R), the force between any other curved surfaces simply scales by R (see Section 10.5). Furthermore, the adhesion or interfacial energy E per unit area between two *flat* surfaces is simply related to F by the Derjaguin approximation: $E = F/2\pi R$. Thus, for $R \approx 1$ cm, and given the measuring sensitivity in F of about 10^{-8} N, the sensitivity in measuring adhesion and interfacial energies is therefore about 10^{-3} mJ m^{-2} (erg cm^{-2}).

Over the past few years the SFA has identified and quantified most of the fundamental interactions occurring between surfaces in both aqueous solutions and nonaqueous liquids. These include the attractive van der Waals and repulsive electrostatic 'double-layer' forces, oscillatory (solvation or structural) forces, repulsive hydration forces, attractive hydrophobic forces, steric interactions involving polymeric systems, and capillary and adhesion forces. These forces are described in the following chapters.

Apart from testing theories of intermolecular forces, direct force measurements have also been useful in explaining or helping to understand more complex phenomena such as the second virial coefficients of colloidal dispersions (Gee et al., 1990), the origin of the lower consulate points in the phase diagrams of certain surfactant–water mixtures (Claesson et al., 1986b), the unexpected stability of certain colloidal dispersions in high salt (see Chapter 13), the crucial role of hydration and ion-correlation forces in clay swelling and ceramic processing (Quirk, 1968; Pashley and Quirk, 1984; Kjellander et al., 1988; Velamakanni et al., 1990; Horn, 1990), and the deformed shapes of adhering particles and vesicles (Bailey et al., 1990).

Though mica, because of its molecularly smooth surface and ease of handling, is the primary surface used in SFA studies, there is currently much interest in developing alternative surfaces with different chemical and physical properties. Thus, the mica surface can be used as a substrate for adsorbing or depositing a thin film of some other material, for example, lipid monolayers or bilayers (see Chapter 18), metal films (Smith et al., 1988; Parker and Christenson, 1988), polymer films (see Chapter 15), or other macromolecules such as proteins (Lee and Belfort, 1989). Alternative materials to mica sheets are also being developed. Thus Horn et al. (1988a, 1989a) have shown how molecularly smooth sapphire and silica sheets can be used in such studies, and Hirz et al. (1991) have studied carbon and metal oxide surfaces, sputtered as thin layers onto mica sheets which now act as substrate supports for these materials.

The scope of phenomena that can be studied using the SFA technique has recently been extended to measurements of dynamic interactions and time-dependent effects, for example, the viscosity of liquids in very thin films (Chan and Horn, 1985; Israelachvili, 1986, 1989), shear and frictional forces (Israelachvili et al., 1988), and the fusion of lipid bilayers (Helm et al., 1989).

Measuring colloidal forces: Total Internal Reflection Microscopy (TIRM)

The forces between two colloidal particles in a liquid can be weaker than 10^{-13} g (10^{-15} N) and yet still be important in determining the properties of the system. Prieve and coworkers introduced a new method for measuring such minute forces, specifically between a colloidal particle and a surface. Consider a particle of radius 1 μm having twice the density of water. When placed in a container of water its effective mass will be 4×10^{-14} N $(4 \times 10^{-12}$ g). The particle will slowly move downwards, but if there is a repulsive force between it and the bottom surface of the container it will come to equilibrium at some finite distance D_0 from the surface. In Chapters 11–14 we shall see that repulsive forces of this magnitude can easily arise from long-range van der Waals forces, electrostatic forces and other types of interactions in liquids. Thus, the equilibrium surface separation D_0 could be many tens of nanometers (see Problem 10.3).

The technique developed by Prieve and Frej (1990) and Prieve et al., (1990) uses TIRM for measuring the distance between an individual colloidal particle of diameter ~ 10 μm hovering over a surface. The surface is usually made of transparent glass, and a laser beam is directed at the particle from below. From the intensity of the reflected beam one can deduce the equilibrium separation D_0. Since the particle is sitting at the bottom of a shallow potential energy well, its Brownian motion causes it to sample positions both smaller and larger than the mean value, D_0. By analysing how the reflected intensity of the light varies with time one can thus determine the distances sampled. From this the force-law can be obtained over a reasonably large range of distances on either side of the equilibrium distance. The TIRM technique promises to provide reliable data on a variety of interparticle interactions under conditions that closely parallel those occurring in colloidal systems.

Measuring atomic forces: the Atomic Force Microscope (AFM)

The AFM is in principle similar to the SFA except that forces are measured not between two macroscopic surfaces but between a fine tip and a surface (Hansma et al., 1988; Rugar and Hansma, 1990). Tip radii can be as small as one atom and larger than 1 μm. Of most current interest are the smaller tips, where in principle one could directly measure the force between an individual atom and a surface, or even between two individual atoms (see Fig. 1.4).

The adhesion force between a molecular-sized tip and a surface should be of order 10^{-9}–10^{-10} N (see Problems 1.3 and 10.8). This would be the force needed to separate the two from contact. At finite distances, the attractive or repulsive force between them would be even less. To measure such small forces one must use not only very sensitive force-measuring springs but also very sensitive ways for measuring their displacements. A whole new technology has arisen devoted to fabricating highly sensitive micron-sized force-sensing devices for AFM work. Spring stiffness as small as 0.5 N m^{-1} can be used and displacements as small as 0.01 nm can now be accurately measured using a variety of laser-optical techniques, thereby allowing forces as small as 10^{-9}–10^{-10} N to be measured (Hansma *et al.*, 1988; Rugar and Hansma, 1990). This sensitivity is sufficient for measuring adhesion and very short-range forces between molecular-sized tips and surfaces, but not longer range forces (see Problem 10.8). Recently, however, Ducker *et al.* (1991) attached a micron-sized quartz sphere to the end of an AFM tip and measured the long-range repulsive electrostatic force (see Chapter 12) between the sphere and a flat surface in aqueous salt solutions out to surface separations of 60 nm.

Interpreting the results of an AFM experiment is not always straightforward (cf. Problem 10.8). The absolute distance between the surfaces is usually not known exactly, and neither is the tip geometry. In addition, the fine tips and the surfaces often deform elastically or plastically during a measurement, which further complicates interpretation of the results. However, the technology is developing rapidly, so that very soon we may expect to see reliable intermolecular, as opposed to intersurface, force laws emerging from AFM measurements.

PROBLEMS AND DISCUSSION TOPICS

10.1 The 'range of an interaction' has always been an ambiguous concept because regardless of whether the interaction potential decays as a power law or exponentially with distance its range is strictly (mathematically) infinite. Propose a practical general definition for 'the range of an interaction'. See if it is readily applicable to the interactions of small molecules, large particles and extended surfaces, both in the absence and presence of other forces.

10.2 Show that the van der Waals interaction energy $W(D)$ per unit length for two parallel cylinders of radii R_1 and R_2 whose surfaces are separated by a distance D (where $R_1 \gg D$ and $R_2 \gg D$) is

$$\frac{W(D)}{\text{length}} = \frac{\pi^2 C \rho_1 \rho_2}{12\sqrt{2} D^{3/2}} \left(\frac{R_1 R_2}{R_1 + R_2} \right)^{1/2}. \qquad (10.21)$$

10.3 The repulsive pressure between two flat surfaces interacting in a liquid medium is given by $P = Ce^{-\kappa D}$ where $C = 100$ Pa and $\kappa^{-1} = 20$ nm are constants, and where D is the separation between the surfaces in the liquid (such long-range repulsive forces can arise from electrostatic or polymer-mediated interactions). Consider a large spherical colloidal particle of radius $R = 1\ \mu\text{m}$ interacting with a flat surface in the liquid medium via the same interaction. When the surface of the particle is at a distance $D_0 = 10$ nm from the flat surface, what is the net repulsive force between the particle and the surface, and what fraction of the total force comes from surface separations greater than $2D_0$?

If the spherical particle is made of iron and is suspended in water, what will be its average distance from the surface at equilibrium?

10.4 The repulsive pressure between two flat surfaces interacting in a particular aqueous solution is described by $P = Ce^{-\kappa D}$, as in Problem 10.3. Consider a spherical colloidal particle of radius R interacting with a flat surface in this solution, where the two surfaces are at a distance D_0 apart.

(i) Calculate their 'effective' area of interaction, i.e. the area at which two flat surfaces, separated by the same distance D_0, would have the same force as that between the sphere and the flat surface. Relate your result to the distance D'_0 between the flat surface and the circular section through the sphere having this area.

(ii) Repeat the above calculations for the case where the interaction is a purely attractive van der Waals force, where the pressure between two flat surfaces is now given by $P = -A/6\pi D^3$.

10.5 Obtain an expression for the repulsive part of the force law between two flat surfaces by summing the repulsive $+B/r^{12}$ term of the Lennard–Jones potential. Is your expression physically meaningful?

10.6 Derive Eq. (10.20).

10.7 For what types of interactions or under what conditions will the Derjaguin approximation not apply?

10.8 The pair potential of two atoms consists of an attractive van der Waals term with a coefficient of $C = -10^{-77}$ J m^6, and a hard-wall repulsion at $r = 0.2$ nm. What is the adhesion force between (i) an atom and a solid surface, and (ii) a solid sphere of radius $R = 100$ nm and a solid surface? Compare the adhesion forces (calculated at $D = 0.2$ nm) with the forces when the surfaces are at $D = 2$ nm from each other. Each solid may be assumed to be composed of the same atoms at their maximum (close-packed) density.

If in each of the above cases the atom (or sphere) were suspended from a spring of stiffness $K = 0.1$ N m^{-1} which is slowly brought down towards the surface, calculate (i) the gap distance, D_J, at which the atom (or sphere) will spontaneously 'jump' into contact with the flat surface, and (ii) the attractive force between the two at $D = D_J$.

In a particular type of AFM experiment to measure surface forces the atom or sphere constitutes the end of the 'tip', and the gap distance D_J is measured from the deflection of the spring (between the start and end of a jump). Is this a reliable way of measuring D_J, or will it overestimate/underestimate the real jump distance between the surfaces?

CHAPTER 11

VAN DER WAALS FORCES BETWEEN SURFACES

11.1 THE FORCE LAWS FOR BODIES OF DIFFERENT GEOMETRIES: THE HAMAKER CONSTANT

As we saw in Part I, van der Waals forces play a central role in all phenomena involving intermolecular forces, for while they are not as strong as Coulombic or H-bonding interactions, they are always present. When we come to consider the long-range interactions between macroscopic particles and surfaces in liquids we shall find that the three most important forces are the van der Waals, electrostatic and steric-polymer forces, and that at shorter distances (below 1 to 3 nm) solvation and other types of steric forces often dominate over both.

Let us begin by deriving the van der Waals interaction energies in vacuum for pairs of bodies of different geometries. Starting at the simplest level we shall assume that the interaction is *non-retarded* and *additive*. In Chapter 10 we saw that for an interatomic van der Waals pair potential of the form $w(r) = -C/r^6$, one may sum (integrate) the energies of all the atoms in one body with all the atoms in the other and thus obtain the 'two-body' potential for an atom near a surface (Eq. (10.2)), for a sphere near a surface (Eq. (10.5)), or for two flat surfaces (Eq. (10.8)). This procedure can be carried out for other geometries as well. The resulting interaction laws for some common geometries are shown in Fig. 11.1, given in terms of the conventional Hamaker constant

$$A = \pi^2 C \rho_1 \rho_2 \tag{11.1}$$

after Hamaker (1937), who together with Bradley (1932), Derjaguin (1934), and de Boer (1936), did much of the earlier work that advanced understanding of the forces between macroscopic bodies.

Typical values for the Hamaker constants of condensed phases, whether solid or liquid, are about 10^{-19} J for interactions across *vacuum*. For example,

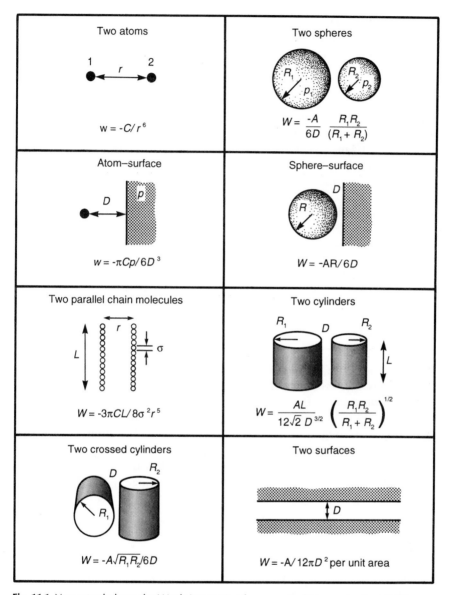

Two atoms	Two spheres
1 $\quad$ r $\quad$ 2	R_1 $\quad$ R_2 $\quad$ p_1 $\quad$ p_2
$w = -C/r^6$	$W = \dfrac{-A}{6D}\dfrac{R_1 R_2}{(R_1 + R_2)}$
Atom–surface	**Sphere–surface**
D $\quad$ p	D $\quad$ R
$w = -\pi C p/6D^3$	$W = -AR/6D$
Two parallel chain molecules	**Two cylinders**
L $\quad$ r $\quad$ σ	R_1 $\quad$ D $\quad$ R_2 $\quad$ L
$W = -3\pi CL/8\sigma^2 r^5$	$W = \dfrac{AL}{12\sqrt{2}\,D^{3/2}}\left(\dfrac{R_1 R_2}{R_1 + R_2}\right)^{1/2}$
Two crossed cylinders	**Two surfaces**
D $\quad$ R_2 $\quad$ R_1	D
$W = -A\sqrt{R_1 R_2}/6D$	$W = -A/12\pi D^2$ per unit area

Fig. 11.1. Non-retarded van der Waals interaction free energies between bodies of different geometries calculated on the basis of pairwise additivity (Hamaker summation method). The *Hamaker constant A* is defined as $A = \pi^2 C \rho_1 \rho_2$ where ρ_1 and ρ_2 are the number of atoms per unit volume in the two bodies and C is the coefficient in the atom–atom pair potential (top left). A more rigorous method of calculating the Hamaker constant in terms of the macroscopic properties of the media is given in Section 11.3. The forces are obtained by differentiating the energies with respect to distance.

if each body is composed of atoms for which $C = 10^{-77}$ J m^6 (cf. Table 6.1) and of number density $\rho = 3 \times 10^{28}$ m^{-3} (corresponding to atoms of radius ~ 0.2 nm), the Hamaker constant is

$$A = \pi^2 10^{-77}(3 \times 10^{28})^2 \approx 10^{-19} \text{ J } (10^{-12} \text{ erg}).$$

Let us consider three cases more specifically. First, for hydrocarbons, treating them as an assembly of CH$_2$ groups, we have $C \approx 5 \times 10^{-78}$ J m^6 and $\rho = 3.3 \times 10^{28}$ m^{-3} per CH$_2$ group, from which we obtain $A \approx 5 \times 10^{-20}$ J. This is shown in Table 11.1 together with similarly calculated estimates for carbon tetrachloride and water.

TABLE 11.1 Hamaker constants determined from pairwise additivity, Eq. (11.1).

Medium	C $(10^{-79}$ J m$^6)$	ρ $(10^{28}$ m$^{-3})$	A $(10^{-19}$ J$)$
Hydrocarbon	50	3.3	0.5
CCl$_4$	1500	0.6	0.5
H$_2$O	140	3.3	1.5

Note that all three Hamaker constants are similar even though the media are composed of molecules differing greatly in polarizability and size. This is not a coincidence. It arises because the coefficient C in the interatomic pair potential is roughly proportional to the square of the polarizability α, which in turn is roughly proportional to the volume v of an atom (Section 5.1). And since $\rho \propto 1/v$ we see that $A \propto C\rho^2 \propto \alpha^2\rho^2 \propto v^2/v^2 \propto constant$. Of course, this is a gross oversimplification; nevertheless, the Hamaker constants of most condensed phases are found to lie in the range $(0.4-4)10^{-19}$ J.

11.2 STRENGTH OF VAN DER WAALS FORCES BETWEEN BODIES IN VACUUM OR AIR

Taking $A = 10^{-19}$ J as a typical value, we can now estimate the strength of the van der Waals interaction between macroscopic bodies in vacuum (or air). Thus, for two spheres of radius $R = 1$ cm $= 10^{-2}$ m in contact at $D \approx 0.2$ nm, their adhesion force will be

$$F = AR/12D^2 = (10^{-19} \times 10^{-2})/12(2 \times 10^{-10})^2$$
$$= 2 \times 10^{-3} \text{ N (or 0.2 g)},$$

while at $D = 10$ nm the force will have fallen by a factor of 2500 to about 10^{-6} N, or 0.1 mg. Note that these forces are easily measurable using conventional methods.

Turning now to the interaction *energy*, at $D = 10$ nm the energy is $-AR/12D \approx -10^{-14}$ J, or about $2 \times 10^6 \, kT$, and even for particles with radii as small as $R = 20$ nm their energy exceeds kT at $D = 10$ nm.

For two planar surfaces in contact ($D \approx 0.2$ nm), the adhesive pressure will be

$$P = A/6\pi D^3 \approx 7 \times 10^8 \, \mathrm{N \, m^{-2}} \approx 7000 \, \mathrm{atm},$$

while at $D = 10$ nm the pressure is reduced by a factor of about 10^5 to a still-significant 0.05 atm. At contact the *adhesion energy* will be $-A/12\pi D^2 \approx -66$ mJ m^{-2}, which corresponds to a surface energy of $\gamma = 33$ mJ m^{-2}. This is exactly of the order expected for the surface energies and tensions of van der Waals solids and liquids, discussed later. We see, therefore, that the van der Waals interaction between macroscopic particles and surfaces is large, and not only when the bodies are in contact. Later we shall see that in a medium the interaction strength is reduced by about an order of magnitude, and that under certain conditions it can become repulsive.

11.3   THE LIFSHITZ THEORY OF VAN DER WAALS FORCES

The assumptions of simple pairwise additivity inherent in the formulae of Fig. 11.1 and the definition of A of Eq. (11.1) ignore the influence of neighbouring atoms on the interaction between any pair of atoms. First, as we saw in Section 5.7 the effective polarizability of an atom changes when it is surrounded by other atoms. Second, recalling our earlier simple model of the dispersion interaction between two Bohr atoms 1 and 2, if a third atom 3 is present, it too will be polarized by the instantaneous field of atom 1, and its induced dipole field will also act on atom 2. Thus, the field from atom 1 reaches atom 2 both directly and by reflection from atom 3. The existence of multiple reflections and the extra force terms to which they give rise is a further instance where straightforward additivity breaks down, and the matter becomes very complicated when many atoms are present (see Problem 6.2). In rarefied media (gases) these effects are small, and the assumptions of additivity hold, but this is not the case for condensed media. Further, the additivity approach cannot be readily extended to bodies interacting in a medium.

The problem of additivity is completely avoided in the *Lifshitz theory*

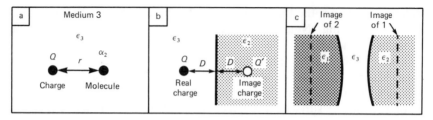

Fig. 11.2. (a) A charge interacts with a neutral molecule because of the field reflected by the molecule on becoming polarized. (b) Likewise, a charge interacts with a surface because of the field reflected by the surface. This reflected field is the same as if there were an 'image' charge Q' at a distance $2D$ from Q. Similarly, a dipole near a surface will see an image of itself reflected by the surface. If $\varepsilon_2 > \varepsilon_3$, the force is attractive; if $\varepsilon_2 < \varepsilon_3$, it is repulsive. (c) Two surfaces will see an image of each other reflected by the other surface that gives rise to the van der Waals force between them. In principle, the reflected or image fields are the same as occur when one looks at a glass surface or a mirror. Metal surfaces reflect most of the light falling on them, and the van der Waals force between metals is much stronger than that between dielectric media.

where the atomic structure is ignored and the forces between large bodies, now treated as continuous media, are derived in terms of such bulk properties as their dielectric constants and refractive indices. However, before we proceed it is well to point out that all the expressions in Fig. 11.1 for the interaction energies remain valid even within the framework of continuum theories. The only thing that changes is the way the Hamaker constant is calculated.

The original Lifshitz theory (Lifshitz, 1956; Dzyaloshinskii *et al.*, 1961) requires a thorough working knowledge of quantum field theory for its understanding, and it is probably due to this that it was initially ignored by most scientists who persisted with the additivity approach of Hamaker. Later, Langbein, Ninham, Parsegian, Van Kampen and others, showed how the essential equations could be derived using much simpler theoretical techniques (for reviews see Israelachvili and Tabor, 1973; Parsegian, 1973; Israelachvili, 1974; Mahanty and Ninham, 1976). Here we shall adopt the simplest of these using a modified additivity approach.

We have already seen that the van der Waals interaction is essentially electrostatic, arising from the dipole field of an atom 'reflected back' by a second atom that has been polarized by this field. Let us first analyse this reflected field when the second atom is replaced by a planar surface and where the first atom is replaced by a charge Q (Fig. 11.2). From Fig. 2.2 we know that the interaction energy of a charge with a molecule (Fig. 11.2a) is given by

$$w(r) = -C/r^n = -Q^2\alpha_2/2(4\pi\varepsilon_0\varepsilon_3)^2 r^4, \qquad (11.2)$$

where α_2 is the excess polarizability of molecule 2 in medium 3. When molecule 2 is replaced by a medium (Fig. 11.2b) the interaction between the charge in medium 3 and the surface of medium 2 may be obtained by the method of additivity, Eq. (10.1), which gives $W(D) = -2\pi C\rho_2/(n-2) \times (n-3)D^{n-3}$. Inserting values for C and n from Eq. (11.2) gives

$$W(D) = -\pi Q^2 \rho_2 \alpha_2 / 2(4\pi\varepsilon_0\varepsilon_3)^2 D. \tag{11.3}$$

However, it is well known that a charge Q in a medium of dielectric constant ε_3 at a distance D from the plane surface of a second medium of dielectric constant ε_2 experiences a force as if there were an 'image' charge of strength $-Q(\varepsilon_2 - \varepsilon_3)/(\varepsilon_2 + \varepsilon_3)$ at a distance D on the other side of the boundary, i.e., at a distance $2D$ away (Landau and Lifshitz, 1984, Ch. II). This force is therefore given by

$$F(D) = \frac{-Q^2}{(4\pi\varepsilon_0\varepsilon_3)(2D)^2}\left(\frac{\varepsilon_2 - \varepsilon_3}{\varepsilon_2 + \varepsilon_3}\right), \tag{11.4}$$

which corresponds to an interaction energy of

$$W(D) = \frac{-Q^2}{4(4\pi\varepsilon_0\varepsilon_3)D}\left(\frac{\varepsilon_2 - \varepsilon_3}{\varepsilon_2 + \varepsilon_3}\right). \tag{11.5}$$

Equating Eq. (11.5) with Eq. (11.3) we immediately find that

$$\rho_2\alpha_2 = 2\varepsilon_0\varepsilon_3(\varepsilon_2 - \varepsilon_3)/(\varepsilon_2 + \varepsilon_3). \tag{11.6}$$

This is an important result, giving the excess bulk or volume polarizability of a planar dielectric medium 2 in medium 3 in terms of the purely macroscopic properties of the media. The Hamaker constant for the interaction of two media 1 and 2 across a third medium 3 (Fig. 11.2c) may now be expressed in terms of McLachlan's equation, Eq. (6.18), for C and Eq. (11.6) for $\rho_1\alpha_1$ and $\rho_2\alpha_2$ as follows:

$$\begin{aligned}
A &= \pi^2 C\rho_1\rho_2 \\
&= \frac{6\pi^2 kT\rho_1\rho_2}{(4\pi\varepsilon_0)^2} \sum_{n=0,1,...}^{\infty}{}' \frac{\alpha_1(iv_n)\alpha_2(iv_n)}{\varepsilon_3^2(iv_n)} \\
&= \frac{3}{2}kT \sum_{n=0,1,...}^{\infty}{}' \left[\frac{\varepsilon_1(iv_n) - \varepsilon_3(iv_n)}{\varepsilon_1(iv_n) + \varepsilon_3(iv_n)}\right]\left[\frac{\varepsilon_2(iv_n) - \varepsilon_3(iv_n)}{\varepsilon_2(iv_n) + \varepsilon_3(iv_n)}\right].
\end{aligned} \tag{11.7}$$

Replacing the sum by the integral of Eq. (6.25) we end up with the expression for the Hamaker constant based on the Lifshitz theory

$$
A \approx \frac{3}{4}kT\left(\frac{\varepsilon_1 - \varepsilon_3}{\varepsilon_1 + \varepsilon_3}\right)\left(\frac{\varepsilon_2 - \varepsilon_3}{\varepsilon_2 + \varepsilon_3}\right)
$$
$$
+ \frac{3h}{4\pi}\int_{\nu_1}^{\infty}\left(\frac{\varepsilon_1(iv) - \varepsilon_3(iv)}{\varepsilon_1(iv) + \varepsilon_3(iv)}\right)\left(\frac{\varepsilon_2(iv) - \varepsilon_3(iv)}{\varepsilon_2(iv) + \varepsilon_3(iv)}\right)dv, \qquad (11.8)
$$

where ε_1, ε_2 and ε_3 are the static dielectric constants of the three media, $\varepsilon(iv)$ are the values of ε at imaginary frequencies, and $\nu_n = (2\pi kT/h)n = 4 \times 10^{13}\,\mathrm{n\,s^{-1}}$ at 300 K. The first term in Eq. (11.8) gives the zero-frequency energy of the van der Waals interaction and includes the Keesom and Debye dipolar contributions. The second term gives the dispersion energy and includes the London energy contribution. Equations (11.7) and (11.8) are not exact but are only the first terms in an infinite series for the nonretarded Hamaker constant. The other terms, however, are small and rarely contribute more than 5%.

The above approach may also be used to calculate the van der Waals interaction between a molecule or small particle 1 in medium 3 with the surface of medium 2 (Fig. 11.3a). The interaction energy may be readily

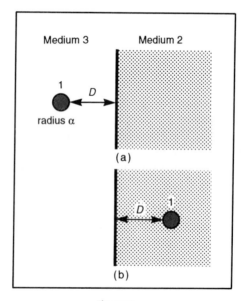

Fig. 11.3.

shown to be given by (Israelachvili, 1974)

$$W(D) = -\pi C\rho_2/6D^3$$

$$= -\frac{3kTa_1^3}{2D^3}\sum'\left[\frac{\varepsilon_1(iv_n) - \varepsilon_3(iv_n)}{\varepsilon_1(iv_n) + 2\varepsilon_3(iv_n)}\right]\left[\frac{\varepsilon_2(iv_n) - \varepsilon_3(iv_n)}{\varepsilon_2(iv_n) + \varepsilon_3(iv_n)}\right]$$

$$\approx -\frac{Aa_1^3}{3D^3}. \tag{11.9}$$

If, on the other hand, molecule 1 is in medium 2 (Fig. 11.3b), we must exchange $\varepsilon_2(iv)$ and $\varepsilon_3(iv)$ in this equation. Equation (11.9) has interesting consequences when the dielectric media 2 and 3 are liquids, or at least permeable to molecule 1, for it predicts that the molecule will behave in one of two ways depending on the relative values of the dielectric permittivities:

(i) The molecule will be attracted to the interface from either side of it (e.g., if ε_1 is intermediate between ε_2 and ε_3).
(ii) The molecule will be attracted towards the interface from one side and then repelled from the other side (e.g., if $\varepsilon_1 > \varepsilon_3 > \varepsilon_2$, the molecule will be driven from right to left in Fig. 11.3).

Thus, in the absence of constraints or other forces the van der Waals interaction alone may promote the migration of small uncharged particles across liquid interfaces. Note that repulsion from both sides of the interface never occurs. These effects are entirely analogous to those previously discussed in Section 9.4 in non-specific terms. The application of the Lifshitz theory to explain the engulfing or rejection of particles by surfaces has been extensively studied by Van Oss et al. (1980).

11.4 HAMAKER CONSTANTS CALCULATED ON THE BASIS OF THE LIFSHITZ THEORY

To obtain the Hamaker constant for any system we first need to know how the dielectric permittivity of the media vary with frequency, after which we can integrate Eq. (11.8) to obtain A. The dielectric permittivity $\varepsilon(iv)$ of a medium varies with frequency v in much the same way as does the atomic polarizability of an atom, discussed in Section 6.6. Thus, $\varepsilon(v)$ and $\varepsilon(iv)$ can usually be represented by a function of the form (Mahanty and Ninham, 1976)

$$\varepsilon(v) = 1 + \frac{\text{constant}}{(1 - iv/v_{rot})} + \frac{\text{constant}}{(1 - v^2/v_e^2)}, \tag{11.10}$$

so that

$$\varepsilon(iv) = 1 + \frac{(\varepsilon - n^2)}{(1 + v/v_{rot})} + \frac{(n^2 - 1)}{(1 + v^2/v_e^2)}, \tag{11.11}$$

$$\varepsilon(0) = \varepsilon,$$

where v_{rot} is the molecular rotational relaxation frequency, which is typically at microwave and lower frequencies ($v_{rot} < 10^{12} \, \text{s}^{-1}$), v_e is the main electronic absorption frequency in the UV typically around $3 \times 10^{15} \, \text{s}^{-1}$, and n is the refractive index of the medium in the visible (i.e., $n^2 = \varepsilon_{vis}(v)$). Note that $\varepsilon(iv)$ of Eq. (11.11) is a real function of v.

Since $v_1 \approx 4 \times 10^{13} \, \text{s}^{-1} \gg v_{rot}$ the dispersion energy is determined solely by the electronic absorption (last term in Eq. (11.11)). We may therefore substitute an expression of the form

$$\varepsilon(iv) = 1 + (n^2 - 1)/(1 + v^2/v_e^2) \tag{11.12}$$

for each medium into Eq. (11.8) and integrate using the definite integral of Eq. (6.27). If the absorption frequencies of all three media are assumed to be the same, we obtain the following approximate expression for the non-retarded Hamaker constant for two macroscopic phases 1 and 2 interacting across a medium 3:

$$A_{\text{Total}} = A_{v=0} + A_{v>0}$$

$$\approx \frac{3}{4}kT\left(\frac{\varepsilon_1 - \varepsilon_3}{\varepsilon_1 + \varepsilon_3}\right)\left(\frac{\varepsilon_2 - \varepsilon_3}{\varepsilon_2 + \varepsilon_3}\right)$$

$$+ \frac{3hv_e}{8\sqrt{2}} \frac{(n_1^2 - n_3^2)(n_2^2 - n_3^2)}{(n_1^2 + n_3^2)^{1/2}(n_2^2 + n_3^2)^{1/2}\{(n_1^2 + n_3^2)^{1/2} + (n_2^2 + n_3^2)^{1/2}\}}. \tag{11.13}$$

For the 'symmetric case' of two identical phases 1 interacting across medium 3, the above reduces to the simple expression

$$A = A_{v=0} + A_{v>0} = \frac{3}{4}kT\left(\frac{\varepsilon_1 - \varepsilon_3}{\varepsilon_1 + \varepsilon_3}\right)^2 + \frac{3hv_e}{16\sqrt{2}} \frac{(n_1^2 - n_3^2)^2}{(n_1^2 + n_3^2)^{3/2}}. \tag{11.14}$$

The above expressions for A apply to any of the macroscopic geometries listed in Fig. 11.1. Four interesting aspects of these equations may be noted:

(1) The van der Waals force between two *identical* bodies in a medium is always attractive (A positive), while that between *different* bodies in a medium can be attractive or repulsive (A negative).

(2) The van der Waals force between any two condensed bodies in vacuum or air ($\varepsilon_3 = 1$ and $n_3 = 1$) is always attractive.

(3) The Hamaker constant for two similar media interacting across another medium remains unchanged if the media are interchanged. Thus, if no other forces are operating, a liquid film in air will always tend to thin under the influence of the attractive van der Waals force between the two surfaces, or in other words, two air phases (or bubbles) attract each other in a liquid.

(4) The purely entropic zero-frequency contribution $A_{v=0}$ can never exceed $\frac{3}{4}kT$, or about 3×10^{-21} J at 300 K. For interactions across vacuum where the dispersion energy contribution $A_{v>0}$ is typically $\sim 10^{-19}$ J, this is always small (about 3%). But, as will be seen, for interactions across a medium the zero-frequency contribution sometimes dominates over the dispersion energy contribution.

A number of authors have given more complex analytic formulae for A when the interacting media have different absorption frequencies (Israelachvili, 1972; Horn and Israelachvili, 1981, Appendix B), while others have described simple numerical procedures for computing Hamaker constants for media with many absorption frequencies when the more exact equation (11.7) must be used (Gregory, 1970; Pashley, 1977; Hough and White, 1980; Prieve and Russel, 1988).

The above analysis applies to what are normally referred to as dielectric or non-conducting materials. For interactions involving conducting media such as metals their static dielectric constant is infinite and the dispersion formula of Eq. (11.11) does not apply. The dielectric permittivity of a metal is given approximately by

$$\varepsilon(v) = 1 - v_e^2/v^2, \qquad (11.15)$$

so that

$$\varepsilon(iv) = 1 + v_e^2/v^2, \quad \text{and} \quad \varepsilon(0) = \infty, \qquad (11.16)$$

where v_e is the so-called plasma frequency of the free electron gas, typically in the range $(3-5) \times 10^{15}$ s^{-1}. Substituting the above equation in Eq. (11.8) and integrating as before, we obtain for two metals interacting across vacuum

$$A \approx (3/16\sqrt{2})hv_e \approx 4 \times 10^{-19} \text{ J}. \qquad (11.17)$$

TABLE 11.2 Non-retarded Hamaker constants for two identical media interacting across vacuum (air)

Medium	Dielectric constant ε	Refractive index n	Absorption frequency[a] ν_e $(10^{15}\,s^{-1})$	Hamaker constant A $(10^{-20}\,J)$		
				Eq. (11.14) $\varepsilon_3 = 1$	Exact solutions[b]	Experimental[c]
Water	80	1.333	3.0	3.7	3.7, 4.0	
n-Pentane	1.84	1.349	3.0	3.8	3.75	
n-Octane	1.95	1.387	3.0	4.5	4.5	
n-Dodecane	2.01	1.411	3.0	5.0	5.0	
n-Tetradecane	2.03	1.418	2.9	5.0	5.1, 5.4	
n-Hexadecane	2.05	1.423	2.9	5.1	5.2	
Hydrocarbon (crystal)	2.25	1.50	3.0	7.1		10
Cyclohexane	2.03	1.426	2.9	5.2		
Benzene	2.28	1.501	2.1	5.0		
Carbon tetrachloride	2.24	1.460	2.7	5.5		
Acetone	21	1.359	2.9	4.1		
Ethanol	26	1.361	3.0	4.2		

Polystyrene	2.55	1.557	2.3	6.5	6.6, 7.9	5–6
Polyvinyl chloride	3.2	1.527	2.9	7.5	7.8	13.5
PTFE	2.1	1.359	2.9	3.8	3.8	
Fused quartz	3.8	1.448	3.2	6.3	6.5	
Mica	7.0	1.60	3.0	10	10	
CaF_2	7.4	1.427	3.8	7.0	7.2	
Liquid He	1.057	1.028	5.9	0.057		
Alumina (Al_2O_3)[d]	11.6	1.75	3.0 est.	14		
Iron oxide (Fe_3O_4)	—	1.97	3.0 est.	21		
Zirconia $(n\text{-}ZrO_2)$[d]	20–40	2.15	3.0 est.	27		
Rutile (TiO_2)[d]	—	2.61	3.0 est.	43		
Silicon carbide[d]	10.2	2.65	3.0 est.	44		
Metals (Au, Ag, Cu)	∞	—	3–5	25–40 Eq. (11.17)		

[a] UV absorption frequencies obtained from 'Cauchy plots' mainly from Hough and White (1980) and H. Christenson (1983, thesis).

[b] Exact solutions computed mainly by Hough and White (1980) and, when in italics, by Parsegian and Weiss (1981).

[c] Experimental values from Israelachvili and Tabor (1972) and Derjaguin et al. (1978).

[d] Velamakanni (1990, thesis).

Due to their high polarizability, as reflected by their high dielectric constants and refractive indices, the Hamaker constants of metals and metal oxides can be up to an order of magnitude higher than those of non-conducting media.

Table 11.2 gives some Hamaker constants for two identical phases interacting across vacuum as calculated using Eq. (11.14) together with more rigorously computed values. Experimentally determined values are also included where available. Note the good agreement between values calculated on the basis of Eq. (11.14) and those computed more rigorously by solving the full Lifshitz equation. Note too the reasonably good agreement between theory and experiment. We may also compare the values of A of Table 11.2 with those obtained from a summation of additive pair potentials given in Table 11.1. Thus, for hydrocarbons and CCl_4, the agreement is surprisingly good, but for highly polar liquids such as water, the additivity approach overestimates the value of A mainly because it overestimates the zero-frequency contribution.

Note that the Lifshitz theory is a continuum theory and so can only be used when the interacting surfaces are farther apart than molecular dimensions, i.e., when $D \gg \sigma$. In Sections 11.10 and 11.11 we shall investigate the applicability of the Lifshitz theory for determining short-range forces, including the adhesion energies of bodies in contact.

11.5 APPLICATIONS OF THE LIFSHITZ THEORY TO INTERACTIONS IN A MEDIUM

The Lifshitz theory is particularly suitable for analysing interactions in a medium. As a graphic example of this, Fig. 11.4 shows how $\varepsilon(iv)$ varies with frequency for water and a typical hydrocarbon such as dodecane. Both of these liquids have roughly the same absorption frequency of about $v_e \approx 3.0 \times 10^{15}\,s^{-1}$. The area under the dashed curve at frequencies above v_1 is roughly proportional to the non-retarded dispersion energy of two hydrocarbon phases across a water film, for which we obtain, using Eq. (11.14),

$$A_{v>0} \approx \frac{3(6.63 \times 10^{-34})(3 \times 10^{15})}{16\sqrt{2}} \frac{(1.41^2 - 1.33^2)^2}{(1.41^2 + 1.33^2)^{3/2}} \approx 0.17 \times 10^{-20}\,J.$$

Concerning the zero-frequency contribution, note that water exhibits strong absorptions at lower frequencies and consequently has a high static dielectric constant ε of 80. In contrast, the dielectric constant of hydrocarbons remains the same right down to zero frequency where $\varepsilon \approx n^2 \approx 2.0$. The large

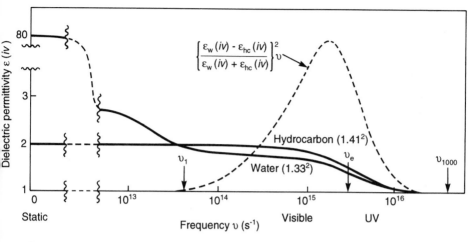

Fig. 11.4. Dielectric permittivity $\varepsilon(iv)$ as a function of frequency v for water and hydrocarbon. In the visible and UV range these are given by

$$\varepsilon_w(iv) = 1 + (n_w^2 - 1)/(1 + v^2/v_e^2), \qquad n_w = 1.33,$$

$$\varepsilon_{hc}(iv) = 1 + (n_{hc}^2 - 1)/(1 + v^2/v_e^2), \qquad n_{hc} = 1.41,$$

where $v_e = 3.0 \times 10^{15} \, \text{s}^{-1}$ for both. The total Hamaker constant for this system at 300 K is about 0.45×10^{-20} J.

difference between ε_{H_2O} and ε_{hc} leads to a large zero-frequency contribution to the Hamaker constant of

$$A_{v=0} = \frac{3}{4}kT\left(\frac{80-2}{80+2}\right)^2 \approx 0.28 \times 10^{-20} \, \text{J at 300 K},$$

giving a total value for A of $(0.17 + 0.28)10^{-20} \approx 0.45 \times 10^{-20}$ J.

Thus, the interaction between hydrocarbon across water is about 10% of that across vacuum (Table 11.2). Note that for interactions across water the zero-frequency contribution $A_{v=0}$ is actually higher than that across vacuum, and that this mainly entropic term therefore dominates over the dispersion contribution.

Ninham and Parsegian (1970), Gingell and Parsegian (1972), and Hough and White (1980) considered the hydrocarbon–water system in great detail and found that the theoretical A for this system lies in the range $(4-7) \times 10^{-21}$ J depending on the refractive index of the hydrocarbon. Note that as far as van der Waals forces are concerned the interaction of hydrocarbon across water is the same as for water across hydrocarbon. Experimental A values for hydrocarbon–water systems have been determined from studies on lipid

TABLE 11.3 Hamaker constants for two media interacting across another medium

Interacting media			Hamaker constant A (10^{-20} J)		
1	3	2	Eq. (11.13)[a]	Exact solutions[b]	Experiment
Air	Water	Air	3.7	3.70	
Pentane	Water	Pentane	0.28	0.34	
Octane	Water	Octane	0.36	0.41	
Dodecane	Water	Dodecane	0.44	0.50	0.5[d]
Hexadecane	Water	Hexadecane	0.49	0.50	0.3–0.6[d,e]
Water	Hydrocarbon	Water	0.3–0.5	0.34–0.54	0.3–0.9[f]
Polystyrene	Water	Polystyrene	1.4	0.95–1.3[c]	
Fused quartz	Water	Fused quartz	0.63	0.83	
Fused quartz	Octane	Fused quartz	0.13		
PTFE	Water	PTFE	0.29	0.33	
Mica	Water	Mica	2.0	2.0	2.2[g]

Alumina (Al$_2$O$_3$)	Alumina (Al$_2$O$_3$)	Water	4.2[i]	5.3[b]	6.7[i]
Zirconia (n-ZrO$_2$)	Zirconia (n-ZrO$_2$)	Water	13[i]		
Rutile (TiO$_2$)	Rutile (TiO$_2$)	Water	26[i]		
Ag, Au, Cu	Ag, Au, Cu	Water	—	30–40[c]	40 (gold)[h]
Water	Water	Pentane	0.08	0.11	
Water	Water	Octane	0.51	0.53	
Octane	Water	Water	−0.24	−0.20	
Fused quartz	Water	Octane	−0.87	−1.0	
Fused quartz	Octane	Air	−0.7		
Fused quartz	Tetradecane	Air	−0.4		
CaF$_2$, SrF$_2$	Liquid He	Vapour	−0.59	−0.59[i]	−0.5[h] / −0.58[i]

[a] Based on dielectric data of Table 11.2, assuming mean values for v_e.
[b] Hough and White (1980).
[c] Parsegian and Weiss (1981).
[d] Lis et al. (1982).
[e] Ohshima et al. (1982).
[f] Requena et al. (1975).
[g] Israelachvili and Adams (1978).
[h] Derjaguin et al. (1978).
[i] Sabisky and Anderson (1973).
[j] Velamakanni (1990, thesis); Horn et al. (1988a).

bilayers. Table 11.3 gives some computed values of A for interactions across various media based on the Lifshitz theory, together with measured values where these are available. As can be seen, the agreement between theory and experiment is good. Note the much smaller Hamaker constants compared to those for interactions across vacuum, especially for media of low refractive index (Table 11.2).

11.6   REPULSIVE VAN DER WAALS FORCES: DISJOINING PRESSURE AND WETTING FILMS

We have already seen in Chapter 6 how repulsive van der Waals forces can arise in a medium and again in Chapter 9 how repulsive forces in general can be understood intuitively. Equation (11.13) shows that the Hamaker constant will be negative, resulting in repulsion, whenever the dielectric properties of the intervening medium are intermediate between those of the two interacting media. Indeed, one of the early successes of the Lifshitz theory was in the quantitative explanation of the wetting properties of liquid helium due to a negative Hamaker constant (Dzyaloshinskii *et al.*, 1961).

It is well known that liquid helium avidly spreads on almost any surface. Thus, if liquid helium is placed in a beaker, it rapidly climbs up the walls and down the other side, and eventually leaves the beaker altogether. The reason for this peculiar behaviour is that the dielectric permittivity of liquid helium, $\varepsilon = n^2 = 1.057$, is lower than that of any other medium (except vapour). Thus, there will be a negative Hamaker constant and a repulsive van der Waals force across an absorbed liquid helium film which will act to thicken the film so as to lower its energy. But when liquid helium climbs up a smooth wall the gain in van der Waals energy is at the expense of gravitational energy, and so the equilibrium film thickness will decrease with height (Fig. 11.5). Sabisky and Anderson (1973) measured the thickness as a function of the height of liquid helium films at 1.38 K on atomically smooth surfaces of CaF_2, SrF_2 and BaF_2. As the height increased the thickness decreased. Let us look at this interesting phenomenon of *wetting films* in more detail (see also Section 12.9).

Consider unit area of a film of mass m, density ρ and thickness D at a height H above the flat liquid–vapour surface (Fig. 11.5). The total free energy of this mass will be

$$G(D) = -\frac{A}{12\pi D^2} + mgH = -\frac{A}{12\pi D^2} + \rho g H D \quad \text{(since } m = \rho D\text{)}.$$

$$(11.18)$$

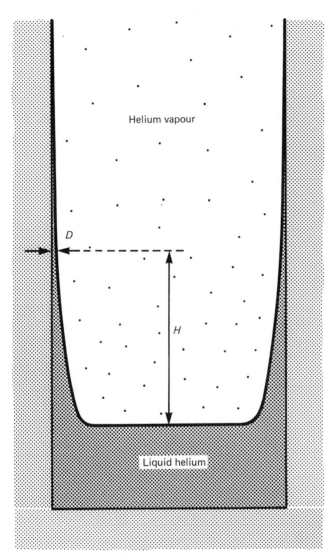

Fig. 11.5. Liquid helium climbs up the walls of containers because of the repulsive van der Waals force across the adsorbed film. At equilibrium the work done against gravity *mgH* is balanced by the work done by the van der Waals force.

The equilibrium film thickness at H will be given when $\partial G/\partial D = 0$, i.e.,

$$A/6\pi D^3 + \rho g H = 0, \qquad (11.19)$$

which gives for the thickness profile as a function of height H:

$$D = (-A/6\pi\rho g H)^{1/3}. \tag{11.20}$$

Equation (11.20) has solutions for finite D only when A is negative. In the case of liquid helium on a CaF_2 surface, from Table 11.3 we have $A = -5.9 \times 10^{-21}$ J, and for helium $\rho = 1.4 \times 10^2$ kg m^{-3}. Thus, we expect for the film profile:

$$D = \frac{28}{\sqrt[3]{H(\text{cm})}} \text{ nm.}$$

Sabisky and Anderson (1973) measured $D = 2.8$ nm at $H = 1000$ cm, exactly as expected. But at $H = 1$ cm they obtained $D = 21.5$ nm instead of $D = 28$ nm. This is due to retardation effects, which were observed for films thicker than 6 nm.

Repulsive van der Waals forces also occur across thin liquid hydrocarbon films on alumina (Blake, 1975) and quartz (Gee et al., 1989). The thicknesses of such films were varied either by pressing a gas bubble in the liquid against the solid surface or by changing the vapour pressure above the film (see below). The measured variation of the disjoining pressure with thickness in the range 0.5–80 nm was found to be in excellent agreement with theory. Once again retardation effects are evident for films thicker than 5 nm.

Returning to Eq. (11.19), the first term is simply the repulsive van der Waals pressure across the film. In fact, we may write that equation in the more general form

$$P(D) = -\rho g H \tag{11.21}$$

which is valid for any repulsive (negative) pressure regardless of the origin of the interaction. Such a repulsive pressure $P(D)$ is often referred to as the *disjoining pressure* of a film.

It is important to appreciate that the equilibrium thickness of the helium film at any point on the surface of the beaker in Fig. 11.5 does not rely on the film being in contact with the bulk liquid reservoir at the bottom of the beaker. If the surface is not in contact with the bulk liquid, equilibrium will simply be established via the vapour (though it may take longer). This fact invites another way of looking at the equilibrium of films. Just above the flat liquid surface the vapour pressure must be at the saturation value, p_0, but at a height H the vapour pressure, p, will be below saturation. Now from Eq. (2.13) we have $p = p_0 e^{-mgH/kT}$, so that the equilibrium condition,

Eq. (11.21), may also be written as

$$P(D) = (kT/v)\log(p/p_0) = (RT/V)\log(p/p_0) \qquad (11.22)$$

where v, V are the molecular and molar volumes respectively. Since the relative vapour pressure, p/p_0, can be controlled in a variety of ways (in addition to controlling the height of the film above the surface of the liquid) we see that Eq. (11.22) provides an alternative way of determining the equilibrium thicknesses of a wetting film in contact with its undersaturated vapour. Indeed, the method of 'vapour pressure control' rather than 'gravity control' of film thicknesses is usually far more practical, as the example below shows.

☐   ● WORKED EXAMPLE ●

Question: (i) At what relative vapour pressure of *n*-octane will a 1.5 nm film adsorb on a horizontal quartz glass surface at 25°C? (ii) If the same film is to be attained by having the surface placed above the bulk liquid level, what height should this be at? (iii) Assuming that a curved meniscus experiences an additional pressure given by the *Laplace equation*, estimate the maximum curvature of the surface asperities that can be tolerated if the film thickness has to be uniform to within 10% of the calculated value. What other experimental parameters are crucial for this?

Answer: (i) For the quartz–octane–vapour system $A = -0.7 \times 10^{-20}$ J (Table 11.3). The molecular weight of octane, C_8H_{18}, is 114×10^{-3} kg mol^{-1}, and the density of the liquid is 0.70×10^3 kg m^{-3}. The molecular volume is obtained from Eq. (5.30) as: $v = M/\rho N_0 = (114 \times 10^{-3})/(0.70 \times 10^3) \times (6.02 \times 10^{23}) = 2.7 \times 10^{-28}$ m^3. Thus, to have an equilibrium film thickness of $D = 1.5 \times 10^{-9}$ m (15 Å) at $T = 298$ K we require

$$p/p_0 = \exp[P(D)v/kT] = \exp[Av/6\pi D^3 kT]$$

$$= \exp[-(0.7 \times 10^{-20})(2.7 \times 10^{-28})/6\pi(1.5 \times 10^{-9})^3(4.12 \times 10^{-21})]$$

$$= 0.993 \text{ (i.e., 99.3\% of the saturated vapour pressure).}$$

(ii) Using equation (11.20), the height is

$$H = \frac{-A}{6\pi\rho g D^3} = \frac{(0.7 \times 10^{-20})}{6\pi(0.70 \times 10^3)(9.81)(1.5 \times 10^{-9})^3} = 16.0 \text{ m}.$$

Such a height is experimentally feasible but difficult (remember that the temperature must be kept exactly the same at both ends). To obtain even thinner films H becomes impractically large, whereas controlling the thickness via the vapour pressure actually becomes easier (why?).

(iii) The van der Waals pressure across a 1.5 nm film of octane is $P(D) = A/6\pi D^3 = -(0.7 \times 10^{-20})/6\pi(1.5 \times 10^{-9})^3 = 1.10 \times 10^5$ Pa (about 1 atm), while that across a film 10% thicker, i.e., with $D = 1.65$ nm, is 0.83×10^5 Pa. The difference in pressure is $\Delta P = 0.27 \times 10^5$ Pa. From the Laplace equation, Eq. (15.31), such a pressure difference is established if the radius of the surface film is $r \approx 2\gamma/\Delta P$, where $\gamma \approx 22$ mJ m^{-2} is the surface tension of octane at 25°C. Thus, if the surface has locally curved regions (asperities) of radii less than $r \approx 2(0.022)/(0.27 \times 10^5) \approx 1.6 \times 10^{-6}$ m, or 1.6 μm, the film thickness will vary from place to place by more than 10% of the mean value.

Other experimental parameters that must be carefully controlled: temperature, purity of liquid and vapour, cleanliness of surface.   □

Negative Hamaker constants are also expected for water films on hydrocarbons and on quartz (Table 11.3). Now, water does indeed wet quartz, but this is also due to the existence of other repulsive forces across the film such as electric double-layer forces (Chapter 12) and hydration forces (Chapter 13). However, water certainly does not wet hydrocarbons. This is because of the attractive *hydrophobic* force across the film, which is stronger than the repulsive van der Waals force (Chapter 13).

Repulsive van der Waals forces often occur between different types of polymers dissolved in organic solvents. Van Oss *et al.* (1980) estimated the Hamaker constants of 31 polymer pairs in different solvents. These varied from $A = +0.75 \times 10^{-20}$ J to $A = -0.38 \times 10^{-20}$ J. It was found that whenever the Hamaker constant exceeds about 0.03×10^{-20} J the polymers are miscible with each other in the solvent, while for $A < -0.03 \times 10^{-20}$ J they are immiscible, and phase separation occurs. Another example of repulsive van der Waals forces is given below after a few words about retardation effects.

11.7   RETARDATION EFFECTS

In Chapter 6 it was pointed out that at distances beyond about 5 nm the dispersion contribution to the total van der Waals force begins to decay more rapidly due to retardation effects. For interactions between molecules, this is of little consequence. However, for interactions between macroscopic

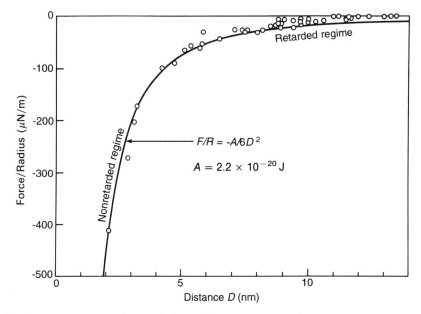

Fig. 11.6. Attractive van der Waals force F between two curved mica surfaces of radius $R \approx 1$ cm measured in water and aqueous electrolyte solutions. The measured non-retarded Hamaker constant is $A = 2.2 \times 10^{-20}$ J. Retardation effects are apparent at distances above 5 nm where the measured forces are weaker than the non-retarded prediction (solid line). (From Israelachvili and Adams, 1978; Israelachvili and Pashley, unpublished.)

bodies, where the forces can still be significant at such large separations, the effects of retardation must sometimes be taken into account.

There is no simple equation for calculating the van der Waals force at all separations. Strictly, the Hamaker constant is never truly constant but decreases progressively as D increases, and it can fall from its full non-retarded value at small D to less than half this value by $D = 10$ nm. Mahanty and Ninham (1976) and Pashley (1977) have described numerical methods for computing the van der Waals force law at all distances by solving the full Lifshitz equation.

As an illustration of retardation effects, Fig. 11.6 shows experimental results obtained for the van der Waals force law between two curved mica surfaces in aqueous electrolyte solutions (in a liquid medium such as water the full force law also involves the electric double-layer force, so that some extrapolation and subtraction of this force must be done in order to obtain the purely van der Waals force contribution to the total interaction). For this system, the non-retarded Hamaker constant was found to be $A = 2.2 \times 10^{-20}$ J, which is about 10 % higher than the theoretical value (Table 11.3). The experimental results clearly show the onset of retardation at separations above about 5 nm—the same distance as in the wetting film experiments described above.

Retarded van der Waals forces have also been measured between mica, glass and metal surfaces in air or vacuum out to distances of 1200 nm, all in good agreement with the full Lifshitz theory. Indeed, the first direct measurements of van der Waals forces by Derjaguin and Abrikossova (1954) were of retarded forces between two quartz glass surfaces in the distance range 100–400 nm.

Since only the dispersion force suffers retardation, and not the zero-frequency term, some very interesting effects can sometimes arise leading to a change in sign of the Hamaker constant at some finite separation. A particularly notable example of this phenomenon concerns the behaviour of liquid hydrocarbons on water. Referring to Table 11.3 we note that the non-retarded Hamaker constant for a pentane film on water is very small, about $A \approx 10^{-21}$ J. This is made up of a negative zero-frequency contribution of $A_{v=0} \approx -0.8 \times 10^{-21}$ J together with a positive dispersion contribution of $A_{v>0} \approx 1.6 \times 10^{-21}$ J which dominates at small separations. At larger distances, however, $A_{v>0}$ becomes retarded and progressively decreases. Thus, at some small but finite distance the value of A changes sign from net positive to net negative. It is for this reason that pentane spreads on water: the van der Waals force across the film is repulsive, so tending to increase its thickness and thus favouring the spreading of the liquid on water. For the higher alkanes, the attractive dispersion force contribution is very much larger (cf. $A_{v>0} \approx 6 \times 10^{-21}$ J for octane), and these hydrocarbons do not spread on water but collect as isolated lenses on the water surface (Fig. 11.7).

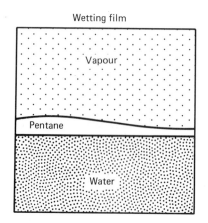 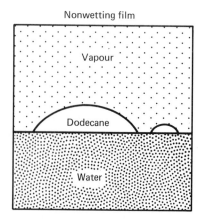

Fig. 11.7.

11.8   SCREENED VAN DER WAALS FORCES IN ELECTROLYTE SOLUTIONS

In Chapter 6 we saw that the zero-frequency contribution to the van der Waals force is essentially an electrostatic interaction. Now, in any medium containing free charges, e.g., water containing free ions in solution or a metal containing free electrons, all electrostatic fields become *screened* due to the polarization (displacement) of these charges. A screened electric field decays roughly exponentially with distance according to $\mathrm{e}^{-\kappa D}$ where the characteristic decay length, κ^{-1}, is known as the *Debye screening length* or *Thomas–Fermi screening length* (see Section 12.15). Typical values for κ^{-1} in aqueous electrolyte solutions are ~ 10 nm in a 10^{-3} M solution and ~ 1 nm in a 0.1 M solution (see Eq. (12.37)). Since both the emanating and reflected fields are screened during a van der Waals interaction, the net effect is that the zero-frequency part is reduced by a factor roughly proportional to $\mathrm{e}^{-2\kappa D}$. This screening affects only the zero-frequency contribution, $A_{v=0}$. The dispersion contribution, $A_{v > v_1}$, remains unscreened because the electrolyte ions cannot respond to, and are therefore not polarized by such high frequencies.

Across an electrolyte solution the screened non-retarded Hamaker constant is given by (Mahanty and Ninham, 1976)

$$A = A_{v=0}(2\kappa D)\mathrm{e}^{-2\kappa D} + A_{v > v_1} \tag{11.23}$$

where the van der Waals screening length is seen to be half the Debye length. The screening of $A_{v=0}$ is analogous to the retardation of $A_{v > v_1}$, but it usually comes in at much smaller separations. For example, in a 0.1 M aqueous NaCl solution the van der Waals screening length is about 0.5 nm, so that by $D = 1$ nm the zero-frequency contribution has already fallen to about 10% of its value at $D = 0$. Thus, for interparticle interactions across such a solution, at separations greater than 1 nm, the attraction is effectively determined solely by the dispersion force.

Marra (1986a) measured the van der Waals force law between two uncharged lipid bilayers in various salt solutions. Over the distance regime from 1 to 4 nm the non-retarded Hamaker constant was found to be $A = 7 \times 10^{-21}$ J in pure water and $A = 3 \times 10^{-21}$ J in 0.1 M NaCl solution. The first value is slightly larger than expected for pure hydrocarbon across water (Section 11.5 and Table 11.3) probably due to the additional contribution from the polarizable headgroups, while the second value is lower than the first by 4×10^{-21} J, consistent with the zero-frequency term being screened by the electrolyte ions.

11.9 COMBINING RELATIONS

Combining relations (or combining laws) are frequently used for obtaining approximate values for unknown Hamaker constants in terms of known ones. Let us define A_{132} as the non-retarded Hamaker constant for media 1 and 2 interacting across medium 3 (Fig. 11.2c). A glance at Eq. (11.7) shows that we may expect A_{132} to be approximately related to A_{131} and A_{232} via

$$A_{132} \approx \pm \sqrt{A_{131} A_{232}} \qquad (11.24)$$

From this we obtain

$$A_{12} \approx \sqrt{A_{11} A_{22}} \qquad (11.25)$$

where A_{12} is for media 1 and 2 interacting across vacuum (i.e., with no medium 3 between them). This is a useful combining relation giving A_{12} in terms of the Hamaker constants of the individual media. Two other useful relations are (Israelachvili, 1972)

$$A_{131} \approx A_{313} \approx A_{11} + A_{33} - 2A_{13} \qquad (11.26)$$

$$\approx (\sqrt{A_{11}} - \sqrt{A_{33}})^2 \qquad (11.27)$$

which when combined with Eq. (11.24) above gives

$$A_{132} \approx (\sqrt{A_{11}} - \sqrt{A_{33}})(\sqrt{A_{22}} - \sqrt{A_{33}}). \qquad (11.28)$$

Note the similarity of these combining relations to those derived in Chapter 9, Eqs (9.1)–(9.11).

As an illustration of the above relations let us consider a few systems whose Hamaker constants are given in Tables 11.2 and 11.3. Thus, for the quartz–octane–air system, Eq. (11.28) would predict for A_{132}:

$$A_{132} \approx (\sqrt{6.3} - \sqrt{4.5})(0 - \sqrt{4.5})10^{-20} = -0.82 \times 10^{-20} \text{ J,}$$

to be compared with the more rigorously computed value of -0.71×10^{-20} J. Likewise for the CaF_2–helium–vapour system, we expect

$$A_{132} \approx (\sqrt{7.2} - \sqrt{0.057})(0 - \sqrt{0.057})10^{-20} = -0.58 \times 10^{-20} \text{ J,}$$

to be compared with -0.59×10^{-20} J.

For the system quartz–octane–quartz, Eq. (11.28) gives

$$A_{132} \approx (\sqrt{6.3} - \sqrt{4.5})^2 \, 10^{-20} = 0.15 \times 10^{-20} \, \text{J},$$

compared with 0.13×10^{-20} J.

Combining relations are applicable only when dispersion forces dominate the interactions as in the above examples, but they break down when applied to media with high dielectric constants such as water or whenever the zero-frequency contribution $A_{\nu=0}$ is large. Thus, for hexadecane across water, Eq. (11.28) would predict a Hamaker constant of $(\sqrt{5.2} - \sqrt{3.7})^2 \, 10^{-20} \approx 0.13 \times 10^{-20}$ J, which is much smaller than the real value of $\sim 0.5 \times 10^{-20}$ J.

In view of the ease with which Hamaker constants may be reliably computed using Eq. (11.13) or more rigorous numerical methods, the use of combining relations is not recommended.

11.10 SURFACE AND ADHESION ENERGIES

Chapter 15 describes various phenomena that arise from the *surface energies* γ of solids and liquids (for a liquid, γ is usually referred to as its *surface tension*). Here we shall see how surface energies are determined from the intermolecular forces between two surfaces. But first, what do we mean by the surface energy of a material? Let us go back to Fig. 10.2 and recall the pairwise summation of energies between all the atoms of one medium with all the atoms of the *other* medium, which, for van der Waals forces, gave the interaction energy between two identical media as $W = -A/12\pi D^2$. Had the summation been carried out between *all* atoms, including atoms in the *same* medium, we should have obtained two additional energy terms:

$$W = -constant + A/12\pi D_0^2 \quad \text{per unit area} \tag{11.29}$$

where the *constant* is simply the bulk cohesive energy, Eq. (6.5), of the atoms with their immediate neighbours at $D = D_0$ (Fig. 11.8). The second (positive) term arises from the unsaturated 'bonds' at the two surfaces. This term is always positive and shows that a free liquid will always tend to minimize its surface energy by minimizing its surface area.

Thus, apart from the bulk energy, the total energy of two planar surfaces at a distance D apart (Fig. 11.8) is given by

$$W = \frac{A}{12\pi} \left(\frac{1}{D_0^2} - \frac{1}{D^2} \right) = \frac{A}{12\pi D_0^2} \left(1 - \frac{D_0^2}{D^2} \right) \quad \text{per unit area.} \tag{11.30}$$

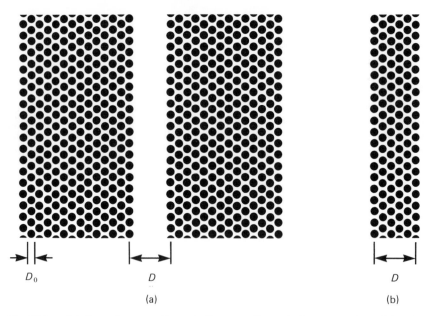

Fig. 11.8.   (a) Two planar media or half-spaces. The pairwise summation of London dispersion energies between all atoms leads to Eq. (11.30). For two surfaces close together, their total surface energy may therefore be written as $2\gamma(1 - D_0^2/D^2)$. The long-range van der Waals interaction energy can be seen to be no more than a perturbation of the surface energy γ. A similar result is obtained for (b), a thin film.

At $D = D_0$ (two surfaces in contact), $W = 0$, while for $D = \infty$ (two isolated surfaces),

$$W = A/12\pi D_0^2 = 2\gamma \qquad (11.31)$$

or

$$\gamma = A/24\pi D_0^2. \qquad (11.32)$$

In other words, the surface energy γ equals half the energy needed to separate two flat surfaces from contact to infinity, viz. it is half the adhesion energy.

We may now test Eq. (11.32) to see how well it predicts the surface energies of materials. Unfortunately it is not at all obvious what value to use for the interfacial contact separation D_0. At first sight one might expect D_0 to be the same as the distance between atomic centres and thus be put equal to σ. However, there are two problems. First, in the process of computing Hamaker constants by the Hamaker summation method the discrete and bumpy surface atoms were artificially 'smeared out' into a continuum by transferring the energy sum to an integral. Our two *structured surfaces* were thereby transformed into two smooth surfaces, and this invalidates the use of Eq. (11.32) at interatomic distances (Tabor, 1982). Second, the continuum Lifshitz theory, which predicts similar values for A as does the Hamaker summation method, is also not expected to apply at atomic-scale distances.

Clearly a molecular approach is called for. Let us therefore first consider the matter in terms of the pairwise additivity of individual atoms or molecules which proved so successful in calculating the cohesive energies of condensed phases (Chapter 6). For an idealized planar close packed solid, each surface atom (of diameter σ) will have only nine nearest neighbours instead of 12. Thus, when it comes into contact with a second surface each surface atom will gain $3w = 3C/\sigma^6$ in binding energy. For a closed-packed solid, each surface atom occupies an area of $\sigma^2 \sin 60°$, and the bulk density of atoms is $\rho = \sqrt{2}/\sigma^3$. Thus, the surface energy should be approximately

$$\gamma \approx \frac{1}{2}\left(\frac{3w}{\sigma^2 \sin 60°}\right) \approx \frac{\sqrt{3}w}{\sigma^2} \approx \frac{\sqrt{3}C\rho^2}{2\sigma^2} \approx \frac{\sqrt{3}A}{2\pi^2\sigma^2}$$

$$\approx \frac{A}{24\pi(\sigma/2.5)^2}, \tag{11.33}$$

where $A = \pi^2 C\rho^2$ is the Hamaker constant as before.

Thus, to use Eq. (11.32) for calculating surface energies we must use a 'cut off' distance D_0 that is substantially less than the interatomic or intermolecular centre-to-centre distance σ (Israelachvili, 1973; Aveyard and Saleem, 1976; Hough and White, 1980; Tabor, 1982). For example, for a typical value of $\sigma \approx 0.4$ nm, we should use $D_0 \approx \sigma/2.5 \approx 0.16$ nm. Table 11.4 gives the predicted surface and adhesion energies of a variety of compounds all based on the same cutoff separation of $D_0 = 0.165$ nm, viz.

$$\gamma \approx A/24\pi(0.165 \text{ nm})^2. \tag{11.34}$$

It is remarkable that this 'universal constant' for D_0 yields values for surface energies γ in such good agreement with those measured, even for very different liquids and solids. Only for highly polar H-bonding liquids, shown in the lower part of Table 11.4, does Eq. (11.34) seriously underestimate their surface energies, which is to be expected (Section 8.4). But for 'ordinary' solids and liquids, even including acetone and ethanol, Eq. (11.34) appears to be reliable to within 10 to 20%, which is anyway within the accuracy that A can be computed. The apparent success of Eq. (11.34) provides us with the following simple formula for estimating the Hamaker constants of non-H-bonding

TABLE 11.4 Comparison of experimental surface energies with those calculated on the basis of the Lifshitz theory

Material (ε) in ordering of increasing ε	Theoretical A $(10^{-20}$ J)	Surface energy, γ (mJ m^{-2})	
		$A/24\pi D_0^2$ ($D_0 = 0.165$ nm)	Experimental[a] (20°C)
Liquid helium (1.057)	0.057	0.28	0.12–0.35 (4–1.6 K)
n-Pentane (1.8)	3.75	18.3	16.1
n-Octane (1.9)	4.5	21.9	21.6
Cyclohexane (2.0)	5.2	25.3	25.5
n-Dodecane (2.0)	5.0	24.4	25.4
n-Hexadecane (2.1)	5.2	25.3	27.5
PTFE (2.1)	3.8	18.5	18.3
CCl$_4$ (2.2)	5.5	26.8	29.7
Benzene (2.3)	5.0	24.4	28.8
Polystyrene (2.6)	6.6	32.1	33
Polyvinyl chloride (3.2)	7.8	38.0	39
Acetone (21)	4.1	20.0	23.7
Ethanol (26)	4.2	20.5	22.8
Methanol (33)	3.6	18	23
Glycol (37)	5.6	28	48
Glycerol (43)	6.7	33	63
Water (80)	3.7	18	73
H$_2$O$_2$ (84)	5.4	26	76
Formamide (109)	6.1	30	58

[a] Note the good agreement between theory and experiment (within 20 %) except for the last six strongly H-bonding liquids.

solids and liquids from their surface energies:

$$A \approx 2.1 \times 10^{-21} \gamma, \tag{11.35}$$

where γ is in mJ m^{-2} (dyn cm^{-1} or erg cm^{-2}) and A is in joules.

11.11　SURFACE ENERGIES OF METALS

From the above analysis we might expect the surface energy of a metal to be given by Eq. (11.34) using a Hamaker constant appropriate for metals, i.e., $A \approx 4 \times 10^{-19}$ J (Table 11.2). This would predict $\gamma \approx 200$ mJ m^{-2}. While this value is higher than for non-metallic compounds, it is still about an order of magnitude lower than typical measured values for metals, which vary from 400 to 4000 mJ m^{-2} (Table 11.5).

TABLE 11.5 Surface energies of metals

Material	Transition temperatures		Surface energy (tension) γ (mJ m^{-2})[a]		
	Boiling point T_B (K)[b]	Melting point T_M (K)[b]	Just above T_M	Just below T_M	At 300 K
Metals					
Aluminium	2543	931	700	800	1100
Silver	2223	1233	1000	1200	1500
Copper	2603	1356	1300	1600	2000
Iron	2773	1803	1500	1800	2400
Tungsten	5273	3653	2500	3600	4400
Semi-conductor					
Silicon	2623	1683	750	$\sim$1100	1400
Weak conductor					
Ice	373	273	75	110	71

[a] Values from Wawra (1975) and other standard references. Values for solids ($T < T_M$) are only approximate, the exact value of γ depends on the crystallographic plane.
[b] At 760 mm Hg.

Clearly, the attractive forces between metal surfaces cannot be accounted for by conventional van der Waals forces even though at larger separations they can (Table 11.3). The strong adhesion is believed to be due to short-range non-additive electron exchange interactions which arise between conducting surfaces at separations below 0.5 nm and give rise to so-called *metallic bonds*. A phenomenological expression for the interaction potential of two similar metallic surfaces is (Banerjee *et al.*, 1991)

$$W(D) = -2\gamma \left[1 - \frac{(D - D_0)}{\lambda_M} \right] e^{-(D - D_0)/\lambda_M} \text{ per unit area} \quad (11.36)$$

where λ_M is some characteristic decay length for metals, similar to the Fermi screening length. Minimum energy occurs at $D = D_0$ where $W(D_0) = -2\gamma$.

Ferrante and Smith (1985) have also computed the adhesion energies of metals when their lattices are not in perfect registry (referred to as 'incommensurate' or 'mismatched' lattices). This can arise between two similar metals if the contacting lattices are at some finite 'twist' angle relative to each other, and it occurs between any two dissimilar metals whose lattice dimensions are different (Table 11.6). As might be expected, the atoms of two incommensurate lattices cannot pack together as closely as two commensurate lattices, and their adhesion energy is often significantly smaller than for commensurate surfaces.

TABLE 11.6 Calculated surface energies of metallic contacts: effects of lattice mismatch

Type of metal–metal interface	Surface energy, γ (mJ m^{-2})[a]	
	Lattices in register (Commensurate)	Lattices out of register (Incommensurate)
Similar materials		
Al(111)–Al(111)	715	490
Zn(0001)–Zn(0001)	545	505
Mg(0001)–Mg(0001)	550	460
Dissimilar materials		
Al(111)–Zn(0001)		520
Zn(0001)–Mg(0001)		490
Al(111)–Mg(0001)		505

[a] As computed by Ferrante and Smith (1985) and Banerjee *et al.* (1991).

11.12   FORCES BETWEEN SURFACES WITH ADSORBED LAYERS

The non-retarded van der Waals force between two surfaces 1 and 1′ with adsorbed layers 2 and 2′ across medium 3 (Fig. 11.9) is given by the approximate expression (Israelachvili, 1972)

$$F(D) = \frac{1}{6\pi}\left[\frac{A_{232'}}{D^3} - \frac{\sqrt{A_{121}A_{32'3}}}{(D+T)^3} - \frac{\sqrt{A_{1'2'1'}A_{323}}}{(D+T')^3} + \frac{\sqrt{A_{1'2'1'}A_{121}}}{(D+T+T')^3} \right]$$

(11.37)

For the 'symmetrical' case when medium 1 = medium 1′, medium 2 = medium 2′, and $T = T'$, Eq. (11.37) can be simplified using combining relations when it reduces to

$$F(D) = \frac{1}{6\pi}\left[\frac{A_{232}}{D^3} - \frac{2A_{123}}{(D+T)^3} + \frac{A_{121}}{(D+2T)^3} \right]$$

(11.38)

At small separations, when $D \ll (T + T')$, Eq. (11.37) becomes

$$F(D) = A_{2'32}/6\pi D^3$$

(11.39)

while at large separations, when $D \gg (T + T')$, we obtain

$$F(D) = A_{1'31}/6\pi D^3$$

(11.40)

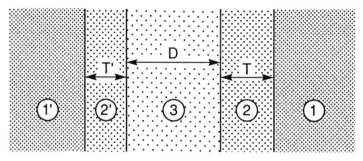

Fig. 11.9.

Thus, as a rule of thumb we may say that the van der Waals interaction is dominated by the properties of the bulk or substrate materials at large separations and by the properties of the adsorbed layers at separations less than the thicknesses of the layers. In particular, this means that the adhesion energies are largely determined by the properties of any adsorbed films even when these are only a monolayer thick. Note that with adsorbed layers the long-range van der Waals forces can change sign over certain distance regimes depending on the properties of the five media. However, for a symmetrical system it can be shown that the interaction is always attractive regardless of the number and properties of the layers.

11.13   EXPERIMENTS ON VAN DER WAALS FORCES

The expression 'experiments on van der Waals forces' is a rather vague one since many important phenomena involve van der Waals interactions in one way or another. Thus, the measurement of the surface tensions of non-polar liquids or the thickness of wetting films may be thought of as experiments on van der Waals forces, and we have seen how the results are in good agreement with theory. Other phenomena involving van der Waals interactions (e.g., physical adsorption, adhesion, the strength of solids) have also been extensively studied, but experiments of this type afford a poor way of rigorously studying van der Waals forces since they usually have to contend with many parameters that themselves are uncertain, so that it becomes difficult to compare the results with theory.

The most direct way to study van der Waals forces is simply to position two bodies close together and measure the force of attraction as a function of the distance between them (Section 10.7). The pioneering measurements carried out by Derjaguin and colleagues in the USSR and Overbeek and colleagues in The Netherlands in the 1950s and 1960s were of this type. The bodies were made of glass, the force was determined by measuring the deflection of a sensitive spring or balance arm, and the distance between the highly polished surfaces was obtained by optical interference (using Newton's rings, for example). In this way the forces between various types of glass in air or vacuum were successfully measured in the range 25–1200 nm, which is in the retarded force regime. In general, experiments with glass and metal-coated glass have yielded results in good agreement with the Lifshitz theory: the expected power law for the force was obtained, and the measured magnitudes of the forces agreed with theory within a factor of about two. For a review of the earlier work on van der Waals forces, see Israelachvili and Tabor (1973), and Israelachvili and Ninham (1977); for more recent work, see Derjaguin et al. (1978) and van Blokland and Overbeek (1978, 1979).

Unfortunately, experiments with glass were unable to provide accurate results for separations less than about 10 nm. For this, a much smoother surface was needed. This problem was resolved by making use of naturally occurring muscovite mica, which may be cleaved to provide molecularly smooth surfaces over large areas, and by employing a multiple beam interferometry technique for measuring surface separations to within ±0.1 nm. In this way (see Section 10.7) Tabor and Winterton (1969) and later Israelachvili and Tabor (1972) and Coakley and Tabor (1978) measured the van der Waals forces between curved mica surfaces or metal-coated surfaces in air in the range 2–130 nm, where the agreement with theory was generally within 30%. These experiments also allowed for the first measurements of the transition from retarded to non-retarded forces. Experiments were also carried out down to separations of 1.4 nm with a surfactant monolayer of thickness 2.5 nm deposited on each mica surface. The results showed that for separations greater than about 5 nm the effective Hamaker constant is as for bulk mica, but that for separations less than 3 nm it is about 25% less and dominated by the properties of the monolayers, a result which is in accord with theoretical expectations (Section 11.12).

After 1975, new experimental techniques were developed to directly measure the van der Waals forces between surfaces in liquids (Israelachvili and Adams, 1978; Derjaguin et al., 1978). Figure 11.6 showed results obtained for the van der Waals force between two mica surfaces in various electrolyte solutions in the distance range 2–15 nm, where again the agreement with theory is within 30%. It has also been verified that in liquids retardation effects come in at smaller separations than in air (above about 5 nm rather than 10 nm), and that the zero frequency contribution is screened in salt solution (Marra, 1986a).

However, in liquids, unlike in air or vacuum, other forces are also usually present, such as long-range electric double-layer forces and—at separations below a few molecular diameters—solvation forces. The major limitation of the Lifshitz theory is that it treats both the surfaces and the intervening solvent medium as structureless continuums and consequently does not encompass molecular effects such as solvation forces and surface structural effects. We have seen in Chapter 7 that at very small separations the solvation force is expected to oscillate with distance with a periodicity equal to the molecular diameter—quite unlike the monotonic force law of the continuum Lifshitz theory. Short-range solvation and structural forces between surfaces are described in Chapter 13.

PROBLEMS AND DISCUSSION TOPICS

11.1 Two identical hard spheres of density $10^4 \, \mathrm{kg \, m^{-3}}$ ($10 \, \mathrm{g \, cm^{-3}}$) and radius R are in contact. Assuming a Hamaker constant of $A = 10^{-19} \, \mathrm{J}$, estimate the value of R at which their van der Waals and gravitational attraction are equal.

11.2 The molecules of two non-polar liquids, A and B, are of equal size but have very different polarizabilities, with $\alpha_A \gg \alpha_B$. Would you expect the liquids to be miscible or immiscible? If the specific solubility of A per unit volume of bulk B is X_s, would you expect the specific solubility to be larger or smaller in (i) small droplets of B (convex meniscus), and (ii) a small volume of capillary condensed liquid in a wedge (concave meniscus)?

 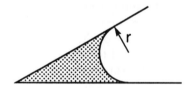

Fig. 11.10.

11.3 (i) Estimate the adhesion force between two perfectly smooth spheres of quartz glass each of radius $R = 1 \, \mu\mathrm{m}$ where the Hamaker constant is given in Table 11.2. Assume that contact occurs at a sharp cutoff distance of $D = 0.25 \, \mathrm{nm}$. How accurately could you measure this force if you had two smooth marbles, a ruler, a protractor and some thread?

(ii) What would be the adhesion force if the glass spheres were immersed in water, and what would be the adhesion *energy* in units of kT at $25°C$ (use Table 11.3).

(iii) If—due to electrostatic or hydration forces or the adsorption of polymers from solution (discussed in Chapters 12–14)—the glass surfaces also experience an additional repulsive force which, between two flat surfaces is given by $F = Be^{-\kappa D}\,\mathrm{N\,m}^{-2}$ per unit area, find the new adhesion force between the spheres in the aqueous solution. Is your answer unique? (Assume that $B = 5 \times 10^5$ and $\kappa^{-1} = 2\,\mathrm{nm}$.)

(iv) What do the force versus distance and energy versus distance curves look like for the above system (iii), and what insights can be obtained from these curves? It is found that if the aqueous solution is added to an initially dry compact powder of spheres the adhesion between the spheres is significantly higher than after the solution is vigorously stirred or sonicated, or if spheres are added individually (e.g. poured in) to the aqueous solution. Explain this phenomenon. Which is the thermodynamically more stable state of this system.

11.4 Two immiscible non-polar liquids A and B are placed in a round container of radius 2 cm exposed to the atmosphere at STP. The volumes of the liquids are the same and equal to 50 ml each. The densities of A and B are 1.0 and $1.1\,\mathrm{g\,ml}^{-1}$, and their refractive indices are 1.45 and 1.40, respectively. Explain what happens at the liquid–air interface and estimate the thickness of the film of liquid A at that interface. Assume that only non-retarded van der Waals forces are involved, and that the UV electronic absorption frequencies of the two liquids is the same and equal to $3 \times 10^{15}\,\mathrm{s}^{-1}$.

Describe, and illustrate with schematic drawings, what can occur at the container–liquid interfaces (with both A and B) depending on the optical properties of the container material relative to those of A and B.

11.5 Two metal surfaces are very close together (as may occur in a crack). The surfaces are not smooth but have asperities of mean radius $R = 10\,\mathrm{nm}$. The elastic properties of the system are such that each asperity may be modelled as being held to its surface by a spring of stiffness $K = 3\,\mathrm{N\,m}^{-1}$. Find the smallest distance that two asperities may remain at equilibrium before the system becomes unstable, and describe the nature of this instability. Assume that the Hamaker constant for the metal–metal van der Waals interaction is $A = 5 \times 10^{-19}\,\mathrm{J}$.

How would you estimate the effective stiffness, K, of the asperity in terms of the Young's modulus of the metal and the asperity dimensions? How would you expect the critical distance for the instability to depend on the radius of the asperities?

11.6 A liquid film on a flat surface exposed to saturated vapour will grow indefinitely if the Hamaker constant across the film A_{SLV} is negative. However, this is not the case for a film on a curved surface. Derive a relation between the equilibrium thickness D of a film on the surface of a cylindrical fibre of radius R in terms of A_{SLV}, R and γ (the surface tension of the liquid).

For a quartz fibre of radius $R = 10\,\mu\mathrm{m}$ in contact with a saturated vapour of octane at $20°C$, calculate the equilibrium thickness of the film on the fibre surface ignoring gravitational effects. Is your calculated thickness likely to be an overestimate or an underestimate of the real value?

(Use Tables 11.3 and 11.4 and Eq. (15.31) for the Laplace pressure. For further reading on this topic see Quéré *et al.* 1989.)

11.7   A liquid is in a beaker of radius R. Show that if $A_{SLV} < 0$, there is a wetting film on the inside of the beaker whose thickness D at a height H above the liquid surface is given by

$$D^3 = \frac{-A_{SLV}}{6\pi\rho gH[1 - \gamma/\rho gH(R - D)]} \qquad (11.41)$$

What happens at $H \approx \gamma/\rho gR$?

11.8   Describe graphically how the Hamaker 'constant' A_{123} might vary with distance D for the liquid–liquid–solid system shown in Fig. 11.11. Assume the film to be thin but the lens to be macroscopic, and ignore gravity.

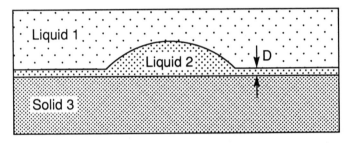

Fig. 11.11. Liquid that does not wet a thin film of itself adsorbed on a surface (known as autophobic). The contact angles in such cases are usually small.

11.9   Two van der Waals liquids A and B are completely miscible. A small quantity of A is dissolved in B. It is found that a very thin surface region of B is depleted of molecules A (below the bulk concentration). Argue whether you expect the surface tension of the mixture to be higher or lower than that of pure liquid B. What if the surface concentration of A is enhanced?

11.10   Colloidal particles often coagulate in non-polar liquids (hydrocarbons, oils) because of the attractive van der Waals forces between them. This is often a nuisance but can be prevented by coating the particles with a surfactant or polymer layer whose refractive index matches that of the liquid. Explain this phenomenon.

At an ACS conference, Dr Chan from Colloids Corp. describes a colloidal dispersion consisting of $0.5 \, \mu m$ diameter smooth silica spheres in oil, where, by coating the spheres with a 'matching layer' of surfactant, the depth of the potential well was reduced by a factor of 10 as ascertained by light scattering measurements. When asked about the thickness of the layer, Dr Chan replies that this is proprietary information.

What was the thickness of the layer?

11.11 Are there realistic conditions for which the effective Hamaker 'constant' (i) increases with distance, and (ii) is higher in a medium than in free space?

11.12 Derive the expression for the non-retarded force between two thin membranes 2 of thickness T whose surfaces are at a distance D apart in medium 1. Show that at a separation of $D = 2.2T$ the force between them is half that of two half-spaces of 2 across 1 at the same separation D. Also show that at larger separations, when $D \gg T$, the non-retarded interaction energy between two sheets approaches

$$W(D) = -A_{121} T^2/2\pi D^4. \tag{11.42}$$

CHAPTER 12

ELECTROSTATIC FORCES BETWEEN SURFACES IN LIQUIDS

12.1 THE CHARGING OF SURFACES IN LIQUIDS: THE ELECTRIC 'DOUBLE LAYER'

Situations in which van der Waals forces alone determine the total interaction are restricted to a few simple systems, for example, to interactions in a vacuum or to non-polar wetting films on surfaces, both of which were discussed in Chapter 11. In more complex, and more interesting, systems long-range electrostatic forces are also involved, and the interplay between these two interactions has many important consequences.

As mentioned earlier the van der Waals force between similar particles in a medium is always attractive, so that if only van der Waals forces were operating, we might expect all dissolved particles to stick together (coagulate) immediately and precipitate out of solution as a mass of solid material. Our own bodies would be subject to the same fate if we remember that we are composed of about 75% water. Fortunately this does not happen, because particles suspended in water and any liquid of high dielectric constant are usually charged and can be prevented from coalescing by repulsive electrostatic forces. Other repulsive forces that can prevent coalescence are solvation and steric forces, described in Chapters 13 and 14. In this chapter we shall concentrate on the electrostatic forces.

The charging of a surface in a liquid can come about in two ways:

(i) by the ionization or dissociation of surface groups (e.g., the dissociation of protons from surface carboxylic groups ($-COOH \rightarrow -COO^- + H^+$), which leaves behind a negatively charged surface) and

(ii) by the adsorption (binding) of ions from solution onto a previously uncharged surface, e.g., the binding of Ca^{2+} onto the zwitterionic headgroups of lipid bilayer surfaces, which charges the surfaces positively. Depending on the ionic conditions, even the air–water and hydrocarbon–water interfaces can become charged in this way. The adsorption of ions from solution can, of course, also occur onto oppositely charged surface sites, e.g., the adsorption

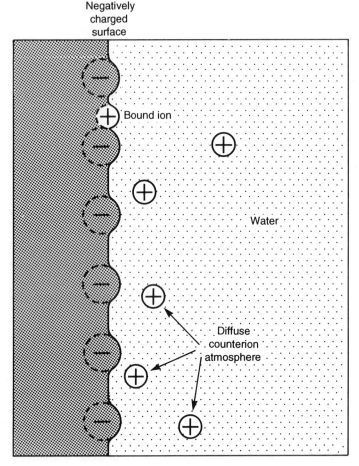

Fig. 12.1. Ions bound to a surface are not rigidly bound but can exchange with other ions in solution; their lifetime on a surface can be as short as 10^{-9} s or as long as many hours.

of cationic Ca^{2+} to anionic COO^- sites vacated by H^+ or Na^+. This is known as *ion exchange*.

Whatever the *charging mechanism*, the final surface charge is balanced by an equal but oppositely charged region of *counterions*, some of which are bound, usually transiently, to the surface within the so-called *Stern* or *Helmholtz* layer, while others form an atmosphere of ions in rapid thermal motion close to the surface, known as the diffuse *electric double layer* (Fig. 12.1).

Two similarly charged surfaces usually repel each other electrostatically in solution, though under certain conditions they may attract at small separations. Zwitterionic surfaces, i.e., those characterized by surface dipoles but no net charge, also interact electrostatically with each other, though here we shall find that the force is usually attractive.

12.2 CHARGED SURFACES IN WATER (NO ADDED ELECTROLYTE)

In the following sections we shall consider the counterion distribution and force between two similarly charged planar surfaces in a pure liquid such as water, where (apart from H_3O^+ and OH^- ions) the only ions in the solution are those that have come off the surfaces. Such systems occur when, for example, colloidal particles, clay sheets, surfactant micelles or bilayers whose surfaces contain ionizable groups interact in water, and also when thick films of water build up (condense) on an ionizable surface such as glass. But first we must consider some fundamental equations that describe the counterion distribution between two charged surfaces in solution.

12.3 THE POISSON–BOLTZMANN (PB) EQUATION

For the case when only counterions are present in solution, the chemical potential of any ion may be written as

$$\mu = ze\psi + kT \log \rho, \tag{12.1}$$

where ψ is the electrostatic potential and ρ the number density of ions of valency z at any point x between two surfaces (Fig. 12.2). Since only differences in potential are ever physically meaningful, we may set $\psi_0 = 0$ at the midplane ($x = 0$), where also $\rho = \rho_0$ and $(d\psi/dx)_0 = 0$ by symmetry.

From the equilibrium requirement that the chemical potential be uniform throughout, Eq. (12.1) gives us the expected Boltzmann distribution of counterions at any point x:

$$\rho = \rho_0 e^{-ze\psi/kT}. \tag{12.2}$$

One further important fundamental equation will be required. This is the well-known _Poisson equation_ for the net excess charge density at x:

$$ze\rho = -\varepsilon\varepsilon_0 (d^2\psi/dx^2) \tag{12.3}$$

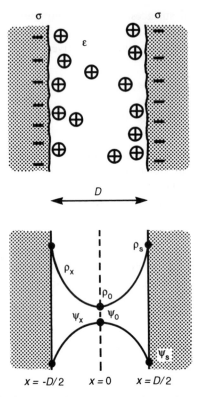

Fig. 12.2. Two negatively charged surfaces of surface charge density σ separated a distance D in water. The only ions in the space between them are the counterions that have dissociated from the surfaces. The counterion density profile ρ_x and electrostatic potential ψ_x are shown schematically in the lower part of the figure. The 'contact' values are ρ_s and ψ_s.

which when combined with the Boltzmann distribution, Eq. (12.2), gives the *Poisson–Boltzmann (PB) equation*:

$$\mathrm{d}^2\psi/\mathrm{d}x^2 = -ze\rho/\varepsilon\varepsilon_0 = -(ze\rho_0/\varepsilon\varepsilon_0)\mathrm{e}^{-ze\psi/kT}. \qquad (12.4)$$

When solved, the PB equation gives the potential ψ, electric field $E = \partial\psi/\partial x$, and counterion density ρ, at any point x in the gap between the two surfaces. Let us first determine these values at the surfaces themselves. These quantities are often referred to as the *contact values*: ψ_s, E_s, ρ_s, etc.

12.4 SURFACE CHARGE, ELECTRIC FIELD AND COUNTERION CONCENTRATION AT A SURFACE

The PB equation is a non-linear second-order differential equation, and to solve for ψ we need two *boundary conditions*, which determine the two integration constants. The first boundary condition follows from the symmetry requirement that the field must vanish at the midplane, i.e., that $E_0 = (d\psi/dx)_0 = 0$. The second boundary condition follows from the requirement of overall *electroneutrality*, i.e., that the total charge of the counterions in the gap must be equal (and opposite) to the charge on the surfaces. If σ is the surface charge density on each surface (in $C\,m^{-2}$) and D is the distance between the surfaces (Fig. 12.2), then the condition of electroneutrality implies that

$$\sigma = - \int_0^{D/2} ze\rho \, dx = +\varepsilon\varepsilon_0 \int_0^{D/2} (d^2\psi/dx^2)\, dx$$

$$= \varepsilon\varepsilon_0 (d\psi/dx)_{D/2} = \varepsilon\varepsilon_0 (d\psi/dx)_s = \varepsilon\varepsilon_0 E_s,$$

i.e.

$$E_s = \sigma/\varepsilon\varepsilon_0. \tag{12.5}$$

Eq. (12.5) gives an important boundary condition relating the surface charge density σ to the electric field E_s at each surface (at $x = \pm D/2$), which we may note is independent of the gap width D.

☐ ● **WORKED EXAMPLE** ●

Question: Is the electric field near a charged surface in water sufficiently intense to immobilize the water molecules adjacent to it?

Answer: Assuming a high charge density of $\sigma = 0.3\,C\,m^{-2}$ (which is one charge per $0.5\,nm^2$—typical of a fully ionized surface), the electric field at the surface, Eq. (12.5), is $E_s = \sigma/\varepsilon\varepsilon_0 = 0.3/80(8.85 \times 10^{-12}) = 4.2 \times 10^8\,V\,m^{-1}$. We may compare this to the field just outside a monovalent ion in water. Using Eq. (3.3), the field at $r = 0.25\,nm$ from the centre of an ion is $E_r = e/4\pi\varepsilon\varepsilon_0 r^2 = 2.9 \times 10^8\,V\,m^{-1}$. Since this is comparable to the field at the charged surface, and since the fields of monovalent ions are usually not strong enough to immobilize water molecules around them (Chapters 3–5), it is unlikely that water molecules will become significantly oriented, immobilized, or 'bound' to, any but the most highly charged surfaces. However, other interactions with the surface, such as H-bonding, may lead to significant effects on the local water structure.   □

Turning now to the ionic distribution, there exists an important general relation between the concentrations of counterions at either surface and at the midplane. Differentiating Eq. (12.2) and then using Eq. (12.4) we obtain

$$\frac{d\rho}{dx} = -\frac{ze\rho_0}{kT} e^{-ze\psi/kT} \left(\frac{d\psi}{dx}\right) = \frac{\varepsilon\varepsilon_0}{kT}\left(\frac{d\psi}{dx}\right)\left(\frac{d^2\psi}{dx^2}\right)$$

$$= \frac{\varepsilon\varepsilon_0}{2kT}\frac{d}{dx}\left(\frac{d\psi}{dx}\right)^2,$$

(12.6)

hence

$$\rho_x - \rho_0 = \int_0^x d\rho = \frac{\varepsilon\varepsilon_0}{2kT}\int_0^x d\left(\frac{d\psi}{dx}\right)^2 = +\frac{\varepsilon\varepsilon_0}{2kT}\left(\frac{d\psi}{dx}\right)_x^2$$

so that

$$\rho_x = \rho_0 + \frac{\varepsilon\varepsilon_0}{2kT}\left(\frac{d\psi}{dx}\right)_x^2,$$

(12.7)

which gives ρ at any point x in terms of ρ_0 at the midplane and $(d\psi/dx)^2$ at x. In particular at the surface, $x = D/2$, we obtain using Eq. (12.5) the contact value of ρ

$$\rho_s = \rho_0 + \sigma^2/2\varepsilon\varepsilon_0 kT.$$

(12.8)

This result shows that the concentration of counterions at the surface depends only on the surface charge density σ and the counterion concentration at the midplane. Note that ρ_s never falls below $\sigma^2/2\varepsilon\varepsilon_0 kT$ even for isolated surfaces, i.e., for two surfaces far apart when $\rho_0 \to 0$. For example, for an isolated surface in water of charge density $\sigma = 0.2$ C m^{-2} (one charge per 0.8 nm^2) at 293 K

$$\rho_s = \sigma^2/2\varepsilon\varepsilon_0 kT = (0.2)^2/2 \times 80 \times 8.85 \times 10^{-12} \times 4.04 \times 10^{-21}$$

$$= 7.0 \times 10^{27} \text{ m}^{-3}$$

which is about 12 M. If these surface counterions are considered to occupy a layer of thickness ~ 0.2 nm, the above value for ρ_s corresponds to a surface counterion density of $(7 \times 10^{27})(0.2 \times 10^{-9}) = 1.4 \times 10^{18}$ ions m^{-2} or one charge per 0.7 nm^2, which is about the same as the surface charge density σ. This is an interesting result, for it shows that regardless of the counterion distribution profile ρ_x away from a surface (Section 12.5), most of the counterions that effectively balance the surface charge are located in the first few ångstroms from the surface (Jönsson et al., 1980). However, for lower surface charge densities, since $\rho_s \propto \sigma^2$, the layer of counterions extends well beyond the surface and becomes much more diffuse (hence the term *diffuse double layer*).

12.5   COUNTERION CONCENTRATION PROFILE AWAY FROM A SURFACE

The above equations are quite general and are the starting point of all theoretical computations of the ionic distributions near planar charged surfaces, even when the solution contains added electrolyte (Section 12.11 onwards). To proceed further for the specific case of counterions only (Fig. 12.2) we must now solve the Poisson–Boltzmann equation, Eq. (12.4), which can be satisfied by

$$\psi = (kT/ze) \log(\cos^2 Kx) \tag{12.9}$$

or

$$e^{-ze\psi/kT} = 1/\cos^2 Kx, \tag{12.10}$$

where K is a constant given by

$$K^2 = (ze)^2 \rho_0/2\varepsilon\varepsilon_0 kT. \tag{12.11}$$

With this form for the potential we see that $\psi = 0$ and $d\psi/dx = 0$ at $x = 0$ for all K, as required. To solve for K we differentiate Eq. (12.9) and then use Eq. (12.5) to obtain for the electric fields

at any point x: $E_x = d\psi/dx = -(2kTK/ze) \tan Kx$,

at the surfaces: $E_s = (d\psi/dx)_s = -(2kTK/ze) \tan(KD/2) = + \sigma/\varepsilon\varepsilon_0$.

$$\tag{12.12}$$

The counterion distribution profile

$$\rho_x = \rho_0 e^{-ze\psi/kT} = \rho_0/\cos^2 Kx \qquad (12.13)$$

is therefore known once K is determined from Eq. (12.12) in terms of σ and D.

□   ● WORKED EXAMPLE ●

Question: Two charged surfaces with $\sigma = 0.2\,\mathrm{C\,m^{-2}}$ are 2 nm apart ($D = 2$ nm). Calculate the field, potential and counterion density at each surface, at 0.2 nm from each surface and at the mid-plane, assuming monovalent counterions.

Answer: From Eq. (12.12) we find that for $z = 1$, $K = 1.3361 \times 10^9\,\mathrm{m^{-1}}$ at 293 K. From Eq. (12.11) this means that $\rho_0 = 0.40 \times 10^{27}\,\mathrm{m^{-3}}$, so that at the surface $\rho_s = \rho_0/\cos^2(KD/2) = 7.4 \times 10^{27}\,\mathrm{m^{-3}}$. The same result is also immediately obtainable from Eq. (12.8) since, as we have previously established, $\sigma^2/2\varepsilon\varepsilon_0 kT = 7.0 \times 10^{27}\,\mathrm{m^{-3}}$. Thus, the counterion concentration at each surface ρ_s is about 18.5 times greater than at the midplane ρ_0, which is only 1 nm away. Putting $K = 1.3661 \times 10^9\,\mathrm{m^{-1}}$, $kT = 4.045 \times 10^{-21}\,\mathrm{J}$, $\sigma = 0.2\,\mathrm{C\,m^{-2}}$, $\varepsilon = 80$, $ze = 1.602 \times 10^{-19}\,\mathrm{C}$, and $D = 2 \times 10^{-9}\,\mathrm{m}$ into Eqs (12.9), (12.12) and (12.13) we obtain:

	ψ (mV)	E (V m^{-1})	ρ (m^{-3})
At $x = 1$ nm ('contact value' at surface)	-74	2.8×10^8	7.4×10^{27}(12 M)
At $x = 0.8$ nm (0.2 nm from surface)	-37	1.2×10^8	1.7×10^{27}(3 M)
At $x = 0$ ('midplane' value 1 nm from surface)	0	0	0.4×10^{27}(0.7 M)

Note the unphysically steep decrease in the ion density ρ near the surface. □

Figure 12.3 shows how the counterion concentration varies with distance for the case of $\sigma = 0.224\,\mathrm{C\,m^{-2}}$, $D = 21$ nm, as calculated on the basis of (i) the Poisson–Boltzmann equation as in the above example, and (ii) a Monte Carlo simulation of the same system. The agreement is quite good though the Monte Carlo result gives a slightly higher counterion concentration

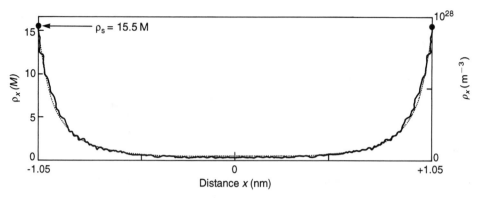

Fig. 12.3. Monovalent counterion concentration profile between two charged surfaces ($\sigma = 0.224\,\mathrm{C\,m^{-2}}$, corresponding to one electronic charge per $0.714\,\mathrm{nm^2}$) a distance 2.1 nm apart in water. The smooth curve is obtained from the Poisson–Boltzmann equation; the other is from a Monte Carlo simulation by Jönsson *et al.* (1980).

very near the surfaces compensated by a lower concentration in the central region between the two surfaces.

12.6   ORIGIN OF THE IONIC DISTRIBUTION, ELECTRIC FIELD, SURFACE POTENTIAL AND PRESSURE

Before we proceed to calculate the force or pressure between two surfaces it is instructive to discuss, in qualitative terms, how the counterion distribution, potential, field and pressure between two surfaces arise. The first thing to notice is that if there were no ions between two similarly charged surfaces, there would be no electric field in the gap between them. This is because the field emanating from a planar charged surface, $E = \sigma/2\varepsilon\varepsilon_0$, is uniform away from the surface (Section 3.3). The two opposing fields emanating from the two plane parallel surfaces therefore cancel out to zero between them. Thus, when the counterions are introduced into the intervening region they do *not* experience an attractive electrostatic force towards each surface. The reason why the counterions build up at each surface is simply because of their mutual repulsion and is similar to the accumulation of mobile charges on the surface of any charged conductor. The repulsive electrostatic interaction between the counterions and their entropy of mixing alone determine their concentration profile ρ_x, the potential profile ψ_x and the field E_x between the surfaces (Jönsson *et al.*, 1980), and we may further note that in all the

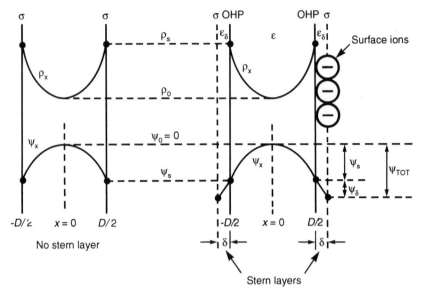

Fig. 12.4. Stern layers of thickness δ at each surface dividing the planes of fixed charge density σ from the boundary of the aqueous solution—the OHP. There is an additional linear drop in potential across the Stern layer given by Eq. (12.14) so that $\psi_{TOT} = \psi_\delta + \psi_s$. However, the counterion density and electrostatic potential within the aqueous region between the two OHPs at $x = D/2$ and $x = -D/2$, and the pressure between the two surfaces, are independent of δ.

theoretical derivations so far the only way the surface charge density σ enters into the picture is through Eq. (12.5), which is simply a statement about the total number of counterions in the gap.

Further, if the locations of the charged surface groups were not at the physical solid–liquid interface (at $x = \pm\frac{1}{2}D$) but at some small distance δ within the surface (Fig. 12.4), the ionic distribution ρ_x, potential ψ_x, field E_x, and the pressure in the medium between $+\frac{1}{2}D$ and $-\frac{1}{2}D$ would not change. But the potential would be different if it were measured at $x = \pm(\frac{1}{2}D + \delta)$. This is the origin of the so called *Stern* and *Helmholtz layers* (Verwey and Overbeek, 1948; Hiemenz, 1977) which separate the charged plane from the *Outer Helmholtz Plane* (OHP) from which the ionic atmosphere begins to obey the Poisson–Boltzmann equation. The combined thickness of the Stern and Helmholtz layers δ is of the order of a few ångstroms and reflects the finite size of the charged surface groups and transiently bound counterions, as illustrated in Fig. 12.4. If the dielectric constant of the Stern–Helmholtz layer is uniform and equal to ε_δ it can be modelled as a capacitor (see Worked Example in Section 3.3) whence the additional drop in potential across the layer is given by

$$\psi_\delta = \sigma\delta/\varepsilon_\delta\varepsilon_0. \tag{12.14}$$

For example, if $\delta = 0.2$ nm, $\sigma = 0.2\,\mathrm{C\,m^{-2}}$ and $\varepsilon_\delta = 40$, we obtain $\psi_\delta = 130$ mV, which is actually higher than the potential drop across the diffuse double layer, calculated in the previous worked example.

We now turn to the origin of the force or pressure between the two surfaces. Contrary to intuition, the origin of the repulsive force between two similarly charged surfaces in a solvent containing counterions and electrolyte ions is entropic (osmotic), not electrostatic. Indeed, the electrostatic contribution to the net force is actually attractive. Consider a surface, initially uncharged, placed in water. When the surface groups dissociate the counterions leave the surface against the attractive Coulombic force pulling them back. What maintains the diffuse double layer is the repulsive osmotic pressure between the counterions which forces them away from the surface and from each other so as to increase their configurational entropy. On bringing two such surfaces together one is therefore forcing the counterions back onto the surfaces against their preferred equilibrium state, i.e., against their osmotic repulsion, but favoured by the electrostatic interaction. The former dominates and the net force is repulsive.

To understand why the purely electrostatic part of the interaction is attractive recall that it involves an equal number of positive (counterion) and negative (surface) charges, i.e., the system is overall electrically neutral. The net Coulombic interaction between a system of charges that are overall neutral always favours their association, as we saw in the case of ionic crystals (e.g., NaCl) in Chapter 3 and dipoles in Chapter 4. This point will be further clarified in the following section.

12.7 THE PRESSURE BETWEEN TWO CHARGED SURFACES IN WATER: THE CONTACT VALUE THEOREM

Using Eq. (2.20) the repulsive pressure P of the counterions at any position x from the centre (Fig. 12.4) is given by $(\partial P/\partial x)_{x,T} = \rho(\partial\mu/\partial x)_{x,T}$, where the chemical potential μ is given by Eq. (12.1). The change in pressure at x on bringing two plates together from infinity ($x' = \infty$, where $P = 0$) to a separation $x' = D/2$ at constant temperature is therefore

$$P_x = - \int_{x' = D/2}^{x' = \infty} [ze\rho(d\psi/dx)_x \, dx' + kT \, d\rho_x].$$ (12.15)

Note that in the above equation the values are computed at the *fixed* point x within the ionic solution which is not the same as the *variable* separation x' between the two surfaces. Replacing $ze\rho$ by the Poisson equation and using the relations

$$\frac{d}{dx}\left(\frac{d\psi}{dx}\right)^2 = 2\left(\frac{d\psi}{dx}\right)\left(\frac{d^2\psi}{dx^2}\right)$$

this becomes

$$P_x(D) - P_x(\infty) = -\tfrac{1}{2}\varepsilon\varepsilon_0\left(\frac{d\psi}{dx}\right)^2_{x(D)} + kT\rho_x(D) + \frac{1}{2}\varepsilon\varepsilon_0\left(\frac{d\psi}{dx}\right)^2_{x(\infty)} - kT\rho_x(\infty)$$

(12.16)

where the subscripts x mean that the values are calculated at x when the surfaces are at a distance D or ∞ apart. In the present case, since there are no electrolyte ions in the bulk solution, $\rho_0(\infty) = 0$, so that by Eq. (12.7) we have $P_x(\infty) = 0$, as expected. The above important equation gives the pressure P at any point x between the two surfaces, and we may notice that it is split into two contributions. The first, being a square, is always negative, i.e., *attractive*. This is the electrostatic field energy contribution, discussed qualitatively in the previous section. The second term is positive and hence repulsive. This is the entropic (osmotic) contribution to the force.

At equilibrium, $P_x(D)$ should be uniform throughout the gap, i.e., independent of x, and it is also the pressure acting on the two surfaces. To verify this we note that using Eq. (12.7) the above may be written as

$$P_x(D) = kT[\rho_0(D) - \rho_0(\infty)]$$ (12.17a)

or

$$P_x(D) = kT\rho_0(D) \qquad \text{since here } \rho_0(\infty) = 0.$$ (12.17b)

which is indeed independent of x and depends only on the increased ionic concentration, or osmotic pressure, at the *midplane*, $\rho_0(D)$, and thus on σ

and D. We may therefore drop the subscript x from $P_x(D)$. It is instructive to insert Eq. (12.8) into the above, whence we obtain

$$P(D) = kT\rho_0(D) = kT[\rho_s(D) - \sigma^2/2\varepsilon\varepsilon_0 kT],$$

that is,

$$P(D) = kT[\rho_s(D) - \rho_s(\infty)]. \tag{12.18}$$

Thus, the pressure is also given by the increase in the ion concentration at the *surfaces* as they approach each other. This important equation, known as the *contact value theorem*, is always valid so long as there is no interaction between the counterions and the surfaces, i.e., so long as there is no counterion adsorption so that the surface charge density remains constant and independent of D.

The contact value theorem is very general and applies to many other types of interactions, for example, to double-layer interactions when electrolyte ions are present in the solution, to solvation interactions where $\rho_s(D)$ is now the surface concentration of solvent molecules (Chapter 13), to polymer-associated steric and depletion interactions where $\rho_s(D)$ is the surface concentration of polymeric groups (Chapters 14 and 18), and to undulation, peristaltic and protrusion forces between fluid membranes (Chapter 18). In the case of overlapping double layers, the resulting force is often referred to as the *electric* or *electrostatic* double-layer repulsion or force, even though, as we have seen, the repulsion is really due to entropic confinement.

Returning to Eq. (12.17b) the pressure may also be expressed in terms of K, as given by Eq. (12.11), by

$$P = kT\rho_0 = 2\varepsilon\varepsilon_0(kT/ze)^2K^2. \tag{12.19}$$

As an example let us apply this result to the worked example of Section 12.5 where for two surfaces with $\sigma = 0.2\,\mathrm{C\,m^{-2}}$ at $D = 2\,\mathrm{nm}$ apart, we found $K = 1.336 \times 10^9\,\mathrm{m^{-1}}$. The repulsive pressure between them is therefore $1.7 \times 10^6\,\mathrm{N\,m^{-2}}$, or about 17 atm. Note that this repulsion exceeds by far any possible van der Waals attraction at this separation; for a typical Hamaker constant of $A \approx 10^{-20}\,\mathrm{J}$ the van der Waals attractive pressure would be only $A/12\pi D^3 \approx 3 \times 10^4\,\mathrm{N\,m^{-2}}$ or about 0.3 atm.

The above equations have been used successfully to account for the equilibrium spacings of ionic surfactant and lipid bilayers in water (Cowley *et al.*, 1978). Figure 12.5 shows experimental results obtained for the repulsive pressure between charged bilayer surfaces in water (cf. Fig. 10.6h), together with the theoretical curve based on Eq. (12.19). The agreement is very good

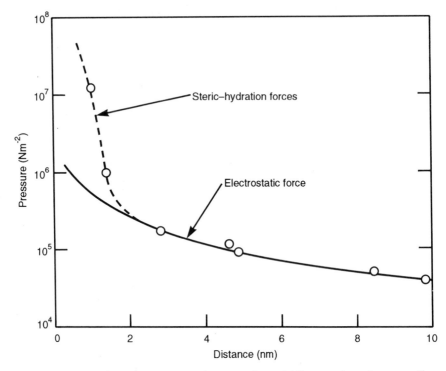

Fig. 12.5. Measured repulsive pressure between charged bilayer surfaces in water. The bilayers were composed of 90% lecithin, a neutral zwitterionic lipid, and 10% phosphatidylglycerol, a negatively charged lipid. For full ionization, the surface charge density should be one electronic charge per 7 nm², whereas the theoretical line through the experimental points suggests one charge per 14 nm² (i.e., about 50% ionization). Below 2 nm there is an additional repulsion due to 'steric-hydration' forces. (From Cowley et al., 1978, ©1978 American Chemical Society.)

down to $D \approx 2$ nm and shows that the effective charge density of the anionic lipid headgroups is about $1e$ per 14 nm². At smaller distances the measured forces are more repulsive than expected due to the steric-hydration interactions between the thermally mobile hydrophilic headgroups that characterize these surfaces (Chapters 14 and 18). Similar methods have been used to measure the repulsive electrostatic forces between biological membranes in salt solutions (Diederichs et al., 1985).

Repulsive electrostatic forces also control the long-range swelling of clays in water. Most naturally occurring clays are composed of lamellar aluminosilicate sheets about 1 to 2 nm thick whose surfaces dissociate in water giving off Na⁺, K⁺, and Ca²⁺ ions, and when placed in water they can swell to more than 10 times their original volume (Norrish, 1954). The

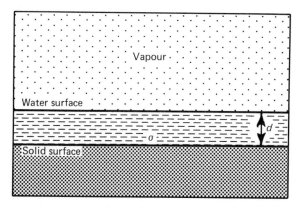

Fig. 12.6. A water film on a charged (ionizable) glass surface will tend to thicken because of the repulsive 'disjoining pressure' of the counterions in the film. If the vapour over the film is saturated, the film will grow indefinitely, but if it is unsaturated, the equilibrium thickness d will be finite as given by Eqs (12.21) and (12.22).

swelling of clays is, however, a complex matter and also involves hydration effects at surface separations below about 3 nm (van Olphen, 1977; Pashley and Quirk, 1984; Kjellander et al., 1988).

In the case of charged spherical particles (e.g., latex particles) in water the long-range electrostatic repulsion between them can result in an ordered lattice of particles even when the distance between them is well in excess of their diameter (Takano and Hachisu, 1978). In such systems (cf. Fig. 10.6g) colloidal particles attempt to get as far apart from each other as possible but, being constrained within a finite volume of solution, are forced to arrange themselves into an ordered lattice (for a review see Forsyth et al., 1978).

Parsegian (1966) and Jönsson and Wennerström (1981) extended the above analysis to the interactions between cylindrical and spherical structures, and the results were used to analyse the relative stability of charged surfactant aggregates which form spontaneously in water. Such micellar structures are soft and fluid-like, and they change from being spherical to cylindrical to sheet-like (bilayers) as the amount of water is reduced (see Part III).

12.8 LIMITATIONS OF THE POISSON–BOLTZMANN EQUATION

Like all continuum and mean-field theories the PB equation breaks down at small distances where it no longer faithfully describes the ionic distribution and forces between two surfaces. The following factors come into play at short separations:

(i) *Ion-correlation effects.* The mobile counterions in the diffuse double-layer constitute a highly polarizable layer at each interface. These two apposing 'conducting' layers therefore experience an attractive van der Waals force which is not included in the PB equation or the Lifshitz theory. Known as the *ion-correlation force* (Guldbrand *et al.*, 1984) this attraction becomes significant at small distances (< 4 nm) and it increases with the surface charge density σ and valency z of the counterions.

In a Monte Carlo study of the ionic density distributions, interaction energies and pressures between planar surfaces, spheres and cylinders, Wennerström *et al.* (1982) concluded that between surfaces of high charge density the attractive ion-correlation force can reduce the effective double-layer repulsion by 10–15% if the counterions are monovalent. However, with divalent counterions (e.g., Ca^{2+}) the ion-correlation attraction was found to *exceed* the double-layer repulsion—the net force becoming overall attractive—below about 2 nm, even in very dilute electrolyte solutions. Such short-range attractive ion-correlation forces have been measured between highly charged anionic bilayer surfaces in $CaCl_2$ solutions (Marra, 1986b, c; Kjellander *et al.*, 1988, 1990), and they are believed to be responsible for the strong adhesion or non-swelling of negatively charged clay and bilayer membrane surfaces in the presence of divalent ions (Kjellander *et al.*, 1988; Khan *et al.*, 1985). Their importance in the interactions of colloidal, amphiphilic and biological surfaces have yet to be fully established.

(ii) *Finite ion size (steric) effects.* This tends to enhance the repulsion between two surfaces. The effect is analogous to the increased (osmotic) pressure of a van der Waals gas due to the finite size (excluded volume) of the gas molecules. The finite sizes of both the counterions and the co-ions contribute to the enhanced repulsion. In the case of the co-ions adsorbed on the surfaces this is simply the steric repulsion between the two overlapping Stern layers (Fig. 12.4).

(iii) *Image forces.* For counterions in water between two surfaces of lower dielectric constant the image force (Fig. 11.2) contributes an additional repulsion.

(iv) *Discreteness of surface charges.* Surface charges are discrete and not 'smeared out' as has been implicitly assumed. Discrete ions generally contribute an attractive force, especially if the surface charges are mobile (thereby providing an additional contribution to the ion-correlation force).

(v) *Solvation forces.* This effect has to do with the solvent rather than the ions (solute). These short-range solvation or hydration forces can be attractive, repulsive or oscillatory, and are fully described in Chapter 13.

Unfortunately, it is not possible to list the relative importance of the above effects. In some cases the short-range forces are dominated by attractive ion-correlation forces, in others by repulsive solvation forces, while in some the forces appear to be well described by continuum theory right down to molecular contact (often because of the fortuitous cancellation of two or more opposing effects).

12.9 THICK WETTING FILMS

At large distances D and high surface charge densities σ the value of $(KD/2)$ in Eq. (12.12) must approach $\pi/2$ (i.e., $K \rightarrow \pi/D$). In this limit the pressure, Eq. (12.19), therefore becomes

$$P(D) = 2\varepsilon\varepsilon_0(\pi kT/ze)^2/D^2, \qquad (12.20)$$

which is known as the *Langmuir equation*. The Langmuir equation has been used to account for the equilibrium thickness of thick wetting films of water on glass surfaces (Figs 10.6f and 12.6). Here the water–air surface replaces the midplane of Fig. 12.2, so that for a film of thickness $d = D/2$, we have

$$P(d) = \varepsilon\varepsilon_0(\pi kT/ze)^2/2d^2, \qquad (12.21)$$

which is sometimes referred to as the *disjoining pressure* of a film. This repulsive pressure is entirely analogous to the repulsive van der Waals force across adsorbed liquid films, such as helium (Section 11.6), that causes them to climb up or spread on surfaces. Note, however, that both the magnitude and range of the double-layer repulsion is usually greater than the van der Waals' ($P \propto 1/d^2$ instead of $P \propto 1/d^3$). In Section 11.6 we saw that the equilibrium thickness d of a wetting film is given by one or other of the following equivalent equations

$$P(d) = -mgH/v = (kT/v)\log(p/p_{\mathrm{sat}}), \qquad (12.22)$$

where H is the height of the film above the surface of the bulk liquid, v and m the molecular volume and mass of the solvent ($\rho = m/v$), and p/p_{sat} the relative vapour pressure. Thus, if water condenses on a charged surface from undersaturated vapour, the film thickness, d will increase to infinity as H approaches zero or, equivalently, as p approaches p_{sat}.

Langmuir (1938) first applied Eq. (12.21) to explain why the rise of water up a capillary tube is higher than expected: since the water also wets the inner surface of the capillary the effective radius is smaller than the dry radius and this leads to a higher capillary rise than expected. Derjaguin and Kusakov (1939) measured how the thickness of a water film on a quartz glass surface decreased when an air bubble was progressively pressed on the film. The results were in rough agreement with the Langmuir equation. Read and Kitchener (1969) repeated these measurements and again found only rough agreement between theory and experiment: in the range 30–130 nm, the measured film thicknesses were 10–20 nm thicker than expected theoretically. This effect could be accounted for if the water–air interface is negatively charged so that for a given pressure P the film thickness would indeed be higher than given by Eq. (12.21), which assumes $\sigma = 0$ and $d\psi/dx = 0$ at that interface. More recently, Derjaguin and Churaev (1974), Pashley and Kitchener (1979), and Gee *et al.* (1990) used the vapour pressure control method to measure the equilibrium film thickness and found that for $d < 30$ nm the films are *much* thicker than expected from Eq. (12.22), an effect that has been attributed to either repulsive hydration forces or to the presence of small amounts of soluble contaminants in the films (Pashley, 1980).

12.10   LIMIT OF SMALL SEPARATIONS: CHARGE REGULATION

At small separations, as $D \to 0$, it is easy to verify from Eq. (12.12) that $K^2 \to -\sigma z e/\varepsilon \varepsilon_0 kTD$ (note that σ and z must have opposite signs). Thus, the repulsive pressure P of Eq. (12.19) approaches infinity according to

$$P(D \to 0) = -2\sigma kT/zeD. \qquad (12.23)$$

From Eqs (12.13) and (12.11) we further find that as $D \to 0$ the counterion density profile between the surfaces becomes uniform and equal to

$$\rho_x \approx \rho_s \approx \rho_0 \approx -2\sigma/zeD \qquad \text{at all } x. \qquad (12.24)$$

Since $-2\sigma/zeD$ is the number density of counterions in the gap this means that the limiting pressure of Eq. (12.23) is simply the osmotic pressure $P = \rho kT$ of an ideal gas at the same density as the trapped counterions.

The infinite pressure as $D \to 0$ implied by Eq. (12.23) is, of course, unrealistic and arises from the assumption that $\sigma = \text{constant}$, i.e., that the surfaces remain fully ionized even when there is a very large pressure pushing the counterions back against the surfaces. In practice when two surfaces are finally forced into molecular contact the counterions are forced to readsorb onto their original surface sites. Thus, as D approaches zero the surface charge density σ also falls, i.e., σ becomes a function of D. This is known as *charge regulation*. The effect of charge regulation is always to reduce the effective repulsion below that calculated on the assumption of constant surface charge and will be discussed again in Section 12.17.

Note, however, that other effects and forces can also come in at small separations (Section 12.8) and that these can be equally important in determining the forces as $D \to 0$.

12.11   CHARGED SURFACES IN ELECTROLYTE SOLUTIONS

It is far more common for charged surfaces or particles to interact across a solution that already contains electrolyte ions (dissociated inorganic salts). In animal fluids, ions are present in concentrations of about 0.2 M, mainly NaCl or KCl with smaller amounts of $MgCl_2$ and $CaCl_2$. The oceans have a similar relative composition of these salts but at a higher total concentration, about 0.6 M. Note that even 'pure water' at pH 7 is strictly an electrolyte solution containing 10^{-7} M of H_3O^+ and OH^- ions, which cannot always be ignored. For example, for a charged isolated surface exposed to a solvent containing no added electrolyte ions (only the counterions), Eqs (12.9) and (12.12) readily show that for the isolated surface, for which $D \to \infty$, we obtain $KD \to \pi$ and $\psi_s \to \infty$. As we shall see, this unrealistic situation is removed as soon as the bulk solvent contains even the minutest concentration of electrolyte ions.

The existence of a bulk 'reservoir' of electrolyte ions has a profound effect not only on the electrostatic potential but also on the force between charged surfaces, and in the rest of this chapter we shall consider this interaction as well as the total interaction potential when the attractive van der Waals force is added. But to understand the double-layer interaction between two surfaces it is necessary first to understand the ionic distribution adjacent to an isolated surface in contact with an electrolyte solution. Consider an isolated surface, or two surfaces far apart, in an aqueous electrolyte (Fig. 12.7). For convenience, we shall put $x = 0$ at the surface rather than at the midplane. Now, all the fundamental equations derived in the previous sections are applicable to solutions containing different types of ions i (of valency $\pm z_i$) so long as this is taken into account by expressing the net charge density at any point x as $\Sigma_i z_i e \rho_{xi}$ and the total ionic concentration (number density) as $\sum_i \rho_{xi}$. Thus, Eq. (12.2) for the Boltzmann distribution of ions i at x now becomes

$$\rho_{xi} = \rho_{\infty i} e^{-z_i e \psi_x / kT} \tag{12.25}$$

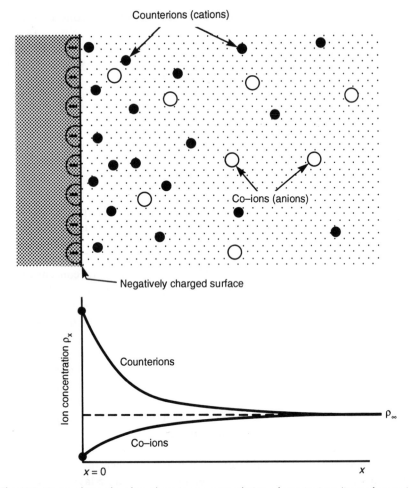

Fig. 12.7. Near a charged surface there is an accumulation of counterions (ions of opposite charge to the surface charge) and a depletion of co-ions, shown graphically below for a 1:1 electrolyte, where ρ_∞ is the electrolyte concentration in the bulk or 'reservoir' at $x = \infty$.

while at the surface, at $x = 0$, the contact values of ρ and ψ are related by

$$\rho_{0i} = \rho_{\infty i} e^{-z_i e \psi_0 / kT} \tag{12.26}$$

where $\rho_{\infty i}$ is the ionic concentration of ions i in the bulk (at $x = \infty$) where $\psi_\infty = 0$. For example, if we have a solution containing $H^+ OH^- +$

$Na^+Cl^- + Ca^{2+}Cl_2^-$, etc., we may write

$$[H^+]_x = [H^+]_\infty e^{-e\psi_x/kT}, \qquad [H^+]_0 = [H^+]_\infty e^{-e\psi_0/kT},$$

$$[Na^+]_x = [Na^+]_\infty e^{-e\psi_x/kT}, \qquad [Na^+]_0 = [Na^+]_\infty e^{-e\psi_0/kT},$$

$$[Ca^{2+}]_x = [Ca^{2+}]_\infty e^{-2e\psi_x/kT}, \qquad [Ca^{2+}]_0 = [Ca^{2+}]_\infty e^{-2e\psi_0/kT},$$

$$[Cl^-]_x = [Cl^-]_\infty e^{+e\psi_x/kT}, \qquad [Cl^-]_0 = [Cl^-]_\infty e^{+e\psi_0/kT},$$

$$(12.27)$$

where $[Na^+]$, etc., are expressed in some convenient concentration unit such as M ($1\,M = 1\,mol\,dm^{-3}$ and corresponds to a number density of $\rho = 6.022 \times 10^{26}\,m^{-3}$).

12.12 THE GRAHAME EQUATION

Let us now find the total concentration of ions at an isolated surface of charge density σ. From Eq. (12.8) this is immediately given by

$$\sum_i \rho_{0i} = \sum_i \rho_{\infty i} + \sigma^2/2\varepsilon\varepsilon_0 kT \qquad \text{(in number per m}^3\text{).} \qquad (12.28)$$

Thus, for $\sigma = 0.2\,C\,m^{-2}$ (corresponding to one electronic charge per $0.8\,nm^2$ or $80\,\text{Å}^2$) at $25°C$, we find $\sigma^2/2\varepsilon\varepsilon_0 kT = 7.0 \times 10^{27}\,m^{-3} = 11.64\,M$. For a $1:1$ electrolyte such as NaCl, the surface concentration of ions in this case is

$$[Na^+]_0 + [Cl^-]_0 = 11.64 + [Na^+]_\infty + [Cl^-]_\infty = 11.64 + 2[Na^+]_\infty$$

$$= 11.64 + 2[NaCl], \qquad (12.29)$$

while for a $2:1$ electrolyte such as $CaCl_2$,

$$[Ca^{2+}]_0 + [Cl^-]_0 = 11.64 + [Ca^{2+}]_\infty + [Cl^-]_\infty = 11.64 + 3[Ca^{2+}]_\infty$$

$$= 11.64 + 3[CaCl_2],$$

where $[NaCl]$ and $[CaCl_2]$ are the bulk molar concentrations of the salts. The ions at the surface are, of course, mainly the counterions (e.g., Na^+ or Ca^{2+} at a negatively charged surface) and their *excess* concentration at the surface over that in the bulk is seen to be

(i) dependent solely on the surface charge density σ (i.e., *independent* of the bulk electrolyte concentration), and

(ii) of magnitude sufficient to balance much of the surface charge (cf. Sections 12.4 and 12.14).

We may now find the relation between the surface charge density σ and the surface potential ψ_0. Incorporating Eq. (12.26) into Eq. (12.28) we obtain for the case of a mixed $NaCl + CaCl_2$ electrolyte:

$$\sigma^2 = 2\varepsilon\varepsilon_0 kT\left(\sum_i \rho_{0i} - \sum_i \rho_{\infty i}\right)$$

$$= 2\varepsilon\varepsilon_0 kT\{[Na^+]_\infty e^{-e\psi_0/kT} + [Ca^{2+}]_\infty e^{-2e\psi_0/kT} + [Cl^-]_\infty e^{+e\psi_0/kT}$$
$$- [Na^+]_\infty - [Ca^{2+}]_\infty - [Cl^-]_\infty\}.$$

On further noting that $[Cl^-]_\infty = [Na^+]_\infty + 2[Ca^{2+}]_\infty$ the above becomes

$$\sigma^2 = 2\varepsilon\varepsilon_0 kT\{[Na^+]_\infty(e^{-e\psi_0/kT} + e^{+e\psi_0/kT} - 2)$$
$$+ [Ca^{2+}]_\infty(e^{-2e\psi_0/kT} + 2e^{+e\psi_0/kT} - 3)\},$$

so that finally

$$\sigma = \sqrt{8\varepsilon\varepsilon_0 kT} \sin h(e\psi_0/2kT)\{[Na^+]_\infty + [Ca^{2+}]_\infty(2 + e^{-e\psi_0/kT})\}^{1/2}$$
$$= 0.117 \sin h(\psi_0/51.4)\{[NaCl] + [CaCl_2](2 + e^{-\psi_0/25.7})\}^{1/2} \quad (12.30)$$

at 25°C, where the concentrations $[NaCl] = [Na^+]_\infty$ and $[CaCl_2] = [Ca^{2+}]_\infty$ are in M, ψ_0 in mV, and σ in $C\,m^{-2}$ (1 $C\,m^{-2}$ corresponds to one electronic charge per 0.16 nm^2 or 16 $Å^2$). Eq. (12.30) allows us to calculate ψ_0 once σ is known, from which the individual ionic concentrations at each surface ρ_{0i} can be obtained using Eqs (12.26) or (12.27). We shall now consider some implications of this important equation, known as the *Grahame equation*.

12.13   SURFACE CHARGE AND POTENTIAL IN THE PRESENCE OF MONOVALENT IONS

For an aqueous 1:1 electrolyte solution such as NaCl against a negatively charged surface of $\sigma = -0.2\ C\,m^{-2}$, we obtain the potentials shown in the middle column of Table 12.1. Note that for no electrolyte we obtain an

TABLE 12.1 Variation of surface potential with aqueous electrolyte concentration for a planar surface of charge density -0.2 C m^{-2} as deduced from the Grahame equation, Eq. (12.30)

1:1 Electrolyte concentration (M)	ψ_0 (mV)	
	Pure 1:1 electrolyte solution	Bulk solution also contains 3×10^{-3} M 2:1 electrolyte
0 (hypothetical)	$-\infty$	-106
10^{-7} (pure water)	-477	-106
10^{-4}	-300	-106
10^{-3}	-241	-106
10^{-2}	-181	-105
10^{-1}	-123	-100
1	-67	-66

infinite potential which is unrealistic; a pure liquid such as water will always contain *some* dissociated ions. It is for this reason that we did not consider an isolated surface in the absence of bulk electrolyte ions in Section 12.5. From Table 12.1 we find that at constant surface charge density the surface potential falls progressively as the electrolyte concentration rises. From the tabulated values of ψ_0 we can determine the ionic concentrations at the surface using Eq. (12.27). For example, in 10^{-7} M 1:1 electrolyte, where $\psi_0 \approx -477.1$ mV, we obtain $10^{-7} \times e^{+477.1/25.69} = 11.64$ M for the counterions, and $10^{-7} \times e^{-477.1/25.69} \approx 10^{-15}$ M for the co-ions. In 1 M, where $\psi_0 = -67.0$ mV, we obtain 13.57 M and 0.07 M for the counterions and co-ions, respectively, which total 13.64 M. As expected, the total concentration of all the ions at the surface agrees exactly with that given by Eq. (12.29).

In most cases neither σ nor ψ_0 remains constant as the solution conditions change. This is because ionizable surface sites are rarely fully dissociated but are partially neutralized by the binding of specific ions from the solution. Such ions are often referred to as *exchangeable* ions, in contrast to those *inert* ions that do not bind to the surface. For example, if only protons can bind to a negatively charged surface, the equilibrium condition at the surface is given by the familiar *mass action equation* (Payens, 1955). Thus, for the reaction

$$S^- + H^+ \rightleftharpoons SH \qquad \text{at the surface,} \qquad (12.31)$$

we may express the proton concentration at the surface as $[H^+]_0$, the concentration or surface density of negative (dissociated) surface sites as $[S^-]_0$, and the density of neutral (undissociated) sites as $[SH]_0$. $[S^-]_0$ is related to σ via $\sigma = -e[S^-]_0$. The surface *dissociation constant* K_d for the above 'reaction' is defined by

$$K_d = \frac{[S^-]_0[H^+]_0}{[SH]_0} \qquad (12.32)$$

$$= \frac{\sigma_0\alpha}{\sigma_0(1-\alpha)}[H^+]_0 = \frac{\alpha}{(1-\alpha)}[H^+]_\infty e^{-e\psi_0/kT}. \qquad (12.33)$$

where σ_0 is the maximum possible charge density (i.e., if all the sites were dissociated) and α is the fraction of sites actually dissociated. Thus, if half the sites are dissociated at $[H^+]_0 = 10^{-4}$ M, we have $K_d = [H^+]_0 = 10^{-4}$ M, which may be quoted in pK units (pK $= -\log_{10}[H^+]_0 = 4.0$ in this case). Some people prefer to describe reactions at surfaces in terms of an *association* or *reaction constant*, K_a, defined by $K_a = 1/K_d$.

For a mixed 1:1 electrolyte of NaCl + HCl, Eq. (12.33) can be combined with the Grahame equation to give

$$\sigma = \sigma_0\alpha = \sigma_0 K_d/(K_d + [HCl]e^{-\psi_0/25.7})$$

$$= 0.117 \sin h(\psi_0/51.4)\sqrt{[NaCl] + [HCl]}, \qquad (12.34)$$

in which both σ and ψ_0 can now be totally determined in terms of the maximum charge density σ_0 and dissociation constant K_d (assuming that there is no binding of Na$^+$ ions). It is clear from the above that if K_d is very large (high surface charge, weak binding of protons) then $\sigma \approx \sigma_0 \approx$ constant, and we obtain the earlier result for the case of fixed surface charge density. However, if K_d takes on a more typical value, the effect can be quite dramatic. For example, if $K_d = 10^{-4}$ M, then for a surface of $\sigma_0 = -0.2$ Cm^{-2} in a 0.1 M NaCl bulk solution at pH 7 we find $\psi_0 = -118$ mV and $\alpha = 0.91$, i.e., the protons have neutralized 9% of the surface site, and ψ_0 is not very different from the value in the absence of protons (see Table 12.1). But at pH 5 we obtain $\psi_0 = -73$ mV and $\alpha = 0.36$, i.e., only 36% of the sites now remain dissociated even though the bulk concentration of HCl is a mere 0.01% of the NaCl concentration. Under such conditions the proton is referred to as a *potential determining* ion. Thus, both ψ_0 and σ will vary as the salt concentration or pH is changed, but the surface will always remain negatively charged.

More generally, a surface may contain both anionic (e.g., acidic) and cationic (e.g., basic) groups to which various cations and anions can bind. Such surfaces are known as *amphoteric*, and the *competitive adsorption* of ions to them can be analysed by assigning a binding constant to each ion type, and then incorporating these into the Grahame equation (Healy and White, 1978; Chan et al., 1980a). The charge density of amphoteric surfaces (e.g., protein surfaces) can be negative or positive depending on the electrolyte conditions. At the *isoelectric point* (iep) or *point of zero charge* (pzc) there are as many negative charges as positive charges so that the mean surface charge density is zero ($\sigma = 0$), though it is well to remember that there may still be patches of high local charge density.

12.14 EFFECT OF DIVALENT IONS

The presence of divalent cations has a dramatic effect on the surface potential and counterion distribution at a negatively charged surface. For example, if all the NaCl solutions of Table 12.1 also contain 3×10^{-3} M $CaCl_2$, the Grahame equation gives the potentials shown in the last column. We see that even at constant surface charge density, relatively small amounts of divalent ions substantially lower the magnitude of ψ_0, in fact, about 100 times more effectively than increasing the concentration of monovalent salt. Indeed, ψ_0 is determined solely by the divalent cations once their concentration is greater than about 3% of the monovalent ion concentration, and for 2:1 electrolyte concentrations above a few mM, typical surface potentials are well below -100 mV irrespective of the 1:1 electrolyte concentration.

Further, even when the bulk concentration of Ca^{2+} is much smaller than that of Na^+ the surface may have a much higher local concentration of Ca^{2+}. For example, in 100 mM NaCl + 3 mM $CaCl_2$ where $\psi_0 = -100$ mV (Table 12.1) the concentration of Ca^{2+} at the surface is $[Ca^{2+}]_0 \approx 3 \times 10^{-3}e^{+200/25.7} \approx 7$ M compared to $[Na^+]_0 \approx 0.1\, e^{+100/25.7} \approx 5$ M.

At such high surface concentrations (of *doubly* charged ions) divalent ions often bind chemically to negative surface sites, thereby lowering σ and reducing ψ_0 even further, and it is not unusual for surfaces to be completely neutralized ($\sigma \to 0$, $\psi_0 \to 0$) in the presence of mM amounts of Ca^{2+}. In the case of trivalent ions such as La^{3+}, bulk concentrations in excess of 10^{-5} M can neutralize a negatively charged surface and even lead to *charge reversal*, wherein the cations continue to adsorb onto a surface that is already net positively charged (see Problem 3.2).

As in the case of monovalent ion binding, the effect of divalent ion binding can be dealt with quantitatively by incorporating the appropriate binding constants into the Grahame equation (Healy and White, 1978; McLaughlin et al., 1981), and when many different ionic species (e.g., Ca^{2+}, H^+) compete for binding sites the variation of ψ_0 and σ with electrolyte concentration and pH can be quite complex. In most cases ion binding tends to lower both σ

and ψ_0 as the concentrations of these ions increase, and we may anticipate that such effects lead to a substantial reduction in the repulsive double-layer forces between surfaces.

12.15   THE DEBYE LENGTH

For low potentials, below about 25 mV, the Grahame equation simplifies to

$$\sigma = \varepsilon\varepsilon_0\kappa\psi_0, \tag{12.35}$$

where

$$\kappa = \left(\sum_i \rho_{\infty i}e^2z_i^2/\varepsilon\varepsilon_0 kT\right)^{1/2} \text{m}^{-1}. \tag{12.36}$$

Thus, the potential becomes proportional to the surface charge density. Equation (12.35) is the same as Eq. (12.14) for a capacitor whose two plates are separated by a distance $1/\kappa$, have charge densities $\pm\sigma$, and potential difference ψ_0. This analogy with a charged capacitor gave rise to the name *diffuse electric double layer* for describing the ionic atmosphere near a charged surface, whose characteristic length or 'thickness' is known as the Debye length, $1/\kappa$.

The magnitude of the Debye length depends solely on the properties of the liquid and not on any property of the surface such as its charge or potential. At 25°C the Debye length of aqueous solutions is

$$1/\kappa = \begin{cases} 0.304/\sqrt{[\text{NaCl}]} \text{ nm} & \text{for 1:1 electrolytes (e.g., NaCl)} \\ 0.176/\sqrt{[\text{CaCl}_2]} \text{ nm} & \text{for 2:1 and 1:2 electrolytes} \\ & \text{(e.g., CaCl}_2 \text{ and Na}_2\text{SO}_4) \\ 0.152/\sqrt{[\text{MgSO}_4]} \text{ nm} & \text{for 2:2 electrolytes (e.g., MgSO}_4) \end{cases}$$

$$\tag{12.37}$$

For example, for NaCl solution, $1/\kappa = 30.4$ nm at 10^{-4} M, 9.6 nm at 1 mM, 0.96 nm at 0.1 M, and 0.3 nm at 1 M. In totally pure water at pH 7, the Debye length is 960 nm, or about 1 μm.

12.16   Variation of Potential and Ionic Concentrations away from a Charged Surface

The potential gradient at any distance x from an isolated surface is given by Eq. (12.7):

$$\sum_i \rho_{xi} = \sum_i \rho_{\infty i} + \frac{\varepsilon \varepsilon_0}{2kT} \left(\frac{d\psi}{dx}\right)^2_x.$$ (12.38)

For a 1:1 electrolyte this gives

$$d\psi/dx = \sqrt{8kT\rho_{\infty i}/\varepsilon\varepsilon_0} \; \sin h(e\psi_x/2kT),$$

which may be readily integrated using the integral $\int \mathrm{csch} X \, dX = \log \tan h(X/2)$ to yield

$$\psi_x = \frac{2kT}{e} \log\left[\frac{1 + \gamma e^{-\kappa x}}{1 - \gamma e^{-\kappa x}}\right] \approx \frac{4kT}{e} \gamma e^{-\kappa x},$$ (12.39)

where

$$\gamma = \tan h(e\psi_0/4kT) \approx \tan h[\psi_0(mV)/103].$$ (12.40)

This is known as the *Gouy–Chapman* theory. For high potentials $\gamma \to 1$, while for low potentials, Eq. (12.39) reduces to the so-called *Debye–Hückel* equation

$$\psi_x \approx \psi_0 e^{-\kappa x},$$ (12.41)

where again the Debye length $1/\kappa$ appears as the characteristic decay length of the potential (see Verwey and Overbeek, 1948, and Hiemenz, 1977, for a fuller discussion of the Gouy–Chapman and Debye–Hückel theories).

The above equations apply to *symmetrical* 1:1 electrolytes, such as NaCl. Equations that apply to *asymmetrical* electrolytes, e.g., 2:1 and 1:2 electrolytes such as $CaCl_2$, have been derived by Grahame (1953). These are more complicated than Eq. (12.39) but for low ψ_0 they all reduce to $\psi_x = \psi_0 e^{-\kappa x}$.

We now have all the equations needed for computing the ionic distributions away from a charged surface. For a 1:1 electrolyte, this is given by inserting Eq. (12.39) into Eq. (12.25) or (12.27). Figure 12.8 shows the variation of ψ_x and ρ_x for a 0.1 M 1:1 electrolyte, together with a Monte Carlo simulation for comparison. Note how the counterion density approaches the bulk value much faster than would be indicated by the Debye length. Indeed, for such

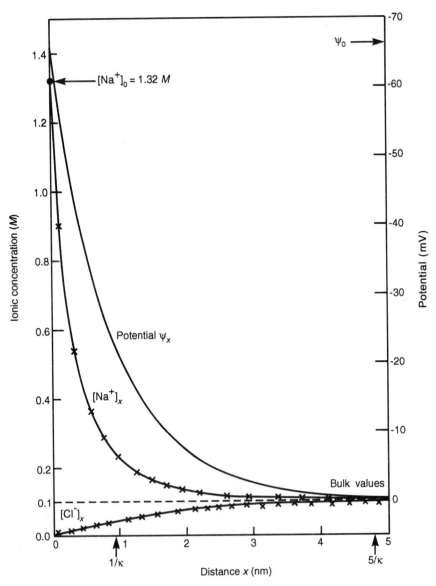

Fig. 12.8. Potential and ionic density profiles for a 0.1 M monovalent electrolyte such as NaCl near a surface of charge density $\sigma = -0.0621\,\mathrm{C\,m^{-2}}$ (about one electronic charge per 2.6 nm²), calculated from Eqs (12.39) and (12.25) with $\psi_0 = -66.2\,\mathrm{mV}$ obtained from the Grahame equation. The crosses are the Monte Carlo results of Torrie and Valleau (1979).

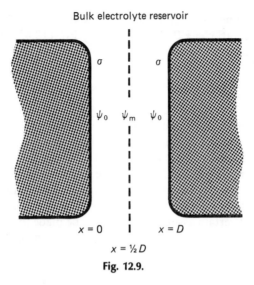

Fig. 12.9.

a high surface charge density and potential the counterion distribution very near the surface is largely independent of the bulk electrolyte concentration, and it is left as an exercise for the reader to verify that even in 10^{-4} M the counterion profile over the first few ångstroms is not much different from that in 0.1 M (so long as σ remains the same).

12.17 THE ELECTROSTATIC DOUBLE-LAYER INTERACTION BETWEEN CHARGED SURFACES IN ELECTROLYTE

The interaction pressure between two identically charged surfaces in an electrolyte solution (Fig. 12.9) can be derived quite simply as follows. First, from Section 12.7 we note that at any point x the pressure $P_x(D)$ is given by

$$P_x(D) - P_x(\infty) = -\tfrac{1}{2}\varepsilon\varepsilon_0\left[\left(\frac{d\psi}{dx}\right)^2_{x(D)} - \left(\frac{d\psi}{dx}\right)^2_{x(\infty)}\right]$$

$$+ kT\left[\sum_i \rho_{xi}(D) - \sum_i \rho_{xi}(\infty)\right]. \qquad (12.42)$$

Second, from Eq. (12.7) we have

$$\sum_i \rho_{xi} = \sum_i \rho_{mi} + \frac{\varepsilon\varepsilon_0}{2kT}\left(\frac{d\psi}{dx}\right)^2_x \qquad (12.43)$$

for any D including $D = \infty$, where $\Sigma\rho_{mi}$ is the total ionic concentration at the midplane, at $x = \frac{1}{2} D$. Incorporating Eq. (12.43) into Eq. (12.42), and again putting $P_x(D = \infty) = 0$, yields

$$P_x(D) = kT\left[\sum_i \rho_{mi}(D) - \sum_i \rho_{mi}(\infty)\right] \qquad (12.44)$$

which as before is the uniform pressure across the gap (independent of position x) acting on the electrolyte ions and on the surfaces. The above result is essentially the same as Eq. (12.17) and shows that P is simply the excess osmotic pressure of the ions in the midplane over the bulk pressure. Since $\Sigma\rho_{mi}(\infty)$ is known from the bulk electrolyte concentration the problem reduces to finding the midplane concentration of ions $\rho_{mi}(D)$ when D is finite, and it is here that certain assumptions have to be made to obtain an analytic result (Verwey and Overbeek, 1948). For a 1:1 electrolyte such as NaCl, Eq. (12.44) may be written as

$$P = kT\rho_\infty[(e^{-e\psi_m/kT} - 1) + (e^{+e\psi_m/kT} - 1)] \approx e^2\psi_m^2\rho_\infty/kT,$$

$$\text{cations} \qquad\qquad \text{anions} \qquad\qquad\qquad (12.45)$$

which assumes that the midplane potential ψ_m (not the surface potential ψ_0) is small. If we further assume that ψ_m is simply the sum of the potentials from each surface at $x = \frac{1}{2} D$ as previously derived for an isolated surface, then Eq. (12.39) gives $\psi_m \approx (8kT\gamma/e)e^{-\kappa D/2}$. Inserting this into Eq. (12.45) gives the final result for the repulsive pressure between two planar surfaces:

$$P = 64kT\rho_\infty\gamma^2 e^{-\kappa D} = (1.59 \times 10^8)[\text{NaCl}]\gamma^2 e^{-\kappa D} \quad \text{N m}^{-2}, \qquad (12.46)$$

where we note that $\gamma = \tan h(ze\psi_o/4kT)$ can never exceed unity. The above equation is known as the *weak overlap approximation* for the interaction between two similar surfaces at constant potential. See Problem 12.3 for the case of two surfaces with unequal surface potentials.

The interaction free energy per unit area corresponding to the above pressure is obtained by a simple integration with respect to D, and gives

$$W = (64kT\rho_\infty\gamma^2/\kappa)e^{-\kappa D} \qquad (12.47)$$

$$= 0.0482[\text{NaCl}]^{1/2}\tan h^2[\psi_0(mV)/103]e^{-\kappa D} \quad \text{J m}^{-2} \quad (\text{for } z = 1)$$

$$= 0.0211[\text{CaCl}_2]^{1/2}\tan h^2[2\psi_0(mV)/103]e^{-\kappa D} \quad \text{J m}^{-2} \quad (\text{for } z = 2)$$

where in the above equations the concentrations [NaCl] and [CaCl$_2$] are in M and the values are for aqueous solutions at 298 K.

Using the Derjaguin approximation, Eq. (10.18), we may immediately write the expression for the force F between two spheres of radius R as $F = \pi R W$, from which the interaction free energy is obtained by a further integration:

$$W = (64\pi k T R \rho_\infty \gamma^2/\kappa^2)e^{-\kappa D} = 4.61 \times 10^{-11} R\gamma^2 e^{-\kappa D} \quad \text{J} \quad \text{(for } z = 1\text{)}.$$

(12.48)

We see therefore that the double-layer interaction between surfaces or particles decays exponentially with distance. The characteristic decay length is the Debye length.

At low surface potentials, below about 25 mV, all the above equations simplify to the following. For two planar surfaces,

$$P \approx 2\varepsilon\varepsilon_0\kappa^2\psi_0^2 e^{-\kappa D} = 2\sigma^2 e^{-\kappa D}/\varepsilon\varepsilon_0 \quad \text{(per unit area)} \qquad (12.49)$$

and

$$W \approx 2\varepsilon\varepsilon_0\kappa\psi_0^2 e^{-\kappa D} = 2\sigma^2 e^{-\kappa D}/\kappa\varepsilon\varepsilon_0 \quad \text{(per unit area)}, \qquad (12.50)$$

while for two spheres of radius R,

$$F \approx 2\pi R\varepsilon\varepsilon_0\kappa\psi_0^2 e^{-\kappa D} = 2\pi R\sigma^2 e^{-\kappa D}/\kappa\varepsilon\varepsilon_0 \qquad (12.51)$$

and

$$W \approx 2\pi R\varepsilon\varepsilon_0\psi_0^2 e^{-\kappa D} = 2\pi R\sigma^2 e^{-\kappa D}/\kappa^2\varepsilon\varepsilon_0. \qquad (12.52)$$

In the above ψ_0 and σ are related by $\sigma = \varepsilon\varepsilon_0\kappa\psi_0$, which, as we have seen, is valid for low potentials. These four equations are quite useful because they are valid for all electrolytes, whether 1:1, 2:1, 2:2, 3:1, or even mixtures, so long as the appropriate Debye lengths are used as given by Eqs (12.36) and (12.37). Thus, they are particularly suitable when divalent ions are present since the surface charge and potential is often low due to ion binding.

All the expressions so far derived for the interactions of two double layers are accurate only for surface separations beyond about one Debye length. At smaller separations one must resort to numerical solutions of the Poisson–Boltzmann equation to obtain the exact interaction potential (Verwey and Overbeek, 1948; Honig and Mul, 1971) for which there are no simple expressions. In addition there is the question of charge regulation at small separations, i.e., does the surface charge density remain constant as two surfaces come close together, or do some of the counterions bind to the surfaces thereby reducing σ. This affects the form of the interaction potential. At large distances, beyond $1/\kappa$, the question does not arise, and the interaction pressures and energies are well described by Eqs (12.46)–(12.48), where ψ_0 and σ are the values appropriate for the isolated surfaces (at $D = \infty$). But at progressively smaller separations, as the counterion concentration at each surface increases, specific ion binding can occur as discussed earlier.

If there is no binding, the surface charge density σ will remain constant, and in the limit of small D the number density of monovalent counterions between the two surfaces will approach a uniform value of $2\sigma/eD$. Thus, from Eq. (12.44) the limiting pressure in this case is

$$P(D \to 0) = kT \sum_i \rho_{mi} = -2\sigma kT/zeD \qquad (12.53)$$

and

$$W(D \to 0) = (-2\sigma kT/ze) \log D + \text{constant} \qquad (12.54)$$

that is, as $D \to 0$ the pressure and the energy become infinite. Note that this is the same limiting pressure as in the case of no bulk electrolyte (counterions only), obtained in Eq. (12.23), and results purely from the limiting osmotic pressure of the 'trapped' counterions. Indeed, this is part of an even more general rule for surfaces of constant charge which states that in the (ideal) limit of small D, all double-layer forces tend towards the osmotic limit, $P \to \rho kT$, where ρ is the number density of counterions remaining in the gap which is independent of the bulk electrolyte concentration.

If there is counterion binding as D decreases, P falls below this limit, and the Poisson–Boltzmann equation must now be solved self-consistently by including the dissociation constants of the adsorbing ions (Section 12.13). The computations have been described by Ninham and Parsegian (1971), and a simple numerical algorithm has been given by Chan et al. (1976, 1980b).

The two main effects of a charge-regulating interaction can be summarized qualitatively as follows (see also Healy et al., 1980):

(i) if counterions adsorb as two surfaces approach each other the strength of the double-layer interaction is always less than that occurring at constant surface charge (Le Chaterlier's principle), and

(ii) in general, the interaction potential will lie between two limits, the upper one corresponding to the interaction at constant surface charge and the lower at constant surface potential.

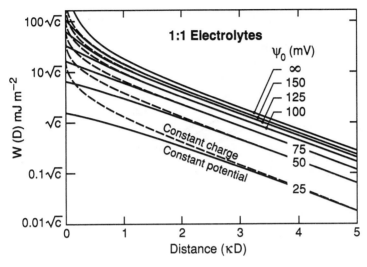

Fig. 12.10. Repulsive double-layer interaction energy for two planar surfaces in a 1:1 electrolyte. (Exact solution kindly computed by M. Sculley, R. Pashley and L. White based on Ninham and Parsegian, 1971.) ψ_0 is the potential of the isolated surfaces and C the electrolyte concentration in M, which is related to the Debye length by $1/\kappa = 0.304/\sqrt{C}$ nm. Theoretically, the double-layer interaction must lie between the constant-charge and constant-potential limits. (– – –, constant charge; ———, constant potential). At separations greater than $1/\kappa$ the forces are well described by Eq. (12.47) for $z = 1$.

This is illustrated in Figs 12.10 and 12.11, which show the double-layer interaction potentials of two planar surfaces in 1:1 and 1:2 electrolytes. The curves are based on exact numerical solutions and show the theoretical limits of the constant charge and constant potential interaction in each case. The figure may be used for reading off the interaction energy of any 1:1 or 2:1 electrolyte at any desired concentration C, and surface separation D. This is because the energy scales with $\sqrt{C}$ and the distance scales with the Debye length, $1/\kappa$. The constant potential curves of Fig. 12.10 compare reasonably well with the approximate expressions of Eq. (12.47) even at small separations, and especially when ψ_0 is between 50 and 100 mV. However, the accurately computed constant charge interaction is always well above that predicted by the approximate expression at small separations. Approximate expressions for interactions at constant surface charge have been derived by Gregory (1973).

For two surfaces of different charge densities or potentials the interaction energy can have a maximum at some finite distance, usually below $1/\kappa$, i.e., the surfaces can *attract* each other at small separations. Approximate equations for the interactions of two surfaces of unequal but constant

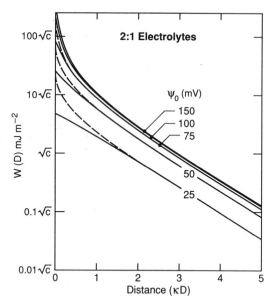

Fig. 12.11. Repulsive double-layer interaction energy for two planar surfaces in a $2:1$ electrolyte (computed as in Fig. 12.10). The electrolyte concentration C is related to the Debye length by $1/\kappa = 0.176/\sqrt{C}$ nm (———, constant charge; ———, constant potential). At separations greater than $1/\kappa$ the forces are well described by Eq. (12.47) for $z = 2$.

potentials were given by Hogg *et al.* (1966) and, in a different form, by Parsegian and Gingell (1972), and for unequal charges by Gregory (1975).

Finally, it is worth again mentioning that at small distances the PB equation often breaks down: the full electrostatic interaction can become more attractive when ion-correlation forces become important (usually at distances below 2–5 nm) or more repulsive when finite ion-size effects become important (usually at even smaller separations).

12.18   VAN DER WAALS AND DOUBLE-LAYER FORCES ACTING TOGETHER: THE DLVO THEORY

The total interaction between any two surfaces must also include the van der Waals attraction. Now, unlike the double-layer interaction, the van der Waals interaction potential is largely insensitive to variations in electrolyte concentration and pH, and so may be considered as fixed in a first approximation. Further, the van der Waals attraction must always exceed the double-layer repulsion at small enough distances since it is a power-law interaction (i.e., $W \propto -1/D^n$), whereas the double-layer interaction energy remains finite or rises much more slowly as $D \to 0$. Figure 12.12 shows schematically the various types of interaction potentials that can occur between two surfaces or colloidal particles under the combined action of these two forces. Depending on the electrolyte concentration and surface charge density or potential one of the following may occur:

(i) For highly charged surfaces in dilute electrolyte (i.e., long Debye length), there is a strong long-range repulsion that peaks at some distance, usually between 1 and 4 nm, at the *energy barrier*. This is illustrated in Fig. 12.12a.

(ii) In more concentrated electrolyte solutions there is a significant *secondary minimum*, usually beyond 3 nm, before the energy barrier (Fig. 12.12, upper inset). The potential energy minimum at contact is known as the *primary minimum*. For a colloidal system, even though the thermodynamically equilibrium state may be with the particles in contact in the deep primary minimum, the energy barrier may be too high for the particles to overcome during any reasonable time period. When this happens, the particles will either sit in the weaker secondary minimum or remain totally dispersed in the solution. In the latter case the colloid is referred to as being *kinetically stable* (as opposed to *thermodynamically stable*).

(iii) For surfaces of low charge density or potential, the energy barrier will always be much lower (Fig. 12.12c). This leads to slow aggregation, known as *coagulation* or *flocculation*. Above some concentration of electrolyte, known as the *critical coagulation concentration*, the energy barrier falls below the $W = 0$ axis (Fig. 12.12d) and the particles then coagulate rapidly. The colloid is now referred to as being *unstable*.

(iv) As the surface charge or potential approaches zero the interaction curve approaches the pure van der Waals curve, and two surfaces now attract each other strongly at all separations (Fig. 12.12e).

The sequence of phenomena described above can be described quantitatively (see Worked Example below) and it forms the basis of the celebrated *DLVO theory* of colloidal stability, after Derjaguin and Landau (1941), and Verwey and Overbeek (1948). See also Shaw (1970), Hiemenz (1977), and Hunter (1989).

The main factor inducing two surfaces to come into adhesive contact in a primary minimum is the lowering of their surface potential or charge, brought about by increased ion binding and/or increased screening of the double-layer repulsion by increasing the salt concentration. However, if the surface charge remains high on raising the salt concentration, two surfaces

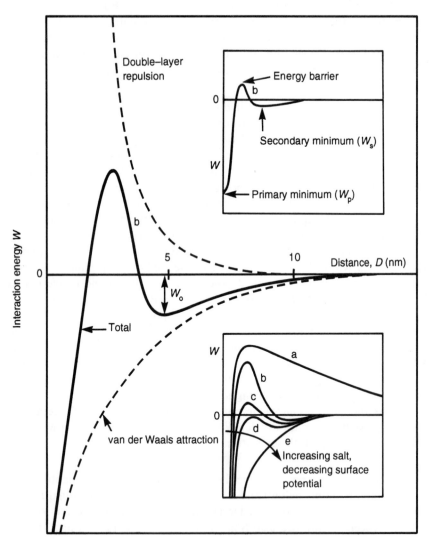

Fig. 12.12. Schematic energy versus distance profiles of DLVO interaction. (a) Surfaces repel strongly; small colloidal particles remain 'stable'. (b) Surfaces come into stable equilibrium at secondary minimum if it is deep enough; colloids remain 'kinetically' stable. (c) Surfaces come into secondary minimum; colloids coagulate slowly. (d) The 'critical coagulation concentration'. Surfaces may remain in secondary minimum or adhere; colloids coagulate rapidly. (e) Surfaces and colloids coalesce rapidly.

can still adhere to each other, but in a secondary minimum, where the adhesion is much weaker and easily reversible.

It is clear that one must have a fairly good idea of the charging process occurring at a surface before attempting to understand its double-layer interactions and the stability of colloidal dispersions, as the following example shows.

□   ● WORKED EXAMPLE ●

Question: For a number of colloidal systems it is found that the critical coagulation concentration varies with the inverse sixth power of the valency z of the electrolyte counterions, i.e., ρ_∞(c.c.c.) $\propto 1/z^6$. Is this empirical observation, known as the *Schultz–Hardy rule*, consistent with the DLVO theory?

Answer: The total DLVO interaction potential between two spherical particles interacting at constant potential is

$$W(D) = (64\pi kTR\rho_\infty \gamma^2/\kappa^2)e^{-\kappa D} - AR/6D. \qquad (12.55)$$

By definition (see Fig. 12.12d), the critical coagulation concentration occurs when both $W = 0$, and $dW/dD = 0$. The first condition leads to

$$\kappa^2/\rho_\infty = 384\pi kTD\gamma^2 e^{-\kappa D}/A,$$

while the second condition leads to $\kappa D = 1$, which shows that the potential maximum occurs at $D = \kappa^{-1}$ (the Debye length). Inserting this into the above equation leads to

$$\kappa^3/\rho_\infty = 768\pi kT\gamma^2 e^{-1}/A,$$

i.e.,

$$\kappa^6/\rho_\infty^2 \propto (T\gamma^2/A)^2.$$

Now, since $\kappa^2 \propto \rho_\infty z^2/\varepsilon T$, the above equation implies that

$$z^6\rho_\infty \propto \varepsilon^3 T^5 \gamma^4/A^2, \qquad (12.56)$$

which is a constant if γ is constant, a condition that holds at high surface potentials ($\psi_0 > 100$ mV) where $\gamma = 1$ (see Eq. (12.40)). In this limit, therefore, the critical coagulation concentrations do indeed scale as $\rho_\infty \propto 1/z^6$. For example, if coagulation occurs at 1 M with a 1:1 electrolyte, it will occur at $\frac{1}{64}$ M with a 2:2 electrolyte (or divalent counterions), and at $\frac{1}{729}$ M with a 3:3 electrolyte (or trivalent counterions). Thus the Schultz–Hardy rule is consistent with the DLVO theory.

But wait. Is it not unreasonable to assume high surface potentials in divalent and trivalent electrolyte solutions? Let us investigate the case of low potentials. Here we have $\gamma \propto z\psi_0/T$, so that Eq. (12.56) now becomes

$$z^2\rho_\infty \propto \varepsilon^3 T\psi_0^4/A^2, \tag{12.57}$$

which is constant if ψ_0 remains constant. Thus for low but constant potentials we obtain a modified form of the Schultz–Hardy rule: $\rho_\infty \propto 1/z^2$.

In real systems the surface potential is neither high nor constant, but usually falls to quite low values as the valancy of the electrolyte counterions increases. For example, if $\psi_0 \propto 1/z$, then for *low* potentials we now obtain: $\rho_\infty \propto \psi_0^4/z^2 \propto 1/z^6$, which brings us back to the Schultz–Hardy rule. Clearly the DLVO theory can be applied in more ways than one to explain the Schultz–Hardy rule.   □

12.19   Experimental measurements of double-layer and DLVO forces

Figure 12.13 shows experimental results of direct force measurements between two mica surfaces in dilute 1:1 and 2:1 electrolyte solutions where the Debye length is large, thereby allowing accurate comparison with theory to be made at distances much smaller than the Debye length. The theoretical DLVO force laws are shown by the continuous curves. The agreement is remarkably good at all separations, even down to 2% of κ^{-1}, and indicates that the DLVO theory is basically sound. One may also conclude that the dielectric constant of water must be the same as the bulk value even at surface separations as small as 2 nm, since otherwise significant deviations from theory would have occurred (Hamnerius *et al.*, 1978, showed that the dielectric constant of water remains unchanged even for 1 nm films). The surface potentials ψ_0 inferred from the magnitude of the double-layer forces agree within 10 mV with those measured independently on isolated mica surfaces by the method of electrophoresis (Lyons *et al.*, 1981). Further, the surface charge density corresponding to these potentials is typically $1e$ per 60 nm². Thus, at separations below about 8 nm the surfaces are actually closer to each other than the mean distance between the surface charges, and yet the

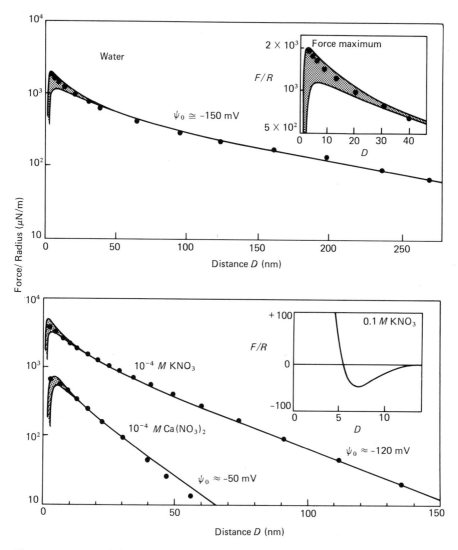

Fig. 12.13. Measured double-layer and van der Waals forces between two curved mica surfaces of radius R (~ 1 cm) in water and in dilute $\sim 10^{-4}$ M KNO_3 and $\sim 10^{-4}$ M $Ca(NO_3)_2$ solutions. The continuous curves are the theoretical DLVO forces (using a Hamaker constant of $A = 2.2 \times 10^{-20}$ J) showing the constant charge and constant potential limits. Theoretically, we expect the interactions to fall between these two limits. (Note that for this geometry the Derjaguin approximation gives $F/R = 2\pi W$, where W is the corresponding interaction energy per unit area between two planar surfaces, as plotted in Figs 12.10 and 12.11.) The inset in the lower part of the figure is the measured force in concentrated 0.1 M KNO_3 showing the emergence of a secondary minimum. (From Israelachvili and Adams, 1978; Pashley, 1981a; Israelachvili, 1982.)

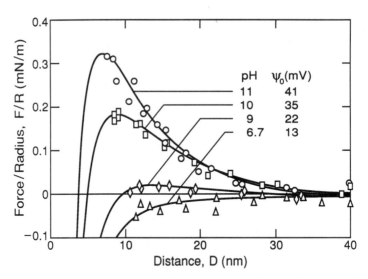

Fig. 12.14. Classic DLVO forces measured between two sapphire surfaces in 10^{-3} M NaCl solutions at different pH. The continuous lines are the theoretical DLVO forces for the potentials shown and a Hamaker constant of $A = 6.7 \times 10^{-20}$ J (from Horn *et al.,* 1988a).

double-layer forces still behave as if the surface charges were smeared out. The reason for this will become clear in the next section.

Other measurements of double-layer or DLVO forces have been carried out in various 1:1 and 2:1 electrolytes (Pashley, 1981a, b; Pashley and Israelachvili, 1984; Horn *et al.,* 1988a), between surfactant and lipid bilayers (Pashley and Israelachvili, 1981; Marra, 1986b, c, Marra and Israelachvili, 1985), across soap films (Derjaguin and Titijevskaia, 1954; Lyklema and Mysels, 1965; Donners *et al.,* 1977), between silica, sapphire and metal surfaces (Horn *et al.,* 1988a, 1989a, b; Smith *et al.,* 1988), as well as in non-aqueous polar liquids (Christenson and Horn, 1983, 1985). The results are invariably in good agreement with the DLVO theory, often down to separations well below the Debye length.

When deviations do occur these can usually be attributed to the presence of other, non-DLVO, forces or to the existence of a Stern layer (Israelachvili, 1985). A Stern layer of thickness δ per surface (2δ for both surfaces) can have a profound effect on the DLVO interaction potential because it pushes the plane of origin of the double-layer interaction (the OHP) out to $D = 2\delta$. It is remarkable that for values of δ as small as 0.2–0.3 nm (corresponding to hydrated sizes of ions) this effect can completely eliminate a deep primary minimum. Instead, the repulsion continues to rise steeply and indefinitely as the surface separation decreases to contact, at $D = 2\delta$. This model was first proposed by Frens and Overbeek (1972) to explain the common

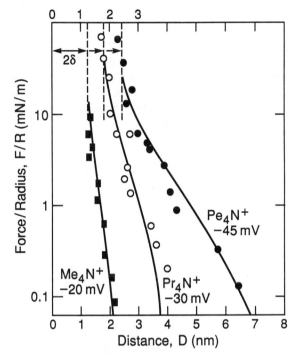

Fig. 12.15. Measured forces between two mica surfaces in various tetra-alkyl ammonium bromide solutions (Claesson et al., 1984). The continuous curves are the expected DLVO interactions assuming potentials as shown and Stern-layer thicknesses of δ per surface equal to the diameters of the adsorbed cations: $\delta = 0.6$ nm for methyl ammonium (Me_4N^+), $\delta = 0.9$ nm for propyl ammonium (Pr_4N^+) and $\delta = 1.2$ nm for pentyl ammonium (Pe_4N^+). Note how the outward shift in the OHP has eliminated the force maximum and primary minimum, resulting in *Stern-layer stabilization*.

phenomenon of colloidal stability in high salt, the spontaneous swelling of certain colloids in water, and *repeptization*—the reversible coagulation of colloidal particles (according to the DLVO theory coagulation in a primary minimum should never be reversible). A direct experimental measurement of Stern-layer stabilization is shown in Fig. 12.15 where the counterions used in that study were unusually large.

It is perhaps surprising that measured double-layer forces are so well described by a theory that, unlike van der Waals force theory, contains a number of fairly drastic assumptions, viz. the assumed smearing out of discrete surface charges, that ions can be considered as point charges, the ignoring of image forces, and that the PB equation remains valid even at fairly high concentrations. One reason for this is that many of these effects act in opposite directions and tend to cancel each other out (Section 12.8). As mentioned above, most experimental deviations in the forces from those expected from the DLVO theory are not due to any breakdown in the DLVO theory, but rather to the existence of a Stern layer or to the presence of other forces such as ion-correlation, solvation, hydrophobic or steric forces. These additional forces, are, of course, very important, especially in more complex colloidal and biological systems where they often dominate the interactions at short range where most of the interesting things happen. Their consideration forms a large part of the rest of this book.

12.20   EFFECTS OF DISCRETE SURFACE CHARGES AND DIPOLES

The charge on a solid surface is obviously not uniformly spread out over the surface, as has been implicit in all the equations derived so far. For a surface with a typical potential of 75 mV in a 1 mM NaCl solution, the surface charge density as given by the Grahame equation is $\sigma = 0.0075$ C m^{-2}, which corresponds to only one charge per 21 nm^2 or 2100 Å^2. In 0.1 M NaCl the same potential implies $1e$ per 2 nm^2. Thus, the charges on real surfaces are typically 1–5 nm apart from each other on average. What effect does this have on the electrostatic interaction between two surfaces?

Let us consider a planar square lattice of like charges q as shown in Fig. 12.16a. If d is the distance between any two neighbouring charges, then the mean surface charge density is $\sigma = q/d^2$, and if this charge were smeared out, the electric field emanating from the surface would be uniform and given by $E_z = \sigma/2\varepsilon\varepsilon_0$. What, then, is the field of a surface lattice of discrete charges having the same mean charge density? To compute this field one must sum the contributions from all the charges. The resulting slowly converging series can be turned into a rapidly converging series by using a mathematical technique known as the Poisson summation formula (Lighthill, 1970). If x and y are the coordinates in the plane relative to any charge as the origin (Fig. 12.16a), the field E_z along the z direction is given by the series (Lennard-Jones and Dent, 1928)

$$E_z = \frac{\sigma}{2\varepsilon\varepsilon_0}\left[1 + 2\left(\cos\frac{2\pi x}{d} + \cos\frac{2\pi y}{d}\right)e^{-2\pi z/d} + \dots\right], \qquad (12.58)$$

where the higher-order terms decay much more rapidly with distance z. The first term is the same as that of a smeared-out surface charge. The second term is interesting, for it shows that the excess field decays away extremely rapidly, with a decay length of $d/2\pi$ (e.g., about 0.3 nm for charges 2 nm

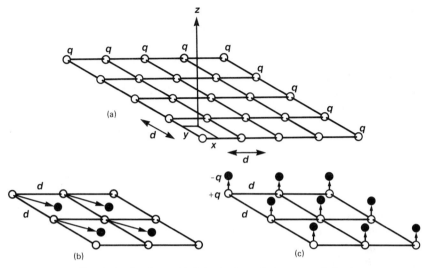

Fig. 12.16. Sections of infinite lattices of charges and dipoles.

apart). Thus, at $z = \frac{1}{2}d$ the electric field is at most 17% different from that of the smeared-out field, while at $z = d$ it has reached 99.3% of the smeared-out value! A similar conclusion is reached for other types of lattices; for example, for a hexagonal lattice where neighbouring ions are separated by a distance d, the mean surface charge density is $\sigma = 2q/\sqrt{3}d^2$ and the decay length is $\sqrt{3}d/4\pi$ which is even smaller than for a square lattice implying that the field decays even faster. It is for these reasons that the smeared-out approximation works so well in considering the electrostatic interactions between charged surfaces.

If charges of opposite sign are now added at the centre of each square, as in Fig. 12.16b, the net surface charge density becomes zero. By superimposing the fields of the positive and negative lattices using Eq. (12.58) it is easy to show that the electric field opposite a positive charge (at $x = 0$, $y = 0$) is

$$E_z = +(4q/\varepsilon\varepsilon_0 d^2)e^{-2\pi z/d} + \cdots, \tag{12.59}$$

while opposite a negative charge (at $x = \frac{1}{2}d$, $y = \frac{1}{2}d$) it is

$$E_z = -(4q/\varepsilon\varepsilon_0 d^2)e^{-2\pi z/d} + \cdots. \tag{12.60}$$

Note that this geometry is equivalent to a dipolar or zwitterionic lattice whose dipoles, of length $d/\sqrt{2}$ and surface density $1/d^2$, are lying *parallel* to the surface.

For dipoles of length l comparable to d arrayed *perpendicular* to the surface, as in Fig. 12.16c, the above two equations become replaced by $E_z \approx \pm (2q/\varepsilon\varepsilon_0 d^2)e^{-2\pi z/d} + \cdots$. This procedure can be readily extended to other lattices including three-dimensional ionic crystals. The end result is always that the field is positive or negative depending on the x, y coordinates and that it decays very rapidly to zero with increasing z.

If a second lattice of vertical dipoles is brought up to the first, the Coulombic interaction pressure between the two dipolar surfaces at a separation D will vary between $\pm(2q^2/\varepsilon\varepsilon_0 d^4)e^{-2\pi D/d}$ depending on whether the approaching dipoles are exactly opposite each other or in register (repulsion) or out of register (attraction). The pressure is anyway very small and in reality, since surface dipoles will not be on a perfect lattice but distributed randomly or moving about (e.g., zwitterionic headgroups on a lipid bilayer surface), the net pressure will average to zero in a first approximation, though a Boltzmann-averaged interaction will yield an overall attractive force.

A similar result is obtained if the dipoles are lying in the plane of the surfaces, as in Fig. 12.16b. This is yet another example where the purely electrostatic interaction between a system of charges or dipoles that are overall electrically neutral produces an attractive force, even though intuitively one might have expected two surfaces with vertical dipoles to always repel each other. In the limit where the surface-bound dipoles are free to rotate in all directions the resulting interaction energy must be the same as the attractive van der Waals–Keesom interaction, which decays as $1/D^4$ (Eq. (11.42)) but is screened if the interaction occurs across electrolyte solution (Section 11.8).

Jönsson and Wennerstrom (1983) pointed out that the total interaction between two dipolar surfaces must also include the image forces of each dipole and its image reflected by the other surface. For two dipolar surfaces of low dielectric constant interacting across water the resulting image force between them can be large and repulsive. They showed that depending on the positional and orientational correlations of the dipoles on each surface the resulting pressure can decay either exponentially or with the inverse fourth power of the separation. Such situations are expected to arise when zwitterionic lipid bilayers and biological membranes ($\varepsilon \approx 2$) interact in water ($\varepsilon \approx 80$), and these purely electrostatic forces have been proposed to be responsible for some of the repulsive short-range 'hydration' forces measured between lipid bilayers in aqueous solutions (cf. Fig. 12.5 and Chapter 18).

PROBLEMS AND DISCUSSION TOPICS

12.1 A glass surface is exposed to water vapour at 96% relative humidity (i.e. $p/p_{sat} = 0.96$). Estimate the equilibrium thickness D of the thin film of water adsorbed on the surface assuming (i) that only electrostatic double-layer forces are operating and that the surface is fully dissociated with a surface charge density of $\sigma = -0.1 \, C \, m^{-2}$, (ii) that the monovalent counterions ($z = 1$) are uniformly distributed throughout the thin water film. With these same assumptions also estimate the repulsive electrostatic pressure between two such planar surfaces immersed in water at a distance $2D$ apart. Is your estimate likely to be too high or too low, and how does it compare with the attractive van der Waals pressure between the surfaces at this separation? Will the van der Waals attraction eventually win out at some smaller, but physically realistic, plate separation?

12.2 Calculate the repulsive pressure between two charged surfaces in pure water where the only ions in the gap are the counterions that have come off from the dissociating surface groups (i.e. no electrolyte present, no bulk reservoir). Assume a surface charge density of one electronic charge per $0.70 \, nm^2$ and $T = 22°C$. Plot your results as pressure against surface separation in the range 0.5–18 nm and compare these with the experimental results of Cowley et al. (1978) where in Fig. 4b on p. 3166 the authors plot their measured values for such a system (Δ points). What conclusions do you arrive at concerning the 'hydration' forces between two pure phosphatidylglycerol bilayers at small separations?

12.3 Show that the double-layer repulsion between two surfaces with low but unequal surface potentials ψ_1 and ψ_2, viz. γ_1 and γ_2, is

$$P = 64kT\rho_\infty \gamma_1 \gamma_2 e^{-\kappa D} \qquad (12.61)$$

which reduces to Eq. (12.46) when $\gamma_1 = \gamma_2$.

12.4 Consider a colloidal dispersion of large spherical particles suspended in a 0.1 M NaCl solution where the Hamaker constant between the particles in the solution is $A = 10^{-20} \, J$, and where experimentally it has been established that for the particle-solution interface the surface potential ψ_0 falls linearly with pH from $\psi_0 = -100 \, mV$ at pH 7, to $\psi_0 = 0$ at pH 5. Assuming that the spheres are quite large, i.e. that $R \gg D$ in the range of interest, calculate the range of pH at which the colloid will become 'unstable', i.e. where rapid coagulation will occur.

12.5 The reason(s) why positively charged divalent counterions are better coagulants or flocculants of negatively charged surfaces or particles than monovalent ions is because of one or more of the following:
   (i) They screen the electrostatic repulsion better.
   (ii) They are more hydrated.
   (iii) They bind more easily and hence lower the surface charge more easily.

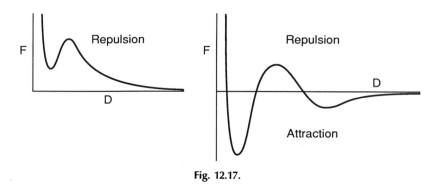

Fig. 12.17.

(iv) The have additional ion-correlation attractive forces.
(v) They can bridge two surfaces by virtue of having two charges.
(vi) They disrupt the water structure more effectively.
(vii) They have a lower kinetic energy.
(viii) They dehydrate surfaces on binding to them.
(ix) They enhance the hydrophobic attraction.

12.6 Two different types of force laws between colloidal particles in an aqueous solution are as shown in Fig. 12.17. Describe how such potentials can arise. If the potentials are assumed to be independent of temperature, sketch how the temperature-composition phase diagrams could look like for these colloidal systems.

12.7 How does the existence of an electric double-layer contribute to the lateral surface pressure of a fully charged surfactant monolayer at a water surface? Consider two limiting cases: (i) the monolayer is totally insoluble so that all the surfactant molecules remain in the monolayer when it is compressed (in this case the lateral pressure varies with area A, and is denoted by Π); and (ii) the monolayer is soluble and so can exchange and equilibrate with the fixed surfactant concentration in the bulk solution (in this case the double-layer contribution to the pressure is independent of area and may be denoted by γ_{el}).
(*Hint*: First obtain an expression for the energy of a single double layer, giving careful thought to the reference state. Refer to Payens (1955), Chan and Mitchell (1983), and Hunter (1989, Chapter 7).)

12.8 When an electric field is applied across an electrolyte solution containing charged particles they are seen to move parallel or antiparallel to the field depending on the sign of their charge. Now, since almost all of the potential drop must occur across the double layer at each electrode surface, there can be no electric field within the conducting electrolyte solution and hence no force on the charged colloidal particles. Why, then, do the particles move?

12.9 Split the double-layer interaction free energy into its enthalpic and entropic components and discuss the implications of your result.

CHAPTER 13

SOLVATION, STRUCTURAL AND HYDRATION FORCES

13.1 NON-DLVO FORCES

When two surfaces or particles approach closer than a few nanometres, continuum theories of attractive van der Waals and repulsive double-layer forces often fail to describe their interaction. This is either because one or both of these continuum theories breaks down at small separations or because other non-DLVO forces come into play. These additional forces can be monotonically repulsive, monotonically attractive or oscillatory, and they can be much stronger than either of the two DLVO forces at small separations.

As we saw in Chapter 7, short-range oscillatory solvation forces arise whenever liquid molecules are induced to order into quasi-discrete layers between two surfaces or within any highly restricted space. Such oscillatory forces have a mainly geometric origin. In addition, surface–solvent interactions can induce positional or orientational order in the adjacent liquid and give rise to a monotonic, rather than oscillatory, solvation force which usually decays exponentially with surface separation. This type of solvation force may be repulsive or attractive, and its range is generally larger than that of oscillatory forces. Additional non-DLVO forces may also arise from the disruption of the liquid hydrogen-bonding network between two surfaces, from electrostatic ion-binding and ion-correlation effects, and from molecular 'bridging' effects.

Solvation forces depend not only on the properties of the intervening medium but also on the chemical and physical properties of the surfaces, for example, whether they are hydrophilic or hydrophobic, whether amorphous or crystalline, smooth or rough, rigid or fluid-like. Such forces can be very strong at short range, and they are therefore particularly important for determining the magnitude of the adhesion between two surfaces or particles in 'contact' (strictly, at their potential energy minimum). We shall start by considering the most general type of solvation force—the oscillatory force arising from the discrete molecular nature of all condensed phases.

13.2   MOLECULAR ORDERING AT SURFACES, INTERFACES AND IN THIN FILMS

The theories of van der Waals and double-layer forces discussed in the previous two chapters are both continuum theories, described in terms of the bulk properties of the intervening solvent such as its refractive index n, dielectric constant ε and density ρ. We have already seen in Chapters 7 and 8 that at small separations, below a few molecular diameters, these values are no longer the same as in the bulk and that the short-distance intermolecular pair potential can be quite different from that expected from continuum theories. In particular we saw that in general the liquid density profiles and interaction potentials in liquids oscillate with distance, with a periodicity close to the molecular size and with a range of a few molecular diameters. These short-distance interactions are usually referred to as *solvation* forces, *structural* forces, or—when the medium is water—*hydration* forces.

To understand how solvation forces arise between two surfaces we must first consider the way solvent molecules order themselves at an isolated surface. We can then consider how this ordering becomes modified in the presence of a second surface, and how this determines the short-range interaction between the two surfaces in the liquid. The solvation (or structuring) of solvent molecules at a surface is in principle no different from that occurring around a small solute molecule, or even around another identical solvent molecule, which—as previously described—is determined primarily by the geometry of molecules and how they can pack around a constraining boundary. (For some mainly theoretical reviews see Nicholson and Parsonage, 1982; Ciccotti *et al.*, 1987; and Evans and Parry, 1990.)

Theoretical work and particularly computer simulations indicate that while liquid density oscillations are not expected to occur at a liquid–vapour or liquid–liquid interface (Fig. 13.1a), a very different situation arises at a solid–liquid interface (Fig. 13.1b). Here, attractive interactions between the wall and liquid molecules and the geometric constraining effect of the 'hard wall' on these molecules force them to order (or structure) into quasi-discrete layers. This layering is reflected in an oscillatory density profile extending several molecular diameters into the liquid (Abraham, 1978; Rao *et al.*, 1979).

The constraining effect of *two* solid surfaces is much more dramatic (Fig. 13.1c). Even in the absence of any attractive wall–liquid interaction, geometric considerations alone dictate that the liquid molecules must reorder themselves so as to be accommodated between the two walls, and the variation of this ordering with separation D gives rise to the solvation force between the two surfaces. For simple spherical molecules between two hard, smooth surfaces the solvation force is usually a decaying oscillatory function of distance. For molecules with asymmetric shapes or whose interaction

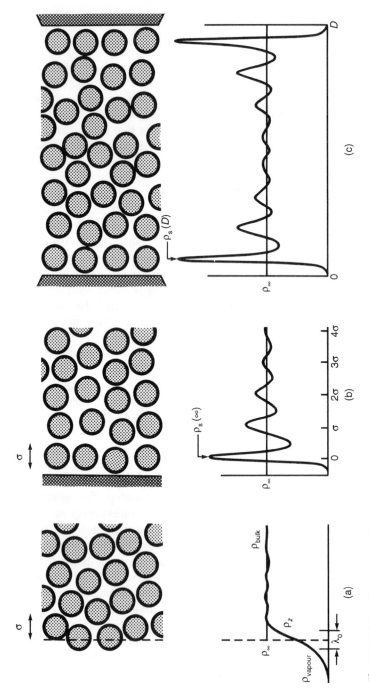

Fig. 13.1. (a) Liquid density profile at a vapour–liquid interface. ρ_∞ is the bulk liquid density and λ_0 is the width or molecular-scale 'roughness' of the interface. (b) Liquid density profile at an isolated solid–liquid interface. $\rho_s(\infty)$ is the 'contact' density at the surface. (c) Liquid density profile between two hard walls a distance D apart. The contact density $\rho_s(D)$ is a function of D as illustrated in Fig. 13.2.

potentials are anisotropic or not pairwise additive, the resulting solvation force may also have a monotonically repulsive or attractive component.

Likewise, if the confining surfaces are themselves not well-ordered but rough or fluid-like, the oscillations will be smoothed out and the resulting solvation force will be monotonic. This occurs, for example, when one or both of the interacting surfaces is a liquid–liquid or liquid–vapour interface where, as illustrated in Fig. 13.1a, little or no structuring is expected. It is not surprising therefore that measurements of the van der Waals forces across thin liquid films of helium and alkanes on solid surfaces are well described by the Lifshitz theory down to film thickness of 1 nm (Section 11.6).

We shall consider the effects of both liquid anisotropy and surface structure on solvation forces after we have first considered the simplest but most general case, viz. that of inert spherical molecules between two smooth 'unstructured' surfaces.

□   ● WORKED EXAMPLE ●

Question: Derive an approximate expression for the molecular-scale thermal roughness of a liquid–vapour interface in terms of the intermolecular bond energy $w(\sigma)$ or surface energy γ, the molecular diameter σ, and the temperature. Estimate the magnitude of this roughness for a van der Waals and a metal liquid–vapour surface, and comment on the different effects these could have on the ordering of the liquid molecules at these surfaces.

Answer: Consider the thermal fluctuations of liquid molecules that causes a certain fraction of them to protrude beyond the surface into the vapour phase (Fig. 13.1a). The additional energy of a molecule that protrudes a small distance z $(z < \sigma)$ may be estimated by multiplying the additional area exposed $\pi\sigma z$ by the surface energy γ (note that the molecular-scale value of γ may not be the same as the macroscopic value; this is discussed later). The *protrusion energy* is therefore proportional to the distance the molecule or group protrudes from the surface. In keeping with our previous notation this may be expressed as

$$\text{Protrusion energy} = (\mu_z^i - \mu_0^i)_{\text{prot}} = \pi\sigma\gamma z = \alpha_p z \qquad (13.1)$$

where α_p is the energy per unit length of a protrusion (in units of J m^{-1}). Proceeding as we did when calculating the density profile of the earth's atmosphere (Section 2.3), we obtain for the density profile of molecular protrusions

$$\rho_z = \rho_\infty e^{-\alpha_p z/kT} = \rho_\infty e^{-\pi\sigma\gamma z/kT} \approx \rho_\infty e^{-\sqrt{3}\pi w(\sigma)z/\sigma kT}, \qquad (13.2)$$

where ρ_∞ ias now the bulk liquid density (Fig. 13.1a) and where we have used $\gamma = \sqrt{3}w(\sigma)/\sigma^2$ from Eq. (11.33). The above shows that the liquid density decays exponentially from the surface according to

$$\rho_z = \rho_\infty e^{-z/\lambda_0},$$

where

$$\lambda_0 = kT/\alpha_p = kT/\pi\sigma\gamma \approx \sigma kT/5w(\sigma) \qquad (13.3)$$

is the *characteristic protrusion decay length*. The above equation cannot apply once $z > \sigma$, since beyond this distance the molecule becomes detached from the surface and is then no longer part of the liquid but becomes part of the vapour (for chain molecules such as alkanes, surfactants and polymers the above equations remain valid out to much larger distances and we shall be considering the consequences of this later).

For small globular molecules we should expect the density to level off at $z \approx \sigma$ to a value corresponding to the saturated vapour pressure. That this is indeed so can be readily checked by putting $z = \sigma$ into Eq. (13.2) which reduces it to the result obtained in Sections 2.1–2.4, viz. $\rho_{vap} = \rho_{bulk}e^{-\mu^i_{liq}/kT} \approx \rho_{bulk}e^{-5w(\sigma)/kT}$.

The protrusion length of Eq. (13.3) is a measure of the molecular-scale dynamic roughness of a surface and is one of a number of contributions to the total 'width' of a surface or interface, often denoted by ξ. Other contributions come from more macroscopic-scale thermal fluctuations such as those arising from capillary waves. For a van der Waals liquid, where typically $\sigma \approx 0.3$ nm and $\gamma \approx 25$ mJ m^{-2} we obtain $\lambda_0 \approx 0.2$ nm.

It is important to appreciate that it is a pure coincidence that the molecular-scale roughness has turned out to be of the same order as the molecular size. For many surfaces it can be much smaller. Thus, for a metal surface, where $\gamma > 300$ mJ m^{-2}, we find $\lambda_0 < 0.02$ nm. However, for metals it is unlikely that γ is so high at the atomic level so that the real value of λ_0 may be larger than 0.02 nm, but still much smaller than the value for a van der Waals liquid. Liquid metal surfaces, by virtue of their high binding energies, are therefore very much smoother than those of other liquids. As a consequence of this, liquid metal molecules (or atoms) behave as if they are packing against a hard wall, as drawn in Fig. 13.1b. It is for this reason that the surfaces of liquid metals are believed to have a layered or 'stratified' structure (Rice, 1987), while van der Waals liquids do not.   □

13.3   ORIGIN OF MAIN TYPE OF SOLVATION FORCE: THE OSCILLATORY FORCE

In Section 12.7 we saw that the repulsive electrostatic double-layer pressure between two charged surfaces separated by a solvent containing the surface counterions is given by the contact value theorem, Eq. (12.18):

$$P(D) = kT[\rho_s(D) - \rho_s(\infty)], \tag{13.4}$$

where ρ_s is the ionic density at each surface. Equation (13.4) also applies to solvation forces (Henderson, 1986; Evans and Parry, 1990) so long as there is no interaction between the walls and liquid molecules, where ρ_s is now the density of liquid molecules at each surface (Fig. 13.1b,c). Thus, a solvation force arises once there is a *change* in the liquid density at the surfaces as they approach each other. For two inert hard walls this is brought about by changes in the molecular packing as D varies, as illustrated in Fig. 13.2(a). Here we see that $\rho_s(D)$ will be high only at surface separations that are multiples of σ but must fall at intermediate separations. At large separations, as $\rho_s(D)$ approaches the value for isolated surfaces $\rho_s(\infty)$, the solvation pressure approaches zero. The resulting variation of the solvation pressure with distance is shown schematically in Fig. 13.2(b). Like the density profile, it is an oscillatory function of distance of periodicity roughly equal to σ and with a range of a few molecular diameters.

In the limit of very small separations, as the last layer of solvent molecules is finally squeezed out, we have $\rho_s(D \to 0) \to 0$. In this limit the solvation pressure approaches a finite value given by

$$P(D \to 0) = -kT\rho_s(\infty) \tag{13.5}$$

which means that the force at contact is negative, i.e., attractive or adhesive. Equations (13.4) and (13.5) are important fundamental equations which crop up in many different systems, and we shall encounter them again when considering other entropic interactions.

Oscillatory forces do not require that there be any attractive liquid–liquid or liquid–wall interaction. All one needs is two hard walls confining molecules whose shapes are not too irregular and that are free to exchange with molecules in the bulk liquid reservoir. In the absence of any attractive forces between the molecules the bulk liquid density may be maintained by an external hydrostatic pressure. In real liquids, attractive intermolecular forces play the role of the external pressure, but the oscillatory forces are much the same.

A number of theoretical studies and computer simulations of various

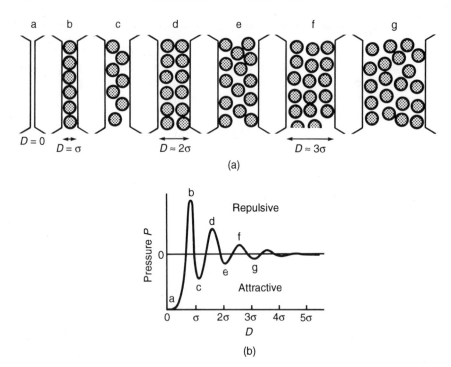

Fig. 13.2. *Top*: Same as in Fig. 13.1c, showing how the molecular ordering changes as the separation D changes. Note that the density of liquid molecules in contact with the surface $\rho_s(D)$ varies between maxima and minima. The molecules in the film are assumed to be free to exchange with those in the bulk. *Bottom*: Corresponding solvation pressure (schematic) as given by Eq. (13.4). The resulting oscillatory solvation force need not be symmetrical about the $P = 0$ axis, but can be superimposed on a monotonic force as shown in Fig. 13.3.

confined liquids, including water, which interact via some form of the Mie potential have invariably led to an oscillatory solvation force at surface separations below a few molecular diameters (van Megen and Snook, 1979, 1981; Snook and van Megen, 1980, 1981; Rickayzen and Richmond, 1985; Kjellander and Marcelja, 1985a,b; Tarazona and Vicente, 1985; Henderson and Lozada-Cassou, 1986; Evans and Parry, 1990). In a first approximation these may be described by an exponentially decaying cos-function of the form

$$P(D) \approx -kT\rho_s(\infty)\cos(2\pi D/\sigma)e^{-D/\sigma} \qquad (13.6)$$

where both the oscillatory period and the characteristic decay length of the envelope are close to σ (Tarazona and Vicente, 1985). By integrating Eq. (13.6) it is a simple matter to show that the solvation force contribution to the interfacial energy of two flat surfaces is

$$\gamma_i = \frac{1}{2} W(0) \approx \frac{kT\rho_s(\infty)\sigma}{8\pi^2}. \tag{13.7}$$

Since $\rho_s(\infty)$ should be approximately equal to the bulk liquid density, we may write $\rho_s(\infty) \approx \sqrt{2}/\sigma^3$, which further simplifies the above to

$$\gamma_i \approx \frac{\sqrt{2}kT}{8\pi^2\sigma^2} \approx \frac{0.02kT}{\sigma^2}. \tag{13.8}$$

Equation (13.8) may be compared with Eq. (11.33) which gives the van der Waals contribution to the interfacial energy as $\gamma_i = \sqrt{3}A/2\pi^2\sigma^2 \approx 0.1A/\sigma^2$. Thus, for Hamaker constants smaller than about $A \approx 0.2\,kT$, i.e., for $A < 10^{-21}$ J, we expect the oscillatory solvation force to dominate the adhesion of two surfaces across a liquid medium (for the opposite case of high A values, see Section 13.4).

It is important to note that once the solvation zones of two surfaces overlap, the mean liquid density ρ in the gap is no longer the same as that of the bulk liquid. And since the van der Waals–Lifshitz interaction depends on both n and ε, which in turn depend on ρ, we must conclude that van der Waals and oscillatory solvation forces are not strictly additive. Indeed, it is more correct to think of the solvation force as *the* van der Waals force at small separations with the molecular properties and density variations of the medium taken into account.

It is also important to appreciate that solvation forces do not arise simply because liquid molecules tend to structure into semi-ordered layers at surfaces. They arise because of the disruption or *change* of this ordering during the approach of a second surface. If there were no change, there would be no solvation force. This is already implicit in Eq. (13.4). The two effects are of course related: the greater the tendency towards structuring at an isolated surface, the greater the solvation force between two such surfaces, but there is a real distinction between the two phenomena that should always be borne in mind.

So far we have only considered the simplest cases of spherical molecules between two smooth surfaces. Such systems are now well understood, both theoretically (see above) and experimentally (see below). Real systems are often much more complex: the liquid molecules are usually non-spherical, interacting via anisotropic orientation-dependent potentials, and the surfaces are not smooth but corrugated or 'structured' at the atomic level. These are the systems we shall be discussing in the rest of this chapter.

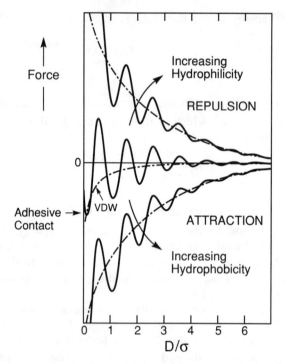

Fig. 13.3. Oscillatory solvation force superimposed on a monotonic solvation force. This type of interaction often arises in aqueous solutions where hydrogen-bond correlation effects can give rise to an additional monotonically decaying 'hydration' force (in addition to any oscillatory and DLVO force). For hydrophilic surfaces the monotonic component is repulsive (upper dashed curve), whereas for hydrophobic surfaces it is attractive (lower dashed curve). For simpler liquids there are no such monotonic components and both theory and experiments show that the oscillations simply decay with distance with the maxima and minima respectively above and below the base line of the van der Waals force (middle curve) or superimposed on the net DLVO interaction (see Figs 13.4–13.6). It is noteworthy that for such liquids the strength of the final adhesion energy or force at molecular contact (the point indicated by the arrow at $D = 0$) is often accurately given by the continuum Lifshitz theory of van der Waals forces, even though this theory fails to describe the force law at finite distances.

Any strongly attractive interaction between a surface and the liquid molecules adjacent to it leads to a denser packing of molecules at the walls (Abraham, 1978; Snook and van Megen, 1979) and thus to higher ρ_s values and a more repulsive but still oscillatory force (Fig. 13.3, upper curve). On the other hand, if the surface–liquid interaction is much weaker than the liquid–liquid interaction the oscillatory force tends to be overall more attractive (Fig. 13.3, lower curve). Such complex force-laws often show both a short-range oscillatory part and long-ranged monotonic part, the latter arising from more complex interactions, for example, involving long-range polarization and H-bond correlation interactions, as is believed to occur in associated liquids such as water (Marcelja and Radic, 1976; Marcelja *et al.*, 1977; Jönsson, 1981; Gruen and Marcelja, 1983; Schiby and Ruckenstein, 1983; Attard and Batchelor, 1988). A theoretical understanding of the interactions of such liquids is not yet available.

To summarize, modern theories of liquids have shown how very complex oscillatory force laws can arise across an assembly of molecules that interact with each other via the simplest possible pair potential. As we shall see, liquids confined within such ultrathin films can take on properties that are both quantitatively and qualitatively different from their bulk properties: the molecules can structure into discrete layers whose properties are 'quantized' with the number of layers, and such films often behave more like a solid or a liquid crystal than a normal liquid, for example, withstanding finite compressive and shear stresses.

13.4 MEASUREMENTS AND PROPERTIES OF SOLVATION FORCES: OSCILLATORY FORCES IN NON-AQUEOUS LIQUIDS

While theoretical work relevant to practical systems is still in its infancy, there is a rapidly growing literature on experimental measurements and other phenomena associated with solvation forces. The simplest systems so far investigated have involved measurements of these forces between molecularly smooth surfaces in organic liquids. Figure 13.4 shows the results obtained by Horn and Israelachvili (1981) for two mica surfaces across an inert liquid of molecular diameter $\sigma \approx 0.9$ nm, together with a plot of the theoretically expected force law. Subsequent measurements of oscillatory forces between different surfaces across both aqueous and non-aqueous liquids have revealed their subtle nature and richness of properties (Christenson and Horn, 1985; Israelachvili, 1987b; Christenson, 1988a). For example, their great sensitivity to the shape and rigidity of the solvent molecules, to the presence of other components, and to surface structure. In particular, the oscillations can be smeared out if the molecules are irregularly shaped, e.g., branched, and therefore unable to pack into ordered layers, or when surfaces are rough even at the ångstrom level. The main features of these forces will now be summarized.

(i) *Inert, spherical, rigid molecules.* In liquids such as CCl_4, benzene, toluene, cyclohexane and OMCTS whose molecules are roughly spherical and fairly rigid, the periodicity of the oscillatory force is equal to the mean molecular diameter σ (within a few percentage points of the diameters obtained from x-ray, gas solubility, and diffusion data).

(ii) *Range of oscillatory forces.* The peak-to-peak amplitudes of the oscillations show a roughly exponential decay with distance with a characteristic decay length of 1.2 to 1.7σ.

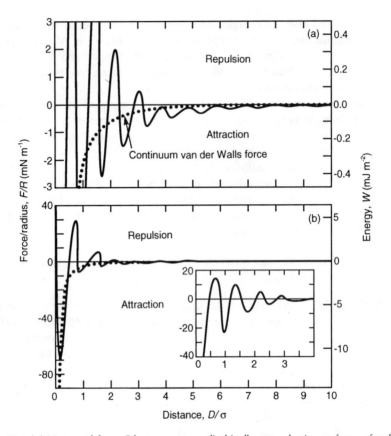

Fig. 13.4. (a) Measured force F between two cylindrically curved mica surfaces of radius $R \approx 1$ cm in octamethylcyclotetrasiloxane (OMCTS), an inert silicone liquid whose nonpolar molecules are quasi-spherical with a mean molecular diameter of $\sigma \approx 0.9$ nm, at 22°C. Dotted line: theoretical continuum van der Waals force computed for this system, given by $F/R = -A/6D^2$ and using a Hamaker constant of $A = 1.35 \times 10^{-20}$ J. The right-hand ordinate gives the corresponding interaction energy per unit area of two flat surfaces according to the Derjaguin approximation: $W = F/2\pi R$. (b) The full experimental force law plotted on a reduced scale. The inset shows a theoretical force law computed for this system based on a molecular theory by Henderson and Lozada-Cassou (1985). (Experimental data from Horn and Israelachvili, 1981.)

(iii) *Magnitude of forces.* The oscillatory force can exceed the van der Waals force at separations below five to 10 molecular diameters, and for simple (non-polymeric) liquids, merges with the continuum van der Waals or DLVO force at larger separations.

(iv) *Effect on adhesion energy.* The depth of the potential well at contact $(D = 0)$ corresponds to an interaction energy that is often surprisingly close to the value expected from the continuum Lifshitz theory of van der Waals forces (Section 11.10). For example, the Lifshitz Hamaker constant for the mica–OMCTS–mica system is about 1.35×10^{-20} J. Using Eqs (11.32)–(11.34) we obtain $W \approx A/12\pi D_0^2 \approx 12$ mJ m^{-2} for the adhesion energy at contact. This may be compared with the value of $W \approx 11$ mJ m^{-2} obtained from the measured adhesion force (Fig. 13.4). Possible reasons for this good agreement were discussed in Section 13.3 and are further investigated in Problem 15.2.

(v) *Temperature dependence.* Oscillatory solvation forces are not strongly temperature dependent and show no change when a liquid is supercooled below its freezing point. They should therefore not be viewed as a surface-induced 'prefreezing' of liquids.

(vi) *Effects of water and other immiscible polar components.* The presence of even trace amounts of water can have a dramatic effect on the solvation force between two hydrophilic surfaces across a non-polar liquid. This is because of the preferential adsorption of water onto such surfaces that disrupts the molecular ordering in the first few layers. This effect usually leads to a shift of the oscillatory force curve to lower, more adhesive, energies (see also Section 15.6).

(vii) *Effects of miscible components (liquid mixtures).* Christenson (1985a) found that the forces between two mica surfaces across a mixture of OMCTS ($\sigma \approx 0.9$ nm) and cyclohexane ($\sigma \approx 0.55$ nm) are essentially the same as that of the dominant component if its volume fraction in the mixture exceeds 90%. However, for a 50–50 mixture the oscillations are not well defined and their range is now *less* than for either of the pure liquids. It appears that a mixture of differently shaped molecules cannot order into coherent layers so that the range of the short-range structure becomes even shorter (note that this is not the case for mixtures of *homologous* molecules, e.g., a mixture of alkanes or short-chained polymers of different lengths but the same molecular width, as discussed in (ix) below).

(viii) *Small flexible (soft) molecules.* Short-chained molecules such as *n*-hexane, and branched chain molecules such as 2,2,4-trimethylpentane, have highly flexible bonds that can rotate freely. Such molecules may be considered as being internally 'liquid-like' and, unlike the more rigid molecules described in (i)–(iii), they have no need to order into discrete layers when confined between two surfaces. Consequently, their short-range structure and oscillatory solvation force does not extend beyond two to four molecules (or some packing dimension of the molecules).

(ix) *Linear chain molecules.* Homologous liquids such as *n*-octane, *n*-tetradecane and *n*-hexadecane exhibit similar oscillatory solvation force laws (Fig. 13.5). For such liquids, the period of the oscillations is about 0.4 nm which corresponds to the molecular width and indicates that the molecular axes are preferentially oriented parallel to the surfaces (Fig. 13.5, inset). Similar results have been obtained with short-chained polymer melts such as polydimethylsiloxanes (Horn and Israelachvili, 1988; Horn *et al.*, 1989b).

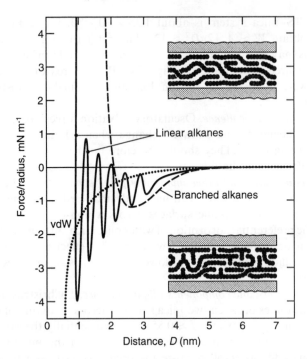

Fig. 13.5. Measured force laws between mica surfaces across straight-chained liquid alkanes such as *n*-tetradecane and *n*-hexadecane (molecular width $\sigma \approx 0.4$ nm), and across the branched alkane (iso-paraffin) 2-methyloctadecane. The dotted line is the theoretical continuum van der Waals interaction. (From Christenson *et al.*, 1987; Gee and Israelachvili, 1990.)

(x) *Non-linear (asymmetric) and branched chain molecules.* Irregularly shaped chain molecules with side groups or branching lack a symmetry axis and so cannot easily order into discrete layers or other ordered structure within a confined space. In such cases the liquid film remains disordered or amorphous and the force law is not oscillatory but monotonic. An example of this is shown in Fig. 13.5 for iso-octadecane where we see how a single methyl side-group on an otherwise linear 18-carbon chain has totally eliminated the oscillations. Similar effects occur with branched polymer melts, such as polybutadienes (Israelachvili and Kott, 1988).

(xi) *Effect of polydispersity.* A small degree of polydispersity appears to have only a small effect on the force law so long as the mixture remains homologous. For example, a polydisperse mixture of *n*-alkanes or a polydisperse polymer melt exhibits similar equilibrium force laws to those of the pure one-component liquids (though the times to reach equilibrium may differ significantly).

(xii) *Effect of molecular polarity (dipole moment and H-bonds)*. The measured oscillatory solvation force laws for polar liquids such as acetone (dipole moment: $u = 2.85D$) are not very different from those of non-polar liquids of similar molecular size and shape. Similarly, in polar and hydrogen-bonding liquids of high dielectric constant such as propylene carbonate ($\varepsilon = 65$, $u = 4.9D$), methanol ($\varepsilon = 33$, $u = 1.7D$), and ethylene glycol ($\varepsilon = 41$, $u = 1.9D$) containing dissolved ions, the force is well described by the continuum DLVO theory at large distances, but at smaller distances the oscillatory solvation force dominates the interaction (Fig. 13.6). It appears, therefore, that dipoles and H-bonds do not have a large effect on the magnitude and range of oscillatory forces (though H-bonds may introduce additional monotonic forces, discussed below).

(xiii) *Effect of surface structure and roughness*. It is now appreciated that the structure of the confining surfaces is just as important as the nature of the liquid for determining the solvation forces. As we have seen, between two surfaces that are completely smooth (or 'unstructured') the liquid molecules will be induced to order into layers, but there will be no lateral ordering within the layers. In other words, there will be positional ordering normal but not parallel to the surfaces. However, if the surfaces have a crystalline (periodic) lattice, this will induce ordering parallel to the surfaces as well (Fig. 13.7), and the oscillatory force will now also depend on the structure of the surface lattices. Further, if the lattices of two opposing structured surfaces are not in register but are at some 'twist angle' relative to each other, or if the two lattices have different dimensions ('mismatched' or 'incommensurate' lattices), the oscillatory force will be further modified as will the adhesion energy (McGuiggan and Israelachvili, 1990).

On the other hand, for surfaces that are *randomly* rough, the oscillatory force becomes smoothed out and disappears altogether, to be replaced by a purely monotonic solvation force (Christenson, 1986). This occurs even if the liquid molecules themselves are perfectly capable of ordering into layers. The situation of *symmetric* liquid molecules confined between *rough* surfaces, is therefore not unlike that of *asymmetric* molecules between *smooth* surfaces, discussed in (x) above.

To summarize, for there to be an oscillatory solvation force, the liquid molecules must be able to be correlated over a reasonably long range. This requires that both the liquid molecules and the surfaces have a high degree of order or symmetry. If either is missing, so will the oscillations. A roughness of only a few ångstroms is often sufficient to eliminate any oscillatory component of a force law (Gee and Israelachvili, 1990).

(xiv) *Effect of surface curvature and geometry*. It is easy to understand how oscillatory forces arise between two flat, plane parallel surfaces (Figs 13.1 and 13.7). Between two curved surfaces, e.g., two spheres, one might imagine

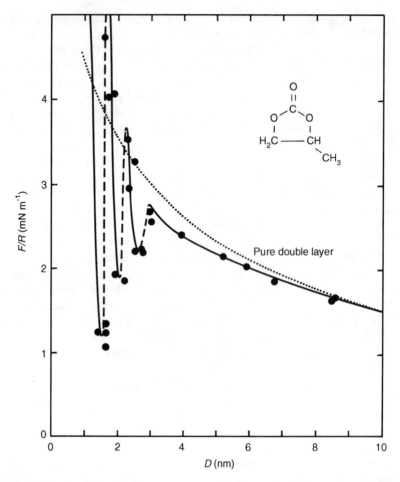

Fig. 13.6. Measured force between two curved mica surfaces in propylene carbonate ($\sigma \approx 0.5$ nm) containing 10^{-4} M electrolyte (tetraethylammonium bromide). Beyond 8 nm and out to 50 nm (not shown) the force law is accurately described by double-layer theory (dotted line). Below 8 nm the van der Waals attraction reduces the net force below the purely double-layer curve, as expected from the DLVO theory, and below 3 nm the force is oscillatory. (From Christenson and Horn, 1983.)

the molecular ordering and oscillatory forces to be smeared out in the same way that they are smeared out between two randomly rough surfaces. However, this is not the case. Ordering can occur so long as the curvature or roughness is itself regular or uniform, i.e., not random. This interesting matter is explored in Problem 13.2.

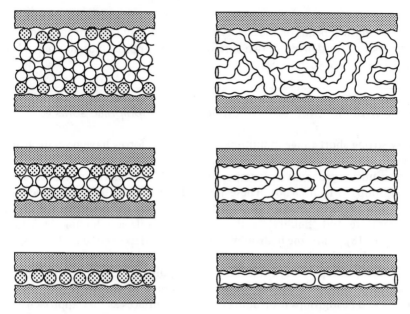

Fig. 13.7. Schematic of how real surfaces, possessing crystalline structure, force the liquid molecules between them to adopt lateral, solid-like ordering commensurate with the atomic corrugations of the surfaces. This ordering is in addition to the layering, which occurs even between smooth unstructured surfaces.

13.5   Solvation forces in aqueous systems: repulsive 'hydration' forces

Certain clays, surfactant soap films, uncharged lipid bilayers and biological membranes swell spontaneously or repel each other in aqueous solutions, and silica dispersions and other colloidal particles sometimes remain 'stable' in very high salt. And yet in all these systems one would expect the surfaces or particles to remain in strong adhesive contact or coagulate in a primary minimum if the only forces operating were DLVO and oscillatory solvation forces.

There are many other aqueous systems where DLVO theory fails and where there is an additional short-range force that is not oscillatory but smoothly varying, i.e., monotonic. Between hydrophilic surfaces this force is exponentially repulsive and is commonly referred to as the *hydration* or *structural* force. The origin and nature of this force has long been controversial, especially in the colloidal and biological literature. Repulsive hydration forces appear to arise whenever water molecules strongly bind to surfaces containing hydrophilic groups, i.e., certain ionic, zwitterionic, or H-bonding groups, and their strength depends on the energy needed to disrupt the hydrogen-bonding network and/or dehydrate two surfaces as they approach each other. There are two types of such forces.

Steric hydration forces between fluid-like amphiphilic surfaces

Very strong short-range monotonically repulsive forces have been measured across soap films composed of various surfactant monolayers as well as across uncharged bilayers composed of lipids with zwitterionic or sugar headgroups (Fig. 13.8a,b). These forces have a range of 1–2 nm and an exponential decay length of 0.1–0.3 nm, both of which increase with the temperature and 'fluidity' of these highly mobile amphiphilic structures (Chapter 18). While the hydrophilicity of such surfaces is due to the presence of strongly hydrophilic groups such as $-N(CH_3)_3^+$, phosphate $-PO_4^-$—, and sugar groups (see Table 8.2), the repulsive force between them is essentially entropic—arising from the overlap of the thermally excited chains and headgroups protruding from these surfaces as they approach each other. Such forces are more akin to the 'steric' forces or 'thermal fluctuation' forces between two polymer-covered or fluid-like interfaces, and they will be considered in Chapters 14 and 18.

Repulsive hydration forces between solid crystalline hydrophilic surfaces

In the case of silica, mica, certain clays and many hydrophilic colloidal particles the hydration forces are believed to arise from strongly H-bonding surface groups, such as hydrated ions or hydroxyl (—OH) groups, which modify the H-bonding network of liquid water adjacent to them. Since this network is quite extensive in range (Stanley and Teixeira, 1980) the resulting interaction force is also of relatively long range.

Repulsive hydration forces were first extensively studied between clay surfaces (van Olphen, 1977). More recently they have been measured in detail between mica and silica surfaces (Figs 13.8c and 13.9) where they have been found to decay exponentially with decay lengths of about 1 nm. Their effective range is about 3–5 nm, which is about twice the range of the oscillatory solvation force in water. Empirically, therefore, the hydration repulsion between two hydrophilic surfaces appears to follow the simple equation

$$W = +W_0 e^{-D/\lambda_0} \tag{13.9}$$

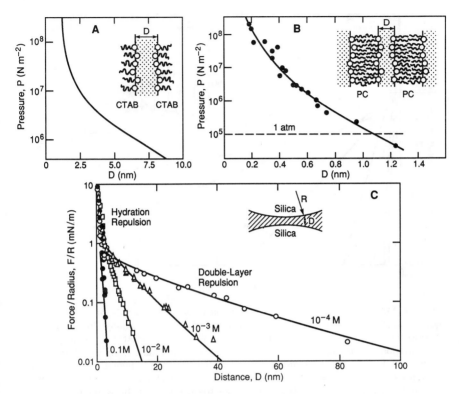

Fig. 13.8. Experimental measurements of short-range exponentially repulsive non-DLVO forces. A. Across a soap film of the cationic surfactant CTAB or HTAB: $C_{16}H_{33}N(CH_3)_3^+ Br^-$. (From Clunie et al., 1967. Reprinted by permission from Nature, Vol. 216, pp. 1203–1204. © 1967 MacMillan Journals Limited.) B. Between fluid bilayers of the uncharged zwitterionic lipid dipalmitoyl-phosphatidylcholine (lecithin or PC) in water. The range of the repulsion is about 1.2 nm, below which it is roughly exponential with a decay length of ~0.14 nm. (From McIntosh and Simon, 1986.) C. Between two silica surfaces in various aqueous NaCl solutions. Here the forces are pure double layer down to ~3 nm below which there is an additional exponentially repulsive hydration force (of decay length 0.5–1.0 nm) instead of the attractive van der Waals force expected from DLVO theory. Similar forces have been measured between glass fibres and other silica surfaces by Rabinovich et al. (1982) and Peschel et al. (1982). (From Horn et al., 1989a.)

where $\lambda_0 \approx 0.6$–1.1 nm for $1:1$ electrolytes (Pashley, 1982), and where W_0 depends on the hydration of the surfaces but is usually below 3–30 mJ m^{-2}— higher W_0 values generally being associated with lower λ_0 values.

In a series of experiments to identify the factors that regulate hydration forces, Pashley (1981a,b, 1982, 1985) and Pashley and Israelachvili (1984) found that the interaction between molecularly smooth mica surfaces in dilute electrolyte solutions obeys the DLVO theory. However, at higher salt

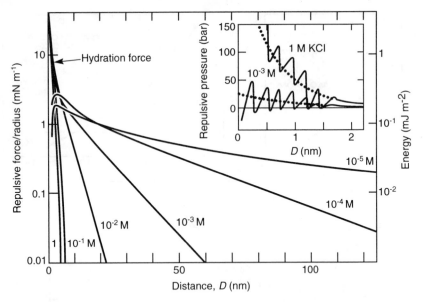

Fig. 13.9. Measured forces between curved mica surfaces in KNO_3 or KCl solutions (qualitatively similar results are obtained in other electrolyte solutions). In 10^{-4} M the force follows the theoretical DLVO force law at all separations. At 10^{-3} M and higher concentrations more cations adsorb (bind) onto the surfaces and bring with them their water of hydration. This gives rise to an additional short-range hydration force below 3–4 nm (see inset and Fig. 13.10). The hydration force is characterized by short-range oscillations of periodicity 0.22–0.26 nm—about the diameter of the water molecule— superimposed on a longer-ranged monotonically repulsive tail of exponential decay length ~1 nm (similar to that observed between glass and silica surfaces, shown in Fig. 13.8C). The right-hand ordinate gives the interaction energy between two flat surfaces according to the Derjaguin approximation. (From Israelachvili and Pashley, 1982a; Pashley, 1981a,b.)

concentrations, specific to each electrolyte, hydrated cations bind to the negatively charged surfaces and give rise to a repulsive hydration force (Fig. 13.9). This is believed to be due to the energy needed to dehydrate the bound cations, which presumably retain some of their water of hydration on binding. This conclusion was arrived at after noting that the strength and range of the hydration forces increase with the hydration number of the cations (cf. Table 4.2) in the order $Mg^{2+} > Ca^{2+} > Li^+ \sim Na^+ > K^+ > Cs^+$. In acid solutions, where only protons bind to the surfaces, no hydration forces were observed (presumably because protons penetrate into the mica lattice) and the measured force laws were close to those expected from DLVO theory at all proton concentrations (pH values).

Israelachvili and Pashley (1983) also found that while the hydration force between two mica surfaces is overall repulsive below about 4 nm, it is not

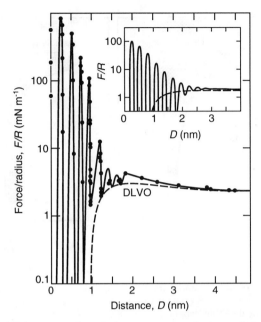

Fig. 13.10. Measured short-range force between two curved mica surfaces of radii $R \approx 1$ cm in 10^{-3} M KCl. The force at distances above 4 nm is given in Fig. 13.9. The dashed line shows the expected DLVO interaction. The values of F/R shown here have been plotted on the assumption of constant surface radius R; however, due to elastic flattening of the glue supporting the mica surfaces, the values of F/R at the force barriers (force maxima) are overestimates. Inset: Theoretical computation for the same system (Henderson and Lozada Cassou, 1986). (Experimental results from Israelachvili and Pashley, 1983.)

always monotonic below about 1.5 nm but exhibits oscillations of mean periodicity 0.25 ± 0.03 nm, roughly equal to the diameter of the water molecule. This is shown in Fig. 13.10. In particular, they observed that the first three minima at $D \approx 0$, 0.28 and 0.56 nm occur at negative energies, a result that rationalizes observations on clay systems. Clay platelets such as motomorillonite repel each other increasingly strongly down to separations of ~ 2 nm (Viani *et al.*, 1984). However, the platelets can also stack into stable aggregates with water interlayers of typical thickness 0.25 and 0.55 nm between them (Del Pennino *et al.*, 1981). In chemistry we would refer to such structures as stable hydrates of fixed stoichiometry, while in physics we may think of them as experiencing an oscillatory force.

These experiments showed that hydration forces can be modified or regulated by exchanging ions of different hydrations on surfaces. Such regulated hydration effects also occur with other surfaces and systems (Israelachvili, 1985). For example, the force between two mercury surfaces

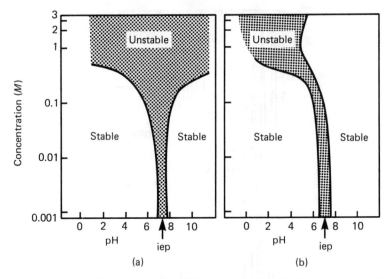

Fig. 13.11. Domains of stability and instability (coagulation) of a dispersion of amphoteric 'latex' particles whose surfaces contain COO⁻ and NH₃⁺ groups. (a) In CsNO₃ solutions the behaviour is 'ideal'—that expected from DLVO theory. (b) In KNO₃, and to an even greater extent in LiNO₃, it is not—the expected coagulation does not occur in high salt above the isoelectric point (iep = 7.2). (From Healy *et al.*, 1978.)

obeys DLVO theory in various electrolyte solutions, but the surfaces fail to coalesce once ionic species (e.g., I^-) specifically bind to the surfaces at higher concentrations (Usui *et al.*, 1967; Usui and Yamasaki, 1969).

Regarding hydration regulation in colloidal dispersions, the effects of different electrolytes on the hydration forces between colloidal particles can determine whether they will coagulate or not. Figure 13.11 shows the experimentally determined regions of stability and instability of amphoteric polystyrene latex particles, composed of —COO⁻— and —NH₃⁺ groups, in CsNO₃ and KNO₃ solutions. In concentrated CsNO₃ solutions, Cs⁺ ions bind to the surfaces at high pH (where there is no competition from protons). But the hydration forces are weak because Cs⁺ is weakly hydrated, and the stability/instability regions are indeed explicable by DLVO theory. However, in concentrated KNO₃ and more so in LiNO₃ solutions the particles remain stable, even at the isoelectric point, because of the stronger hydration forces arising from the binding of the more hydrated K⁺ and Li⁺ ions.

Similar effects occur in other colloidal systems. Thus, the effectiveness of monovalent cations as coagulants usually decreases according to the so-called 'lyotropic series', viz., $Cs^+ > K^+ > Na^+ > Li^+$ for monovalent ions, and $Ca^{2+} > Mg^{2+}$ for divalent ions (Hiemenz, 1977), consistent with the model experiments on mica and exactly as expected from the increasing hydration of these ions. Computer simulations of the interactions between alkali metal and chloride ions in water (Pettitt and Rossky, 1986) also show that the depth of the primary potential minimum becomes *deeper* on going from Li^+Cl^- to Na^+Cl^- to K^+Cl^- (rather than the opposite as would be expected from the unhydrated bare ion radii of these ions). Likewise, between similar ions, the range of the repulsion in water appears to *increase* on going from K^+-K^+ to Na^+-Na^+ to Li^+-Li^+.

Repulsive hydration forces are important in many phenomena. For example, they are believed to be responsible for the unexpectedly thick wetting films of water on silica, discussed in Section 12.9. The fact that these forces can often be regulated by ion exchange makes them useful for controlling various technological processes such as clay swelling (Quirk, 1968), ceramic processing and rheology (Velamakanni et al., 1990), and colloidal and bubble coalescence (Frens and Overbeek, 1972; Healy et al., 1978; Elimelech, 1990; Lessard and Zieminski, 1971).

From the previous discussions we can infer that the hydration force is not of a simple nature, and it may be fair to say that it is probably the most important yet the least understood of all the forces in liquids. Clearly, the almost unique properties of water are implicated, but the nature of the surfaces is equally important. Some particle surfaces can have their hydrophilicity regulated, for example, by ion exchange. These can be coagulated by simply adding more salt or changing the pH. Other surfaces appear to be intrinsically hydrophilic (e.g., silica and lecithin) and cannot be coagulated by changing the ionic conditions. However, such surfaces can often be rendered hydrophobic by chemically modifying their surface groups. For example, by heating silica to above 600°C, two surface —OH groups release a water molecule and combine to form a hydrophobic —O— group, whence the exponentially repulsive hydration force changes into an exponentially attractive hydrophobic force of similar decay length (see next section).

How do exponentially decaying hydration forces arise? Theoretical work and computer simulations by Jönsson (1981), Christou et al. (1981), Kjellander and Marcelja (1985a,b), and Henderson and Lozada-Cassou (1986) suggest that the solvation forces in water should be purely oscillatory, while other theoretical studies (Marcelja and Radic, 1976; Marcelja et al., 1977; Gruen and Marcelja, 1983; Jönsson and Wennerström, 1983; Schiby and Ruckenstein, 1983; Luzar et al., 1987; Attard and Batchelor, 1988) suggest a monotonic exponential repulsion or attraction, possibly superimposed on an oscillatory profile. The latter is consistent with experimental findings, as shown in the inset to Fig. 13.9 where it appears that the oscillatory force is simply additive with the monotonic hydration and DLVO forces, suggesting that these arise from essentially different mechanisms.

It is probable that the short-range hydration force between all smooth, rigid or crystalline surfaces (e.g., mineral surfaces such as mica) has an oscillatory component. This may or may not be superimposed on a monotonically repulsive profile due to image interactions (Jönsson and Wennerström, 1983) and/or to structural or H-bonding polarization interactions (Marcelja and Radic, 1976; Gruen and Marcelja, 1983).

It also appears that between rough surfaces (e.g., of silica) and especially between fluid surfaces (e.g., of lipid bilayers), the oscillations are smeared out and that any longer-ranged structural force collapses. What remains is a much shorter-ranged monotonic hydration repulsion, as has so far always been observed with such systems (Fig. 13.8). This would be in addition to any monotonically repulsive 'steric' interaction which arises between fluid-like surfaces, as discussed in Chapters 14 and 18.

It is clear that the situation in water is governed by much more than the simple molecular packing effects that seem to dominate the interactions in non-aqueous liquids. In Part III we continue to consider the role of hydration, hydrophobic and steric forces in the interactions between micelles, bilayers, biological membranes and other macromolecular structures.

13.6 SOLVATION FORCES IN AQUEOUS SYSTEMS: ATTRACTIVE 'HYDROPHOBIC' FORCES

A hydrophobic surface is one that is inert to water in the sense that it cannot bind to water molecules via ionic or hydrogen bonds. In Chapter 8 we saw that the orientation of water molecules in contact with a hydrophobic molecule is entropically unfavourable and that two such molecules therefore attract each other, since by coming together the entropically unfavoured water is ejected into the bulk thereby reducing the total free energy of the system. Similar effects occur between hydrophobic surfaces in water. Hydrocarbons and fluorocarbons are hydrophobic, as is air, and the strongly attractive hydrophobic force between such surfaces has many important manifestations and consequences, some of which are illustrated in Fig. 13.12.

In recent years there has been a steady accumulation of experimental data on the force laws between various hydrophobic surfaces in aqueous solutions. These surfaces include mica surfaces coated with surfactant monolayers exposing hydrocarbon or fluorocarbon groups (Israelachvili and Pashley, 1982b; Pashley et al., 1985; Claesson et al., 1986a; Claesson and Christenson, 1988; Parker et al., 1989b; Kurihara et al., 1990), and silica and mica surfaces that had been rendered hydrophobic by chemical methylation or plasma etching (Rabinovich and Derjaguin, 1988; Parker et al., 1989b; Christenson

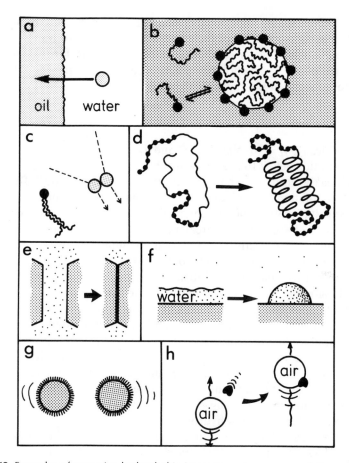

Fig. 13.12. Examples of attractive hydrophobic interactions in aqueous solutions. (a) Low solubility/immiscibility (see Chapters 15 and 16); (b) micellization (see Chapters 16 and 17); (c) dimerization and association of hydrocarbon chains; (d) protein folding (Fischer and Schmid, 1990); (e) strong adhesion; (f) non-wetting of water on hydrophobic surfaces; (g) rapid coagulation of hydrophobic or surfactant-coated surfaces; (h) hydrophobic particle attachment to rising air bubbles (basic mechanism of 'froth flotation' used to separate hydrophobic and hydrophilic particles).

et al., 1990). These studies have found that the hydrophobic force law between two macroscopic surfaces is of surprisingly long range, decaying exponentially with a characteristic decay length of 1–2 nm in the range 0–10 nm, and then more gradually farther out. The hydrophobic force can be far stronger than the van der Waals attraction, especially between hydrocarbon surfaces for which the Hamaker constant is quite small.

As might be expected, the magnitude of the hydrophobic attraction falls with the decreasing hydrophobicity (increasing hydrophilicity) of surfaces as judged from measurements of their interfacial energy with water γ_i or from the contact angle of water on them (Chapter 15). Thus, for two surfaces in water, their purely hydrophobic interaction energy (i.e., ignoring DLVO and oscillatory forces) in the range 0–10 nm is given by

$$W = -2\gamma_i e^{-D/\lambda_0}, \tag{13.10}$$

where, typically, $\gamma_i = 10$–$50\ \text{mJ m}^{-2}$, and $\lambda_0 = 1$–$2\ \text{nm}$.

At separation below 10 nm the hydrophobic force appears to be insensitive or only weakly sensitive to changes in the type and concentration of electrolyte ions in the solution. The absence of a 'screening' effect by ions attests to the non-electrostatic origin of this interaction. In contrast, some experiments have shown that at separations greater than 10 nm the attraction does depend on the intervening electrolyte, and that in dilute solutions, or solutions containing divalent ions, it can continue to exceed the van der Waals attraction out to separations of 80 nm (Christenson et al., 1989, 1990).

The long-range nature of the hydrophobic interaction has a number of important consequences (Fig. 13.12). It accounts for the rapid coagulation of hydrophobic particles in water (Xu and Yoon, 1990) and may also account for the rapid folding of proteins. It also explains the ease with which water films rupture on hydrophobic surfaces (Tchaliovska et al., 1990). In this the van der Waals force across the water film is repulsive (Section 11.6) and therefore favours wetting, but this is more than offset by the attractive hydrophobic interaction acting between the two hydrophobic phases across water (remember that air is hydrophobic!).

Like the repulsive hydration force, the origin of the hydrophobic force is still unknown. Luzar et al. (1987) carried out a Monte Carlo simulation of the interaction between two hydrophobic surfaces across water at separations below 1.5 nm. They obtained a decaying oscillatory force superimposed on a monotonically attractive curve, i.e., similar but of opposite sign to the upper curve in the inset to Fig. 13.9. Figure 13.3 illustrates the way the hydration force between two solid surfaces in water is believed to vary as the surfaces change progressively from being strongly hydrophilic (top curve) to strongly hydrophobic (bottom curve). At each extreme the hydration force can be far stronger than any of the DLVO forces, thereby dominating the short-range interaction and adhesion. However, for surfaces that have no net hydrophilic or hydrophobic character the force law is of the DLVO type, with oscillations superimposed at small separations, and with the adhesion energy at contact often well described by the Lifshitz theory (Fig. 13.3, arrow).

Between amphiphilic surfaces (e.g., of lipid bilayers) the attraction appears to be well described by the van der Waals force, but only when the hydrophilic

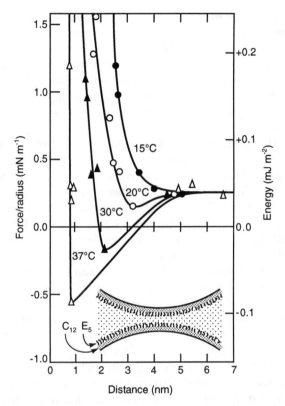

Fig. 13.13. Example of increasing attraction with temperature characteristic of hydrophobic forces: measured forces between monolayers of the nonionic surfactant $C_{12}E_5$ (see Table 16.1) in water whose hydrocarbon/ethylene-oxide interface is believed to become increasingly more hydrophobic at higher temperatures. (From Claesson *et al.*, 1986b.)

headgroups completely screen the hydrophobic hydrocarbon chains from the water phase (Marra and Israelachvili, 1985; Marra, 1985, 1986a). However, the attraction becomes progressively more attractive and longer ranged, i.e., more hydrophobic, as these elastic bilayers are stressed to expose more of their hydrophobic groups to the water (Helm *et al.*, 1989). Some headgroups become more hydrophobic at higher temperatures (see Chapter 8). In such cases the attraction also increases with temperature, as shown in Fig. 13.13. We shall return to consider other aspects of the complex interactions of amphiphilic molecules and surfaces in Chapter 14 and in Part III.

As already mentioned, the origins of both the monotonically repulsive and attractive hydration forces are as yet unclear. Their long range makes it difficult to carry out Monte Carlo or Molecular Dynamics simulations because—unlike the case with the relatively short-range oscillatory solvation forces—these much longer ranged forces would require many hundreds or thousands of water molecules for a successful simulation. Since both these forces have been found to be exponential with similar decay lengths, the repulsive force may simply be a negative form of the attractive force, with the sign of the interaction merely depending on how the water molecules or H-bonds are oriented at each surface.

It is likely that both the orientation and density of water molecules at a surface are important in determining the overall hydration force. As we have seen in Section 13.3 simple theoretical considerations show that a repulsive force arises when the density of water between the two surfaces increases as they approach each other, while an attractive interaction arises if the density falls, i.e., if the region between the two surfaces becomes depleted of water. For sufficiently hydrophobic surfaces the latter should result in the spontaneous nucleation of a vapour phase, or cavity, between the two surfaces, an effect that has actually been observed (Christenson and Claesson, 1988).

It is questionable whether the hydration or hydrophobic force should be viewed as an ordinary type of solvation or structural force—simply reflecting the packing of the water molecules. It is important to note that for any given positional arrangement of water molecules, whether in the liquid or solid state, there is an almost infinite variety of ways the H-bonds can be interconnected over three-dimensional space while satisfying the 'Bernal–Fowler' rules requiring two donors and two acceptors per water molecule (Hobbs, 1974; Stanley and Teixeira, 1980). In other words, the 'H-bonding structure' is actually quite distinct from the 'molecular structure'. It is the energy (or entropy) associated with the H-bonding network and proton-hopping defects, which extends over a much larger region of space than the molecular correlations, that is probably at the root of the long-range solvation interactions of water.

PROBLEMS AND DISCUSSION TOPICS

13.1 From a consideration of the molecular size contribution to the solvation interaction between two surfaces, estimate the adhesive force per unit area (i.e., the pressure) between two flat surfaces in an inert liquid whose molecules occupy a volume of 40 Å^3. Compare this value with what would be expected from the van der Waals force, assuming a typical Hamaker constant of 10^{-20} J.

13.2 Starting from the approximate expression for the decaying oscillatory force (or pressure) between two flat but structured surfaces, Eq. (13.6), obtain an expression for the force law between a large sphere of radius R and a flat surface. What does your answer indicate concerning the 'smearing out' of oscillations between two curved solid surfaces? Will the same apply to (i) two solid surfaces that are randomly rough, and (ii) two fluid-like surfaces that are dynamically rough at any instant but smooth when averaged over time?

13.3 How do you envisage that the last layer of molecules are removed (squeezed out) between two *real*, i.e., structured and elastically deformable, surfaces as they come into final molecular contact with no solvent molecules between them? Make atomic-scale drawings of all the crucial intermediate stages to illustrate your model.

13.4 A liquid droplet of a thermotropic liquid crystal in the nematic phase is placed on a solid surface. The molecules of the nematic can be considered as short rigid rods (cigar-shaped) aligned in the same direction but having no long-range positional order. Discuss the factors that determine whether the molecules align parallel or perpendicular to the surface and how this affects the solvation force between two surfaces across the nematic liquid.

13.5 A colloidal dispersion is stable (i.e., the particles remain dispersed) when the solvent is pure water. On addition of a small amount of a certain electrolyte it becomes unstable (i.e., the particles coagulate). On progressive addition of the same electrolyte it becomes stable again, then unstable, then finally stable again at very high electrolyte concentration. Explain this phenomenon in qualitative terms, and suggest what is the likely nature of the surface of the colloidal particles and the type of electrolyte used.

CHAPTER 14

STERIC AND FLUCTUATION FORCES

14.1 DIFFUSE INTERFACES

So far, we have assumed that all our interacting surfaces are smooth and rigid, possessing sharp well-defined boundaries. There are many instances where this is not the case, where interfaces are spatially diffuse, and where the forces between them depend on how their diffuse boundaries overlap. By a diffuse surface or interface we do not mean that it is simply 'rough', but rather that it has thermally mobile surface groups, i.e., that it is dynamically rough, not statically rough. There are two common types of such diffuse interfaces.

First, there is the interface that is inherently mobile or fluid-like, as occurs at liquid–liquid, liquid–vapour and some amphiphile–water interfaces. We have already seen in Section 13.1 (Fig. 13.1a) that a simple liquid–vapour surface has molecular-scale thermal fluctuations or protrusions. Even though the scale of these fluctuations may be no more than a few ångstroms, this is sufficient to affect the molecular structure of the surfaces significantly. As we shall see, different types of fluctuations can arise depending on the shapes of the molecules and their specific interactions at an interface. In some cases the resulting protrusions can have quite large amplitudes. When two such surfaces or interfaces approach each other their protrusions become increasingly confined into a smaller region of space and, in the absence of any other interaction, a repulsive force arises associated with the unfavourable entropy of this confinement. Such forces are essentially entropic or osmotic in origin and are referred to as 'thermal fluctuation', 'entropically driven' or 'protrusion' forces.

The second type of a thermally diffuse interface occurs when chain molecules, attached at some point to a surface, dangle out into the solution where they are thermally mobile (like seaweed on the sea floor). On approach of another surface the entropy of confining these dangling chains again results in a repulsive entropic force which, for overlapping polymer molecules, is known as the 'steric' or 'overlap' repulsion.

For both of the above examples, complex molecular rearrangements and other interactions can lead to quite complex interaction potentials. For example, the force may be attractive before it becomes repulsive. We shall start by considering the second of the above cases—that associated with the interactions of polymer-covered surfaces.

14.2 POLYMERS AT SURFACES

A polymer is a macromolecule composed of many monomer units or *segments*. If all the monomer units are the same, it is called a *homopolymer*, if different, a *copolymer*. Proteins are copolymers of amino acids. Some common polymers are listed in Table 14.1.

The molecular weight M_0 of a monomer unit is typically between 50 and 100, while the total molecular weight $M = nM_0$ can range from 1000 to above 10^6. When in solution, a polymer chain can adopt a number of configurations depending on the net segment–segment forces in the liquid. If these are weak the polymer assumes the shape of an 'unperturbed' random coil, shown schematically in Fig. 14.1a. An important length scale is the root mean square radius of the polymer coil in solution. For an unperturbed coil this is known as the *unperturbed radius of gyration*, R_g (Flory, 1953, 1969, Ch. 1) and is given by

$$R_g = \frac{l\sqrt{n}}{\sqrt{6}} = \frac{l\sqrt{M/M_0}}{\sqrt{6}}, \qquad (14.1)$$

where n is the number of segments and l the effective segment length. As an example, if $l = 1.0$ nm and the segment molecular weight is $M_0 = 200$, then for a polymer of $M = 10^6$ we obtain $R_g \approx 29$ nm. However, the real volume of the chain is only a small fraction of the volume encompassed by R_g. Thus, if the segment width is about the same as the segment length l, then the molecular volume is $\pi(l/2)^2 nl \approx nl^3$, while the volume encompassed by R_g is $\frac{4}{3}\pi R_g^3 \approx 0.3n^{3/2}l^3$. The ratio of these volumes is $\sim\sqrt{n/10}$. Thus, for a polymer with $n = 1000$ segments, only about 1% of the random coil volume is actually occupied.

Equation (14.1) is valid so long as the solvent is 'ideal' for the polymer, i.e., so long as there are no interactions—either attractive, repulsive or excluded volume—between the segments in the solvent. In real (non-ideal) solvents the effective size of a coil can be larger or smaller than the unperturbed radius R_g, and is sometimes referred to as the *Flory radius*, R_F, where

TABLE 14.1 Some common polymers and polymer groups

Polymer	Characteristic linkage or monomer unit	Uses
Polymer families		
Polypeptides	—NH-CO— (amino acids)	Proteins, wool, silk, steric stabilizers, colloidal additives
Polyvinyls	—CH$_2$-CH(X)—	Plastics
Polyesters	—CO-O—	Clothing, containers
Polysiloxanes, silicones	—Si-O—	Lubricants, rubbers, paints
Polyamides	—NH-CO—	'Nylon', fabrics, auto parts
Polyurethanes	—NH-CO-O—	Adhesives, flexible furniture
Cellulose	—C-O—	Paper, photographic film
Polycarbonates	—O-CO-O—	Optical equipment, CDs
Fluoropolymers	H replaced by F	Non-stick surfaces, lubricants
Common hydrophilic polymers[a]		
Polyethylene oxide (PEO, PEG)	—CH$_2$-O-CH$_2$—	Detergents, cosmetics
Polyacrylamide (PAA)	—CH$_2$-CH(CONH$_2$)—	Plastics, textiles, diapers
Polyvinyl alcohol (PVA)	—CH$_2$-CH(OH)—	Fibres, adhesives, textiles
Common hydrophobic polymers[a]		
Polyethylene (PE)	—CH$_2$-CH$_2$—	Coatings, containers, films
Polystyrene (PS)	—CH$_2$-CH(C$_6$H$_5$)—	Packaging, housewares
Polybutadiene (PBD)	—CH$_2$=CH-CH=CH$_2$—	Latex paints, rubbers, tyres
Polydimethylsiloxane (PDMS)	—Si(CH$_3$)$_2$-O—	Silicone oil, lubricants
Polypropylene (PP)	—CH$_2$-CH(CH$_3$)—	Carpets, bottles, wrap films
Polymethylmethacrylate (PMMA)	—CH$_2$-CCH$_3$(CO$_2$CH$_3$)—	Transparent windows, 'plexiglass', 'perspex'
Polytetrafluoroethylene (PTFE)	—CF$_2$-CF$_2$—	'Teflon', inert, non-wetting, low adhesion–low friction surfaces, lubricants
Polyvinyl chloride (PVC)	—CH$_2$-CH(Cl)—	Plastic sheet, insulation, pipes

[a] Hydrophilic polymers are soluble in aqueous solutions, whereas hydrophobic polymers are soluble in organic solvents such as liquid hydrocarbons. In spite of their high molecular weight some polymers such as PBD and PDMS are liquid at room temperature. These are known as *polymer melts.*

$R_F = \alpha R_g$. The *intramolecular expansion factor* α is unity in an ideal solvent (Flory, 1969, Ch. 2). In a 'good' solvent there is a repulsion between the segments, α exceeds unity, the coil swells and becomes more 'expanded', and its Flory radius is given by

$$R_F \approx ln^{3/5}. \tag{14.2}$$

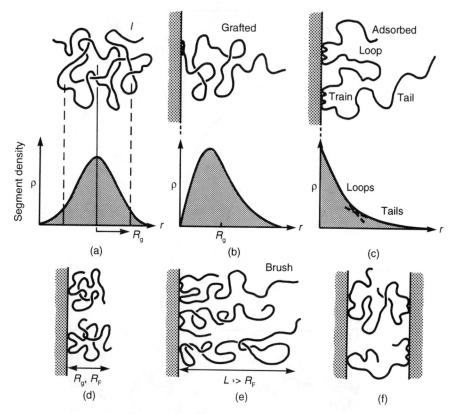

Fig. 14.1. (a) Polymer in solution; (b) chemisorbed (end-grafted) copolymer; (c) physisorbed homopolymer; (d) adsorption at low surface coverage with no nearest neighbour overlap ('mushrooms'); (e) adsorption at high coverage ('brush'); (f) bridging.

Polymers are completely soluble (miscible) in good solvents. In a 'poor' solvent the segments attract each other, α is less than unity, and the coil shrinks. If the segment–segment attractions are very strong, due to intra-ionic, van der Waals or hydrophobic interactions, the coil loses all semblance of randomness and collapses or 'folds' into a compact structure (e.g., proteins and DNA).

A poor solvent can often be made into a good one by adding certain solutes or by raising (or lowering) the temperature above (or below) some critical value known as the *theta temperature* (T_θ or θ) at which point $\alpha = 1$. A solvent at $T = T_\theta$, is known as an *ideal solvent* or a *theta solvent* (θ-solvent).

Polymers can adsorb avidly on surfaces, often reaching saturation adsorption at very low concentrations of a few parts per million in solution. If the volume of an adsorbed coil is the same as its volume in solution (Fig. 14.1d), its

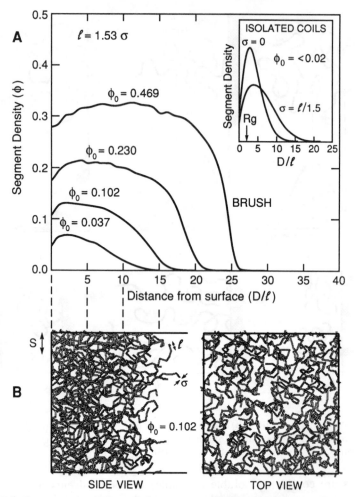

Fig. 14.2. Computer simulations of the structure of end-grafted polymer layers exposed to a theta solvent showing the transition from a low coverage layer to a brush layer with increasing grafting density, ϕ_0. Each polymer has $n = 30$ segments of diameter σ and length $l = 1.53\sigma$. From Eq. (14.1) we therefore obtain $R_g = 2.2l$. (A) Segment density profiles $\phi(D)$ for four different grafting densities, ϕ_0, defined by the area fraction occupied by the grafted end-segments: $\phi_0 = n\pi(\sigma/2)^2$, where n is the number of grafting sites per unit area (the plotted segment densities do not include the density of the grafted end segments themselves). Note that the grafted coils begin to overlap with each other on each surface once $n\pi R_g^2 > 1$, i.e., for $\phi_0 > 0.02$. *Inset*: Density profile of an isolated coil (effectively for $\phi_0 < 0.02$) showing how excluded volume effects (finite σ) blows it up. (B) Side and top views of coils for $\phi_0 = 0.102$, drawn on the same scale as above. (Computed by Young-Hwa Kim at Corporate Research Laboratories, 3M Corporation, St Paul, Minnesota (1989) using a grand canonical MC simulation. Similar results have been obtained by Murat and Grest, 1989.)

radius will be R_g (or R_F) and the surface area covered per coil is approximately $\pi R_g^2 \approx \frac{1}{2}nl^2$. If again the segment width is assumed to be close to l, the total projected area of the *fully extended* coil is nl^2. Thus, as a rough guide, the amount of polymer adsorbed at full surface coverage but without overlap of the unperturbed coils (Fig. 14.1d) is quantitatively similar to that which would occur if all the coils were to lie flat and close-packed on the surface.

In practice the situation is far more complex and often much more polymer is adsorbed resulting in a layer much thicker than R_g (Fig. 14.1e). Various types of adsorptions are possible, depending on the bulk polymer concentration, on whether the polymer is a homopolymer or a copolymer, and on whether the adsorption is via physical forces (physisorption) or by the 'grafting' or 'anchoring' of specific groups via chemical bonds (chemisorption). Some of these are depicted schematically in Fig. 14.1b–e, as is the segment distribution or density profile expected for the different types of adsorptions. Figure 14.2 shows a computer simulation of the density profiles and segment configurations as we go from low coverage (isolated coils) to high coverage (brush layer) of a 30 segment end-grafted polymer in a θ-solvent.

There are many ways of experimentally studying the structure and dynamics of adsorbed polymer layers. These include Small Angle Neutron Scattering (SANS), neutron and x-ray reflectivity, NMR, ellipsometry, internal reflection spectroscopy and various other scattering, reflectivity and spectroscopic techniques (for reviews see Takahashi and Kawaguchi, 1982; Cosgrove, 1990).

The forces between surfaces with adsorbed polymer layers or across polymer solutions are usually measured using one of the standard force-measuring techniques, such as the SFA technique, described in Section 10.7. For reviews of all aspects of intersurface and interparticle forces in polymer fluids see Patel and Tirrell (1989), and Ploehn and Russel (1990).

Theoretically, polymer adsorption and the interactions between polymer-covered surfaces are examined using either scaling or mean-field theories (de Gennes, 1979, 1981, 1982, 1985, 1987a; Scheutjens and Fleer, 1980, 1985; Cohen-Stuart *et al.*, 1986; Fleer, 1988; Ingersent *et al.*, 1986, 1990) or via computer simulations as shown in Fig. 14.2.

14.3   REPULSIVE 'STERIC' OR 'OVERLAP' FORCES BETWEEN POLYMER-COVERED SURFACES

When two polymer-covered surfaces approach each other they experience a force once the outer segments begin to overlap, i.e., once the separation is below a few R_g. This interaction usually leads to a repulsive osmotic force due to the unfavourable entropy associated with compressing (confining) the chains between the surfaces. In the case of polymers this repulsion is usually referred to as the *steric* or *overlap* repulsion and it plays an important role in many industrial processes. This is because colloidal particles that normally coagulate in a solvent can often be stabilized by adding a small amount of polymer to the dispersion. Such *polymer additives* are known as *protectives against coagulation* and they lead to the *steric stabilization* of a colloid. Both synthetic polymers (Table 14.1) and biopolymers (proteins, gelatin) are commonly used in both non-polar and polar solvents (e.g., in paints, toners, emulsions, cosmetics, pharmaceuticals, processed food, soils, lubricants).

In this section we shall consider the purely repulsive steric force between two polymer-covered surfaces interacting in a liquid. Other polymer-associated interactions that can be attractive or that modify steric interactions will be considered in the following sections.

Theories of steric interactions are complex (Hesselink, 1971; Hesselink *et al.*, 1971; Vrij, 1976; Scheutjens and Fleer, 1982; de Gennes, 1987a). The forces depend on the quantity or coverage of polymer on each surface, on whether the polymer is simply *adsorbed* from solution (a reversible process) or irreversibly *grafted* onto the surfaces, and finally on the *quality* of the solvent. Only for interactions in poor and theta solvents are the theories sufficiently well developed (Ingersent *et al.*, 1986, 1990). Two limiting situations, corresponding to low and high surface coverages, will now be described.

We first consider the simplest case of the repulsive steric interaction between surfaces containing an adsorbed polymer layer where each molecule is grafted at one end to the surface but is otherwise inert (Fig. 14.1b). In the limit of low surface coverage, i.e., where there is no overlap or entanglement of neighbouring chains, then each chain interacts with the opposite surface independently of the other chains. For two such surfaces in a theta solvent, the repulsive energy per unit area is a complex series, but over the distance regime from $D = 8R_g$ down to $D = 2R_g$ it is roughly exponential (see Fig. 1 of Dolan and Edwards, 1974) and is adequately given by

$$W(D) = 2\Gamma kTe^{-D^2/4R_g^2} + \ldots \approx 36\Gamma kTe^{-D/R_g} \qquad (14.3a)$$

or

$$W(D) \approx 36kTe^{-D/R_g} \text{ per molecule} \qquad (14.3b)$$

where Γ is the number of grafted chains per unit area, which is related to the mean distance between attachment points, s, by $\Gamma = 1/s^2$ (Fig. 14.2). Equation (14.3) is valid for low coverages ($s > R_g$) when the layer thickness is roughly equal to R_g, and therefore varies as $M^{0.5}$ in a θ-solvent. In good solvents the coils swell, the layer thickness varies as $M^{0.6}$ rather than $M^{0.5}$ (cf. Eq. (14.2)), and the range of the repulsion is therefore greater than that given by Eq. (14.3).

As we go from low coverage to high coverage the adsorbed or grafted chains are now so close to each other that they are forced to extend away from the surface much farther than R_g. In the case of end-grafted chains, as might be expected intuitively, the thickness of the 'brush' layer L now increases linearly with the length of the polymer molecules, i.e., L is proportional to M rather than to $M^{0.5}$ or $M^{0.6}$ as occurs at low coverage. More generally, for a brush in a θ-solvent its thickness, L, scales according to $L \propto M^\nu \propto n^\nu$, where ν varies from 0.5 to 1 as we go from low coverage ($s > R_g$) to high coverage ($s < R_g$). For a brush in a *good* solvent its thickness has been given by Alexander (1977) as

$$L = \frac{nl^{5/3}}{s^{2/3}} = \Gamma^{1/2} R_F^{5/3}. \tag{14.4}$$

Once two brush-bearing surfaces are closer than $2L$ from each other there is a repulsive pressure between them given by the Alexander–de Gennes theory (de Gennes, 1985, 1987a) as

$$P(D) \approx \frac{kT}{s^3}[(2L/D)^{9/4} - (D/2L)^{3/4}] \text{ for } D < 2L. \tag{14.5}$$

For $D/2L$ in the range 0.2 to 0.9 the above pressure is roughly exponential and is adequately given by

$$P(D) \approx \frac{100kT}{s^3}e^{-\pi D/L} = \frac{100kT}{s}\Gamma e^{-\pi D/L} \tag{14.6a}$$

so that

$$W(D) \approx \frac{100L}{\pi s^3}kTe^{-\pi D/L} = \frac{100L}{\pi s}\Gamma kTe^{-\pi D/L} \tag{14.6b}$$

The first term in Eq. (14.5) comes from the osmotic repulsion between the coils which favours their stretching and so acts to increase D, while the second term comes from the elastic energy of the chains which opposes stretching and so acts to decrease D. A different and more complex expression for the force law has been derived by Milner et al. (1988) which nevertheless predicts a very similar force law to the Alexander–de Gennes equation.

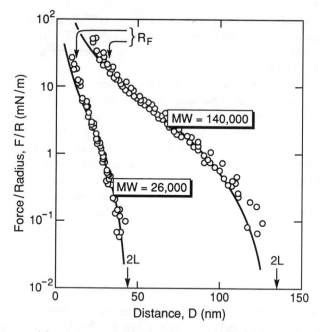

Fig. 14.3. Measured forces between two polystyrene brush layers end-grafted onto mica surfaces in toluene (a good solvent for PS). Left curve: $MW = 26\,000$, $R_F = 12$ nm; right curve: $MW = 140\,000$, $R_F = 32$ nm. Both force curves were reversible on approach and separation. *Solid lines*: theoretical fits using Eq. (14.5) with the following (measured) parameters: spacing between attachment sites: $s = 8.5$ nm, brush thicknesses: $L = 22.5$ nm and 65 nm, respectively. (Adapted from Taunton *et al.*, 1990.)

Figure 14.3 shows results obtained by Taunton *et al.* (1990) for the forces between two end-grafted polystyrene brushes in toluene, together with theoretical fits based on the full Alexander–de Gennes equation. The agreement is surprisingly good. Such forces also contribute to the repulsion between micelles and bilayers in water especially when the amphiphilic molecules at the surfaces have long hydrophilic headgroups (Chapter 18).

The steric forces between surfaces with end-grafted chains are now fairly well understood both theoretically and experimentally. This is because they are reasonably well defined: each molecule is permanently attached to the surface at one end, the coverage is fixed, and the molecules do not interact either with each other or with the two surfaces. Di-block copolymers are usually employed for such purposes, where one of the blocks acts as the anchoring group while the other protrudes into the solvent to form the polymer layer.

But things are usually more complicated than in the example given in

Fig. 14.3. First, many polymers such as homopolymers do not have specific anchoring groups that chemisorb irreversibly to a surface. Instead, each segment can bind to the surfaces via much weaker physical forces (Fig. 14.1c). Such adsorbed—as opposed to grafted—layers are highly dynamic, with individual segments continually attaching and detaching from the surfaces, and where whole molecules slowly exchange with those in the bulk solution.

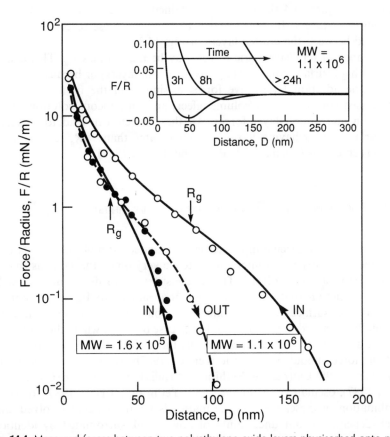

Fig. 14.4. Measured forces between two polyethylene oxide layers physisorbed onto mica from 150 μg/ml PEO solutions in aqueous 0.1 M KNO$_3$ (a good solvent for PEO). *Main figure*: Equilibrium forces at full coverage after ~16 h adsorption time. Left curve: $MW = 160\,000$, $R_g = 32$ nm; right curve: $MW = 1\,100\,000$, $R_g = 86$ nm. Note the hysteresis (irreversibility) on approach and separation for this *physisorbed* polymer, in contrast to the absence of hysteresis with *grafted* chains (Fig. 14.3). *Solid lines*: Theoretical curves based on a modified form of the Alexander–de Gennes equation, Eq. (14.5). *Inset*: Evolution of the forces with the time allowed for the higher MW polymer to adsorb from solution. Note the gradual reduction in the attractive bridging component. (Adapted from Klein and Luckham, 1982, 1984b; Luckham and Klein, 1990.)

The steric forces between such surfaces are more difficult to formulate because neither the amount of adsorbed polymer nor the number of binding sites per molecule remain constant as two surfaces approach each other. Further, bridging can now occur whereby different segments from the same coil can be simultaneously bound to both surfaces. Indeed, the force between two such surfaces at any particular separation can take a long time (many hours) to reach its equilibrium value, and measured force profiles are often very hysteretic. Figure 14.4 shows results obtained for the steric forces between two *adsorbed* layers. It is instructive to compare these forces with those measured between the *grafted* layers of Fig. 14.3.

The range of repulsive steric forces may be many times R_g. This can be due to a high surface coverage (e.g., as for brush layers) or because $R_F \gg R_g$ due to the additional repulsive forces arising from the finite size of the compressed coils (excluded volume effect) and, in aqueous solutions, from hydration and double-layer forces. All these effects can lead to repulsive intersurface forces having a range even greater than $10R_g$, especially in aqueous solutions (Klein, 1988; Patel and Tirrell, 1989).

14.4 FORCES IN PURE POLYMER LIQUIDS (POLYMER MELTS)

Since a polymer molecule in a melt is surrounded by molecules of its own kind one might expect its interactions to be very much the same as that of a polymer in a theta solvent. This is consistent with de Gennes' (1987b) scaling prediction that a repulsive potential similar to Eq. (14.3) applies between two surfaces in a polymer melt, but only if the molecules are terminally anchored to the surfaces. In the case of a melt where the molecules are not attached to the surfaces, various theoretical studies have predicted that the force should be zero (de Gennes, 1987b; Kumar et al., 1988; ten Brinke et al., 1988), attractive (Yethiraj and Hall, 1990), repulsive (Christenson et al., 1987), oscillatory (Madden, 1987; Yethiraj and Hall, 1989), or some combination of these. Clearly, the matter is far from being resolved, and much depends on our uncertainty of how a coil, surrounded by identical coils, interacts with two confining surfaces.

On the experimental side, the forces between mica surfaces across pure polymer melts such as polydimethylsiloxane (PDMS), polybutadiene (PBD) and fluoropolymers have been measured by Horn and Israelachvili (1988), Horn et al. (1989b), Israelachvili and Kott (1988), and Montfort and Hadziioannou (1988). The measured forces generally exhibit oscillations at small distances, with a periodicity equal to the segment width, and a probably non-equilibrium monotonically decaying repulsion farther out extending over a distance of up to $10R_g$. The oscillations are consistent with computer simulations (Yethiraj and Hall, 1989), while the smoothly decaying repulsive tail of these interactions appear to be well described by Eq. (14.3) for anchored chains, probably reflecting the strong binding or effective immobilization of these polymers at the surfaces during the time course of the measurements. In contrast, the forces between inert hydrocarbon surfaces across chain-like hydrocarbon liquids exhibit an attractive rather than repulsive tail (Gee and Israelachvili, 1990) probably reflecting the very much weaker binding of these molecules to the surfaces.

Concerning the short-range oscillatory forces in melts, it has been found that irregularly shaped polymers, e.g., those with large bumpy segments or with randomly branched side groups, show no short-range oscillations, indicative of their inability to order into discrete, well-defined layers (Israelachvili and Kott, 1988; Montfort and Hadziioannou, 1988; Gee and Israelachvili, 1990). Instead, the oscillations are replaced by a smooth monotonic repulsion.

It is worth noting that when polymer molecules are concentrated within an adsorbed surface layer, or confined within a thin film between two surfaces, their molecular relaxation times can be many orders of magnitude higher than in the bulk. In some cases the molecules, which may be liquid in the bulk, freeze into an amorphous glassy state at the surfaces (Van Alsten and Granick, 1990). Consequently, it is unlikely that the measured force laws, even though they often appear to be reversible and reproducible, are ever at true thermodynamic equilibrium.

14.5   ATTRACTIVE 'INTERSEGMENT', 'BRIDGING' AND 'DEPLETION' FORCES

We now turn our attention to attractive polymer-mediated interactions, bearing in mind that the attraction may be operating over only a small distance regime and that the full force law may have both attractive and repulsive regimes.

We have already noted that segments attract each other in a poor solvent. The attraction may be due to van der Waals or some solvation force, and if it is not too strong the main effect on an isolated coil in solution is that the coil radius shrinks below R_g. However, the coil does not totally collapse because the osmotic repulsion still wins out at some smaller radius. Understanding this scenario helps understand the interactions between polymer-covered surfaces in poor solvents.

As two polymer-coated surfaces come together in a poor solvent the attraction between the outermost segments is felt as an initial 'intersegment'

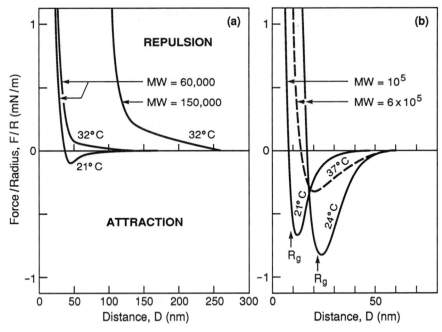

Fig. 14.5. Measured forces between mica surfaces covered with polystyrene below and above the theta temperature, θ, corresponding to poor and good solvent conditions. (a) End-grafted polystyrene brushes in toluene ($\theta = 35°C$). Left curves: $MW = 60\,000$, $R_g = 7$ nm; right curve: $MW = 150\,000$, $R_g = 11$ nm. At $T = 21°C$ (poor solvent conditions) the attraction is due to intersegment forces. At $T = 32°C$ (good solvent conditions) there is no intersegment nor a bridging force and the attraction vanishes. Note that the forces scale roughly as M or R_g^2—a characteristic feature of brush layers. (Adapted from Hadziioannou et al., 1986; Patel, 1986.) (b) Physisorbed polystyrene in cyclohexane ($\theta = 34.5°C$). Left curve: $MW = 100\,000$, $R_g = 8.5$ nm. Right curves: $MW = 600\,000$, $R_g = 21$ nm. At $T = 24°C$ (poor solvent conditions) the attraction is due to both intersegment and bridging forces; at $T = 37°C$ (good solvent conditions) only to bridging forces. The reduced range of the interaction at 37°C is due to the reduced adsorption of physisorbed polymer at the higher temperature. Note that the forces scale roughly as $\sqrt{M}$ or R_g—indicative of low coverage rather than brush layers. (Adapted from Klein, 1980, 1982, 1983; Israelachvili et al., 1984.)

attraction between the surfaces. Closer in, the steric overlap repulsion wins out and the force becomes overall repulsive. Figure 14.5 shows that this is what is observed in the interactions between both grafted and adsorbed polystyrene layers at temperatures below the theta temperature. Such force laws are expected in all poor solvent conditions regardless of whether the polymer is a homopolymer or a copolymer, physisorbed or chemisorbed, or at low or high coverage.

As we go from poor to good solvent conditions, e.g., by raising the temperature, the intersegment attraction vanishes and if no other attractive force is operating the force law becomes purely repulsive at all separations. This is what is observed in the case of grafted polystyrene on mica (Fig. 14.5a). However, in the case of physisorbed polystyrene (Fig. 14.5b), even though the attraction has diminished, it has not completely disappeared above T_θ. This is because of the residual attractive 'bridging' force that is still operative in the latter but not the former case. These attractive bridging forces will now be described.

We have seen that segment–segment forces can be attractive or repulsive depending on the polymer and solvent conditions. Likewise, there are segment–surface forces that can also be attractive or repulsive. The net interaction of one polymer-covered surface with another depends not only on the segment–segment forces, but also on the segment–surface forces and the availability of binding sites for these segments on the opposite surface. If the segment–surface interaction is attractive *and* the opposing surface has unoccupied or exchangeable binding sites, some polymer coils will form bridges between two surfaces (Fig. 14.1f) and give rise to an attractive bridging force (Almog and Klein, 1985; Hu *et al.*, 1989; Ingersent *et al.*, 1986, 1990; Ji *et al.*, 1990). Clearly, any polymer that naturally adsorbs to a surface from solution has the potential to form bridges between two such surfaces. On the other hand, a polymer that is specifically grafted to a surface may or may not be attracted to an opposite surface.

Situations which favour a strong bridging attraction are therefore those where the polymer is attracted to the surfaces, but where the coverage is not too high nor too low. If the coverage is too high, as in the case of a brush, there will be few free binding sites for bridges to form, whereas if it is too low the density of bridges will also be low (Ji *et al.*, 1990). All this is nicely exemplified in Figs 14.4 and 14.5 where the attractive forces above T_θ are due to bridging. In Fig. 14.4 (inset), as the adsorption gradually increased to form brush layers much thicker than R_F the bridging attraction disappeared. In contrast, in Fig. 14.5b (dashed line) the low coverage in that system ensured that the layer thickness never exceeded R_F so that there was a strongly attractive bridging force even above T_θ.

Under suitable conditions attractive bridging forces can be strong and of long range, far exceeding any van der Waals attraction between two surfaces. Both experiment and theory indicate that the bridging force decays roughly exponentially with distance, with a decay length that is close to the value of R_g of the tails and/or loops on the surfaces (Israelachvili *et al.*, 1984; Ji *et al.*, 1990). It has still not been established whether bridging forces occur in a polymer melt, that is, in a system where the 'physisorbed polymer' and the 'solvent' are identical.

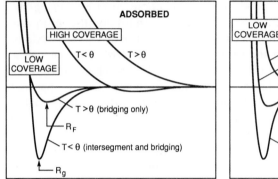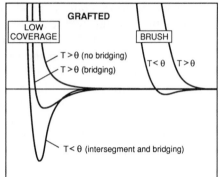

Fig. 14.6. Effects of surface coverage and solvent quality on force profiles of adsorbed and grafted chains (schematic). (Adapted from Klein, 1988; Ingersent et al., 1986, 1990.)

Figure 14.6 summarizes everything that has been discussed so far on polymer mediated forces, especially on how they depend on molecular weight, solvent quality and coverage.

There is a third type of attractive interaction that remains to be mentioned — that associated with polymers that are repelled from surfaces so that there is no adsorption from solution at all. One might expect that under these circumstances there would be no interaction either, but this is not so. The interaction is weak, attractive, and in certain cases can be quite important (Asakura and Oosawa, 1954, 1958; Vrij, 1976; Joanny et al., 1979).

Consider two surfaces in a dilute solution of coils of average radius R_g that do not interact with the surfaces. When the surfaces are closer than R_g the coils will be pushed out from the gap resulting in a reduced polymer concentration between the surfaces. If the bulk polymer concentration is ρ, then applying the contact value theorem, Eq. (13.4), we immediately obtain the attractive 'depletion' force per unit area between the surfaces at contact: $P(D \to 0) = -\rho kT$. If this force is assumed to act uniformly over a distance of about R_g, then we obtain for the depletion free energy per unit area

$$W(D \to 0) \approx -\rho R_g kT. \tag{14.7}$$

For example, if $\rho = 10^{24} \, \text{m}^{-3}$ and $R_g = 5 \, \text{nm}$, then at $25\,^{\circ}\text{C}$ the interaction energy between two surfaces will decrease by $0.2 \, \text{mJ} \, \text{m}^{-2}$ due to depletion. This is small compared to typical van der Waals and solvation adhesion energies, but can be important in cases where there is no other attractive force.

Clearly, for depletion forces to be significant we need a high bulk concentration of polymer molecules (high ρ) as well as a large R_g (high MW). But since in practice it is not possible to have both, the best practical way to attain a strong depletion attraction is to have as high a polymer concentration as possible which in turn requires that the molecular weight and hence R_g be low. In the limit of very high ρ and low R_g, the adhesive minimum becomes deeper, but the range of the depletion force decreases, and by the time the polymer mole fraction has reached unity the monotonic depletion attraction will have developed ripples transforming it into the decaying oscillatory force function characteristic of a pure liquid or polymer melt (see Sections 13.4 and 14.4).

Theories of depletion and bridging forces are not as well advanced as those of other polymer-mediated forces. It is usually difficult to distinguish experimentally between depletion and bridging attraction, and depletion forces have often been invoked to explain colloidal particle coagulation where bridging forces are equally likely. Evans and Needham (1988) were the first to measure unambiguously the depletion energy of two interacting bilayer surfaces in a concentrated dextran solution, and they successfully verified Eq. (14.7).

14.6   NON-EQUILIBRIUM ASPECTS OF POLYMER INTERACTIONS

So far, all theories of polymer-mediated interactions have been equilibrium theories, similar to theories of DLVO forces discussed earlier. This is fine for the DLVO forces, since when two surfaces or colloidal particles approach each other both the electronic and ionic distributions can usually respond sufficiently rapidly to ensure that the van der Waals and double-layer forces they experience *will* be the equilibrium forces. Even the short-range oscillatory solvation forces in liquids are likely to attain equilibrium quickly if the molecules are small and spherical. But one cannot always be sure that the equilibrium force law is operating between two surfaces interacting across a complex polymer system or even a pure polymer melt. Indeed, a distinctive feature of polymer interactions—one that has often been noted by experimentalists—is the extreme sluggishness with which equilibrium is attained once polymer molecules are confined within a narrow space (Israelachvili et al., 1979, 1980c; Van Alsten and Granick, 1990). This leads to time-dependent effects in force measurements and to 'ageing' effects in colloidal systems.

At least four different molecular relaxation mechanisms can be taking place when polymers are confined between two surfaces, each having its own relaxation time. First, solvent has to flow out through the network of entangled polymer coils; second, the coils themselves must reorder as they become compressed; third, new binding sites and bridges have to be formed, and fourth, a certain fraction of polymer molecules may have to enter or leave the gap region altogether (requiring them to diffuse through the network of entangled coils on their way to or from the bulk solution). Most of these processes involve the concerted motions of many entangled molecules which may require many hours or days even though the rate of similar molecular motions of isolated coils in the bulk may take less than 10^{-6} s (a difference of a factor of 10^{10}).

Because of this, the interactions between compressed polymer layers are often far from equilibrium, exhibiting hysteresis, time-dependent, and history-dependent effects (cf. Fig. 14.4). This is probably the most important factor that distinguishes polymer-mediated interactions from other interactions and one that must always be borne in mind when comparing theory with experiment.

14.7   THERMAL FLUCTUATION FORCES BETWEEN FLUID-LIKE SURFACES

Not all surfaces or interfaces are rigid. Some structures such as micelles, bilayers, microemulsion droplets and biological membranes are aggregates of weakly held amphiphilic or co-polymer molecules (Chapter 16). These structures are thermally mobile or 'fluid-like', their shape is constantly changing as their molecules twist, turn and bob in and out of the surfaces.

As two such surfaces come together they experience a number of repulsive 'thermal fluctuation' forces associated with the entropic confinement (overlap) of their various fluctuation modes. These can be either collective molecular motions such as the undulating ripples of a thin membrane, or they can be molecular-scale protrusions as discussed in Section 13.1 and measured by Pfieffer et al. (1989). We shall proceed by first considering the protrusion interaction.

14.8   PROTRUSION FORCES

A repulsive protrusion force arises when two amphiphilic surfaces come close enough together that their molecular-scale protrusions overlap (Fig. 14.7). This force is analogous to the steric or overlap force between surfaces with adsorbed polymer layers. Here, however, as two surfaces approach each other their protruding segments are forced back *into* the surfaces, whereas with polymers the molecules are compressed but remain *between* the surfaces (for

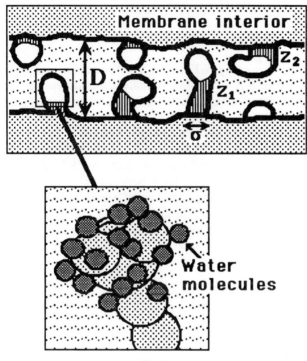

Fig. 14.7.

grafted chains) or they may be forced out *laterally* into the bulk liquid (for adsorbed chains).

Protrusion forces are particularly important between amphiphilic sufaces interacting in aqueous and highly polar liquids (Israelachvili and Wennerström, 1990). To calculate the protrusion force between two amphiphilic surfaces in water, we first note that each protrusion is associated with a positive hydrophobic energy due to the increased molecular hydrocarbon–water contact area. In a first approximation, this energy increases linearly with the distance, z, that the molecules protrude into the water (Fig. 14.7). We may thus define a 'protrusion' potential as in Eq. (13.1)

$$v(z_i) = \alpha_p z_i \tag{14.8}$$

where the interaction parameter, α_p, is in units of $J\,m^{-1}$. The density of protrusions extending a distance z from the surface is therefore expected to decay exponentially according to

$$\rho(z) = \rho(0)e^{-\alpha_p z_i/kT} = \rho(0)e^{-z_i/\lambda},$$

where

$$\lambda = kT/\alpha_p \tag{14.9}$$

is the protrusion decay length. Equation (14.9) was first used by Aniansson (1978) and Aniansson et al. (1976) to analyse the protrusion dynamics of surfactant molecules in and out of micelles and their exchange rates with the monomers in the bulk solution. Values for the interaction parameter, α_p, of single-chained and double-chained amphiphiles in water range from $1.5 \times 10^{-11}\,\mathrm{J\,m^{-1}}$ to $5 \times 10^{-11}\,\mathrm{J\,m^{-1}}$ at 25°C, which corresponds to decay lengths, λ, in the range 0.08 to 0.3 nm (see Sections 18.4–18.6).

Turning now to the protrusion force between two amphiphilic surfaces, let each surface have molecular protrusions of lateral dimensions σ, extending a distance z_i into the solution, and let there be Γ protrusion sites per unit area ($\Gamma \approx 1/\sigma^2$). For two surfaces facing each other, whose protrusions are not allowed to overlap, the potential distribution theorem (cf. Eq. (4.9)) gives for the interaction free energy

$$W(D) = -\Gamma kT \ln \left\{ \int_0^D dz_2 \int_0^{D-z_2} \exp[-\alpha_p(z_1 + z_2)/kT]\,dz_1 \right\}$$

$$= -\Gamma kT \ln\{(kT/\alpha_p)^2 [1 - (1 + D\alpha_p/kT)e^{-\alpha_p D/kT}]\} \tag{14.10}$$

which gives the force per unit area (the protrusion pressure) as

$$P(D) = -\partial W/\partial D = \frac{(\Gamma \alpha_p^2 D/kT)e^{-\alpha_p D/kT}}{[1 - (1 + \alpha_p D/kT)e^{-\alpha_p D/kT}]} = \frac{\Gamma \alpha_p (D/\lambda)e^{-D/\lambda}}{[1 - (1 + D/\lambda)e^{-D/\lambda}]}. \tag{14.11}$$

In the distance regime between 1 and 10 decay lengths, the protrusion force as given by Eq. (14.11) varies roughly exponentially and is adequately given by

$$P(D) = 2.7\Gamma \alpha_p e^{-D/\lambda} \qquad \text{for } \lambda < D < 10\lambda \tag{14.12}$$

where $\lambda \approx kT/\alpha_p$ is the protrusion decay length as before. Equation (14.11) also predicts a steep upturn in the force once D falls below λ, when it diverges according to

$$P(D \to 0) = 2\Gamma kT/D. \tag{14.13}$$

Note that this is the expected osmotic limit for an ideal gas of protrusions (or any non-interacting particles) of density 2Γ per unit area (Γ per surface) confined uniformly within a gap of thickness D, i.e. of number density $2\Gamma/D$. We have previously noted a similar small-distance limit for the double-layer repulsion between two surfaces of constant surface charge density (cf. Equations (12.23) and (12.53)).

The protrusion pressure of Eq. (14.12) corresponds to an energy per unit area of

$$W(D) = 2.7\Gamma\alpha_p\lambda e^{-D/\lambda} \approx 3\Gamma kTe^{-D/\lambda}. \tag{14.14}$$

This may be compared with $W(D) \approx 36\Gamma kTe^{-D/R_s}$ for the interaction energy between two surfaces with end-grafted chains, Eq. (14.3a), and with $W(D) \approx 30\Gamma kT(L/s)e^{-\pi D/L}$ for the energy between two brush layers, Eq. (14.6b).

In Chapter 18 we shall use the above equations as a basis for analysing the short-range repulsive forces measured between surfactant and lipid bilayers in water and non-aqueous solvents.

14.9 UNDULATION AND PERISTALTIC FORCES

In addition to their molecular-scale protrusions, all fluid-like structures also undergo collective thermal fluctuations at the macroscopic level which can be analysed within a continuum framework. Fluid membranes or bilayers, for example, can be considered as elastic sheets which have two characteristic types of wave-like motions: *undulatory* and *peristaltic* (Fig. 14.8). The first is associated with the membrane's bending modulus, k_b, the second with its area expansion modulus, k_a.

Repulsive undulation and peristaltic forces arise from the entropic confinement of their undulation and peristaltic waves as two membranes approach each other. Both these forces can be easily derived from the contact

A. Undulation forces   B. Peristaltic forces

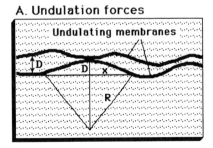 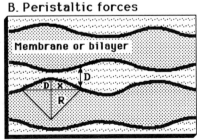

Fig. 14.8.

value theorem which gives the entropic force per unit area between two surfaces as $P(D) = kT[\rho_s(D) - \rho_s(\infty)]$ where $\rho_s(D)$ is the volume density of molecules or molecular groups in contact with the surfaces when the distance between them is D. In the case of undulation forces these contacts can be associated with the thermally excited waves (undulation modes) of amplitude D as shown in Fig. 14.8A. Ignoring numerical factors, the density of contacts (or modes) at $D = D$ and at $D = \infty$ are

$$\rho_s(D) = 1/(\text{volume per mode}) = 1/\pi x^2 D,$$

and

$$\rho_s(\infty) = 0. \tag{14.15}$$

For the spherical geometry of an undulation mode, the 'chord theorem', Eq. (9.7), gives $x^2 \approx 2RD$. The undulation pressure is therefore

$$P(D) = \frac{kT}{\pi x^2 D} \approx \frac{kT}{2\pi RD^2}. \tag{14.16}$$

Now, by definition, the elastic bending (or curvature) energy E_b of a curved membrane with local radii R_1 and R_2 is

$$E_b = \frac{1}{2}k_b\left(\frac{1}{R_1} + \frac{1}{R_2}\right)^2 \quad \text{per unit area}$$

$$= 2k_b/R^2 \quad \text{for } R_1 = R_2 = R. \tag{14.17}$$

At temperature T we expect each mode, which occupies an area πx^2, to have energy $\sim kT$. Thus $kT \approx 2\pi x^2 k_b/R^2 = 4\pi D k_b/R$. Substituting this into Eq. (14.16) and ignoring numerical factors we obtain $P(D) \approx (kT)^2/k_b D^3$, which is the desired (approximate) expression for the repulsive undulation force. The more exact expression, first derived by Helfrich (Helfrich, 1978; Servuss and Helfrich, 1989) is

$$P(D) = \frac{3\pi^2(kT)^2}{64k_bD^3} \approx \frac{(kT)^2}{2k_bD^3}. \tag{14.18}$$

The undulation force has been measured and the inverse third distance dependence verified experimentally (Safinya et al., 1986; Abillon and Perez, 1990). Note that the undulation force has the same form as the non-retarded van der Waals force (force $\propto 1/D^3$) but is of opposite sign. However, the undulation force can be drastically reduced or even eliminated when a membrane carries a surface charge of when it is in tension, since this suppresses the undulations. Tension can be brought about by mechanical or osmotic stresses and has the effect of increasing the adhesion between stressed membranes or bilayers (Servuss and Helfrich, 1989; Bailey *et al.*, 1990). In contrast, van der Waals forces do not change when the interacting surfaces are subjected to a tensile or compressive stress.

In addition to bending undulations, fluid membranes can also undergo peristaltic (or squeezing) fluctuations wherein the thickness of the membrane fluctuates locally about the mean thickness, but without bending (Fig. 14.8B). The same approximate analysis as was used above to derive the undulation force can now be used to derive the peristaltic force associated with the entropic confinement of the peristaltic modes as two membranes approach each other. Referring to Fig. 14.8B, consider a bulged region, or mode, of area $a = \pi x^2$ where the local membrane thickness is greater than the mean. Adjacent regions will have thicknesses smaller than the mean, the total volume of the membrane being assumed constant. Using straightforward geometry, it can be shown that the surface area of the membrane exceeds the mean area, a, by $\Delta a = \pi D^2$ per mode. Now, by definition, the elastic energy E_a of a membrane whose equilibrium area a is stretched by Δa is given by

$$E_a = \frac{1}{2} K_a \frac{(\Delta a)^2}{a} = \frac{\pi k_a D^4}{2x^2} \qquad (14.19)$$

where k_a is the area expansion or compressibility modulus. Equating this energy per mode with kT as before, and again using Eq. (14.16), we obtain

$$P(D) = \frac{kT}{\pi x^2 D} \approx \frac{2(kT)^2}{\pi^2 k_a D^5}, \qquad (14.20)$$

which gives an estimate for the peristaltic pressure between two membranes.

Note that in contrast to the undulation force, which is in terms of a membrane's bending modulus k_b, the peristaltic force depends on the expansion modulus, k_a. These two elastic properties are quite different and have different dimensions. The protrusion force depends on yet another property—the molecular density profile away from the interface or, for amphiphilic molecules in micelles or bilayers, the hydrophobic energy increment per CH_2 group added to the hydrocarbon chain. While these three properties are different, they are not necessarily totally independent of each other.

In Chapter 18 we shall make a quantitative assessment of the contributions of protrusion, undulation, peristaltic, hydration, van der Waals and other forces to the net interaction between surfactant and lipid bilayer membranes.

PROBLEMS AND DISCUSSION TOPICS

14.1 Five scientists are arguing about the force between colloidal particles dispersed in a particular polymer solution at the theta temperature. The polymers are known not to adsorb on the particle surfaces in this solvent.

Dr A: When the particle surfaces are closer than R_g there will be a depletion zone of reduced polymer density in the narrow gap between them. This leads to a repulsive force between the particles, because the polymer molecules in the more concentrated bulk solution want to get back into the gap, thereby pushing the surfaces apart.

Dr B: I disagree. The depletion must lead to an attractive force, because if you imagine the solutions in the gap and the bulk as two distinct phases (as in an osmotic pressure cell), solvent will want to diffuse out from the gap into the more concentrated bulk region, and this will act to pull the surfaces together. Thus the force should be attractive, not repulsive.

Dr C: I agree that the force should be attractive, but for a different and much simpler reason. Since the concentration of polymer is less in the gap, the contact value theorem immediately tells us that the force must be attractive.

Dr D: But you have overlooked that when the polymer molecules move out of the gap, solvent molecules (which are presumably smaller) must come in to replace them. Thus there is actually a net *increase* in the overall number density of molecules in the gap, and according to the contact value theorem this will lead to a repulsion.

Dr E: This is really part of a more general phenomenon that applies to all systems composed of non-interacting particles (but of different sizes) where segregation effects can arise due entirely to entropic effects. If we look at the *whole* system it becomes apparent that as the larger particles come closer together (on the average) their entropy will decrease, but this is more than compensated by the increased entropy of the smaller particles that now fill the space vacated by the larger particles. In the present case the polymers act as the smaller particles and so there will be an effective attraction between the surfaces (of the larger particles).

Critically analyse the above five arguments.

14.2 A colloidal system is stable in the absence of polymer. A small amount of polymer is added and the colloidal particles coagulate, but if a large amount of polymer is added they redisperse. In another system no coagulation occurs when a small amount of polymer is added, but does occur at high polymer concentrations. Discuss the likely nature of the polymer, particle surfaces and solvent in each case.

14.3 Show that the depletion contribution to the adhesion energy of two large spherical colloidal particles of radius R immersed in a liquid containing a volume fraction ϕ of non-adsorbing polymers is approximately $kT\phi(R/R_F)$. Under what conditions will this dominate over other adhesive forces such as van der Waals and solvation forces? How could one experimentally distinguish between depletion and bridging attraction?

14.4 Compute the force-law in water between two bilayers of the poly[ethylene–oxide] surfactant $C_{12}EO_4$ (see Table 16.1). Assume that the interaction is determined mainly by headgroup overlap forces as modelled by the Alexander–de Gennes theory for interacting brush layers. Use $L = 1.6$ nm (0.4 nm per EO group), $s = 0.93$ nm (mean spacing between groups), and $T = 25°C$. Are these values reasonable? Compare your computed force profile in the range $D = 3.2–1.5$ nm with the measured force (see Fig. 2b of Lyle and Tiddy, 1986). Comment on whether the Alexander–de Gennes theory really applies to this interaction or whether any agreement between theory and experiment is fortuitous.

CHAPTER 15

ADHESION

15.1 SURFACE AND INTERFACIAL ENERGIES

In Part I we saw how various interaction potentials between molecules arise, and we considered the implications of the energy minimum, or the 'adhesion' energy, of molecules in contact. Here we shall look at phenomena involving surfaces and particles in adhesive contact, and it is best to begin by defining some commonly used terms and deriving some useful thermodynamic relations.

Work of adhesion and cohesion in vacuum

This is the free energy change, or reversible work done, to separate unit areas of two media 1 and 2 from contact to infinity in vacuum (Fig. 15.1a,b). For two different media ($1 \neq 2$), this energy is referred to as the *work of adhesion* W_{12}, while for two identical media ($1 = 2$), it becomes the *work of cohesion* W_{11}. If 1 is a solid and 2 a liquid, W_{12} is often denoted by W_{SL}. Note that since all media attract each other in vacuum W_{11} and W_{12} are always positive.

Surface energy and surface tension

This is the free energy change γ when the surface area of a medium is increased by unit area. Now the process of creating unit area of surface is equivalent to separating two half-unit areas from contact (Fig. 15.1b,c), so that we may write

$$\gamma_1 = \tfrac{1}{2} W_{11} \tag{15.1}$$

For solids γ_1 is commonly denoted by γ_s and is given in units of energy per

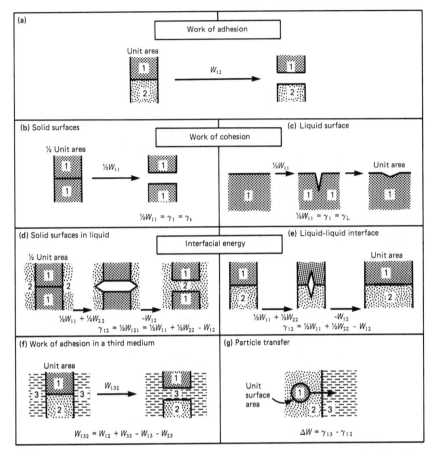

Fig. 15.1. Definition of various energy terms associated with the adhesion of solid surfaces and the surface area changes of liquids. Note that while W and γ are supposed to be well-defined thermodynamic free energies, in practice only for liquids can the area be changed gradually and reversibly (as in (c)), whereas for solids the adhesion, cohesion and debonding process (as in (a) or (b)) is usually irreversible and involves the dissipation of heat (see Sections 15.4 and 15.5).

unit area: $mJ\,m^{-2}$ (the same as $erg\,cm^{-2}$). For liquids, γ_1 is commonly denoted by γ_L and is usually given in units of tension per unit length: $mN\,m^{-1}$ (the same as $dyn\,cm^{-1}$), which is numerically and dimensionally the same as the surface free energy.

It is evident that the intermolecular forces that determine the surface energy of a substance are the same as those that determine its latent heat and boiling point (Section 2.5). As might be expected, substances such as metals with high boiling points ($T_B > 2000°C$) usually have high surface energies

($\gamma > 1000 \, \text{mJ m}^{-2}$), while lower boiling point substances have progressively lower surface energies. For example, for mercury: $\gamma = 485 \, \text{mJ m}^{-2}$, $T_B = 357°C$; for water: $\gamma = 73 \, \text{mJ m}^{-2}$, $T_B = 100°C$; for argon: $\gamma = 13.2 \, \text{mJ m}^{-2}$, $T_B = -186°C$, and for hydrogen: $\gamma = 2.3 \, \text{mJ m}^{-2}$, $T_B = -253°C$. In Sections 11.10 and 11.11 we saw how the surface energies of simple liquids and solids, but not metals or highly polar compounds can be calculated reasonably accurately on the basis of current theories of van der Waals forces.

It is important to appreciate that when the process of increasing the surface area of a medium takes place in a foreign vapour, such as laboratory air, some adsorption of vapour molecules (e.g., water, hydrocarbons) may take place on the newly created surface. This has the effect of lowering γ_S and γ_L from their values in vacuum, and the surface energies in vapour are denoted by γ_{SV} and γ_{LV} (Adamson, 1976, Ch. 7). For example, when mica is cleaved in high vacuum the surface energy is $\gamma_S \approx 4500 \, \text{mJ m}^{-2}$, but when cleaved in humid laboratory air it falls to $\gamma_{SV} \approx 300 \, \text{mJ m}^{-2}$ (Bailey et al., 1970).

Interfacial energy

When two immiscible liquids 1 and 2 are in contact, the free energy change in expanding their 'interfacial' area by unit area is known as their *interfacial energy* or *interfacial tension* γ_{12}. The energetics associated with this expansion process may be understood by splitting it into two hypothetical steps (Fig. 15.1e): first, unit areas of media 1 and 2 are created, and are then brought into contact. The total free energy change γ_{12} is therefore

$$\gamma_{12} = \tfrac{1}{2}W_{11} + \tfrac{1}{2}W_{22} - W_{12} = \gamma_1 + \gamma_2 - W_{12}, \qquad (15.2)$$

which is often referred to as the *Dupré equation*. As shown in Fig. 15.1d, this energy is formally the same as that expended on separating two media 1 in medium 2 (W_{121}) or, conversely, of separating two media 2 in medium 1 (W_{212}). We may therefore also write

$$\gamma_{12} = \tfrac{1}{2}W_{121} = \tfrac{1}{2}W_{212}. \qquad (15.3)$$

For a solid–liquid interface, γ_{12} is commonly denoted by γ_{SL}, so that the Dupré equation may be written as

$$\gamma_{SL} = \gamma_S + \gamma_L - W_{SL}. \qquad (15.4)$$

Table 15.1 gives the surface and interfacial energies of some common substances.

TABLE 15.1 Surface and interfacial energies (mJ m^{-2})[a]

Liquid 1	Surface energy γ_1	Interfacial energy γ_{12}
		With water at 20°C ($\gamma_2 = 73$)
Cyclohexanol	32	4
Undecanol	29	9
Diethyl ether	17	11
Chloroform	27	28
Benzene	29	35
Carbon tetrachloride	27	45
n-Octane (saturated)	21.8[b]	51
1-Octene (unsaturated)	21.8[b]	—
iso-Octanes (branched)	18.8–22.0[b]	—
n-Tetradecane	27	52
Paraffin wax	25	~50
Cyclohexane	25	51
PTFE	19	~50
Octadecane ($C_{18}H_{38}$)	28	52
Octadecene ($C_{18}H_{36}$)	—	19[b]
Octadecadiene ($C_{18}H_{34}$)	—	15[b]
		With tetradecane at 25°C ($\gamma_2 = 26$)
Water	72	53
Glycerol	64	36
Formamide	58	32
1,3 Propanediol	49	21
Ethylene glycol	48	20
Methylformamide	40	12
1,2 Propanediol	38	13
Dimethylformamide	37	5
		With glycol at 20°C ($\gamma_2 = 48$)
Cyclohexane	25.5	14
n-Hexane	18.5	16

[a] Values compiled from standard references, especially TRC Thermodynamic Tables for Hydrocarbons (1990), Landolt-Börnstein (1982), and Zografi and Yalkowsky (1974).
[b] Note that $C=C$ double bonds and branching have only a small effect on the *surface* energies of hydrocarbons, but $C=C$ bonds do have a dramatic effect on reducing their *interfacial* energies with water. Unsaturated hydrocarbons are also much more soluble in water than their saturated homologues.

If only dispersion forces are responsible for the interaction between media 1 and 2, then we have previously seen that to a good approximation

$$W_{12} \approx \sqrt{W_{11}^d W_{22}^d} \approx 2\sqrt{\gamma_1^d \gamma_2^d}$$

so that Eq. (15.2) now becomes (Adamson, 1976, Chs 3, 7)

$$\gamma_{12} \approx \gamma_1 + \gamma_2 - 2\sqrt{\gamma_1^d \gamma_2^d} \tag{15.5}$$

where γ_1^d and γ_2^d are the dispersion force contributions to the surface tensions. Fowkes (1964) and Good and Elbing (1970) estimated that for water, the dispersion contribution to the total surface tension is 20 ± 2 mN m^{-1} or about 27% of the total,[1] the remaining 53 mN m^{-1} arising from non-dispersion (i.e., polar and H-bonding) interactions. Since water and hydrocarbon attract each other mainly via dispersion forces the interfacial tension of a hydrocarbon–water interface should therefore be given by Eq. (15.5). Thus, for octane–water, putting $\gamma_1^d = 21.8$ mN m^{-1} and $\gamma_2^d = 20$ mN m^{-1}, we calculate

$$\gamma_{12} \approx 21.8 + 72.75 - 2\sqrt{21.8 \times 20} = 52.8 \text{ mN m}^{-1}$$

which is very close to the measured value of 50.8 mN m^{-1}. This good agreement is obtained for many hydrocarbon–water interfaces, but the agreement is not so good for unsaturated hydrocarbons and for aromatic molecules such as benzene and toluene (Fowkes, 1964; Good and Elbing, 1970). Indeed, it is interesting to note that only one double bond can lower the interfacial energy of octadecane with water from 52 to 19 mJ m^{-2} (Table 15.1). Clearly, the C=C bond appears to have an unexpectedly strong interaction with water, a phenomenon that is also reflected in the much higher solubilities of unsaturated alkanes in water compared to their saturated counterparts.

Work of adhesion in a third medium

It is left as an exercise for the reader to establish that the energy change on separating two media 1 and 2 in medium 3 (Fig. 15.1f) is given by

$$W_{132} = W_{12} + W_{33} - W_{13} - W_{23} = \gamma_{13} + \gamma_{23} - \gamma_{12}. \tag{15.6}$$

---

[1] Note the near agreement between this value and the probably fortuitous theoretical estimates of 24 and 25% obtained in Tables 6.3 and 11.4.

Note that W_{132} can be positive (attraction between 1 and 2) or negative (repulsion between 1 and 2). If medium 3 is vacuum, $W_{132} \rightarrow W_{12}$, $\gamma_{13} \rightarrow \gamma_1$, $\gamma_{23} \rightarrow \gamma_2$, and the above reduces to Eq. (15.2) as expected.

Surface energy of transfer

When a macroscopic particle 1 moves from medium 2 into medium 3 (Fig. 15.1g) the change in energy per unit area of the particle's surface is

$$\Delta W = (W_{12} - \tfrac{1}{2}W_{22}) - (W_{13} - \tfrac{1}{2}W_{33}) = \gamma_{13} - \gamma_{12} \qquad (15.7)$$

where $(W_{12} - \tfrac{1}{2}W_{22})$ is the energy required to first separate unit areas of media 1 and 2 and then bring into contact the newly created free surfaces of medium 2, and where $-(W_{13} - \tfrac{1}{2}W_{33})$ is the reverse operation for medium 1 with medium 3.

15.2 SURFACE ENERGIES OF SMALL CLUSTERS AND HIGHLY CURVED SURFACES

Here we shall briefly investigate the validity of applying the concept of surface energy to a very small droplet or even an isolated molecule. Clearly the idea of a surface *tension* cannot apply to a single molecule or even to a cluster of a few molecules. But the concept of a surface or interfacial *energy* remains valid, even for isolated molecules, since there is always an energy change associated with transferring a molecule from one medium to another, and this can always be expressed in terms of $4\pi a^2 \gamma$ or $4\pi a^2 (\gamma_{13} - \gamma_{12})$, where a is the molecular radius. The question is, how close will the surface energy of a small cluster correspond to that of the bulk, planar interface? There is at present no simple answer to this problem (Sinanoglu, 1981), but we may gain a feel for the size limit by comparing the pair interactions between simple spherical molecules at a planar surface with those occurring at a highly curved surface. If $-w$ is the pair energy at molecular contact, then for a planar close-packed surface lattice, with three unsaturated bonds per surface molecule, the surface energy is given by Eq. (11.33):

$$\gamma \approx \sqrt{3}w/\sigma^2 \approx 1.7w/\sigma^2.$$

For an isolated molecule, with 12 unsaturated bonds, its effective surface energy will be

$$\gamma \approx 12w/2[4\pi(\sigma/2)^2] \approx 1.9w/\sigma^2, \qquad (15.8)$$

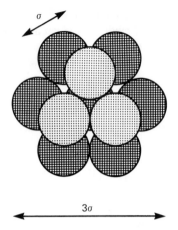

Fig. 15.2. Cluster of 13 molecules: one central molecule surrounded by 12 close-packed neighbours (six in the plane of the page with three above and three below). Note that each of the 12 surrounding molecules has five contact points with its neighbours.

while for a cluster of 13 molecules, with seven unsaturated bonds per each of the 12 surface molecules (Fig. 15.2), we find

$$\gamma \approx 12 \times 7w/2[4\pi(3\sigma/2)^2] \approx 1.5w/\sigma^2. \tag{15.9}$$

By symmetry, it is also clear that cavities of one or 13 (missing) molecules must also have the same surface energies as above.

Thus, we arrive at the remarkable conclusion that the magnitude of the effective surface energy γ of a very small cluster or hole, or even an isolated molecule, is similar to that of a planar macroscopic surface (Sinanoglu, 1981). We encounter various experimental manifestations of this phenomenon in Sections 8.5 and 15.6, in Problem 15.14, and elsewhere in this book.

But the above conclusion is strictly true only for molecules whose pair potentials are additive and where long-range forces and many-body effects are not important. While this condition holds for van der Waals substances, it does not apply to metallic, ionic and hydrogen-bonding compounds. For example, the high latent heats, melting points, surface energies and electronic conductivities of metals is believed to depend on the correlated (cooperative) interactions of many atoms. In a very small droplet this cannot occur, and metal clusters with less than 15–30 atoms lose their bulk 'metallic' properties and become indistinguishable from van der Waals substances. For example, the melting points of small droplets of gold are significantly lower than the bulk value of 1336 K, falling to 1000 K for a cluster of diameter 4 nm and to about 500 K for a diameter of 2.5 nm (Buffat and Borel, 1976).

Similarly, Buffey *et al.* (1990) have found that small water clusters with about 20 molecules or less are probably in the liquid state already at 200 K. Note that these modified properties apply not only to isolated droplets but to any highly curved surfaces such as protruding asperities on a rough surface. It is also likely that the properties of associated and H-bonding liquids become more van der Waals-like in the first layer or two of a surface. This effect is believed to cause the surface of ice to have a thin liquid layer on it (known as 'surface melting') which is responsible for the low friction of ice (Dash, 1989).

15.3   CONTACT ANGLES AND WETTING FILMS

Surface and interfacial energies determine how macroscopic liquid droplets deform when they adhere to a surface. In Fig. 15.3a (top) a large initially spherical droplet 2 in medium 3 approaches and then settles on the rigid flat surface of medium 1. The final total surface energy of the system is therefore

$$W_{tot} = \gamma_{23}(A_c + A_f) - W_{132}A_f, \tag{15.10}$$

where A_c and A_f are the curved and flat areas of the droplet. At equilibrium: $\gamma_{23}(dA_c + dA_f) - W_{132}\,dA_f = 0$. For a droplet of constant volume, it is easy to show using straightforward geometry that $dA_c/dA_f = \cos\theta$. Thus, the equilibrium condition ($\theta = \theta_0$) is

$$\gamma_{23}(1 + \cos\theta_0) = W_{132} = \gamma_{13} + \gamma_{23} - \gamma_{12}, \tag{15.11}$$

or

$$\gamma_{12} + \gamma_{23}\cos\theta = \gamma_{13}, \tag{15.12}$$

which is often derived by balancing the resolved interfacial tensions in the plane of the surface (Fig. 15.3a). If media 2 and 3 are interchanged, as in Fig. 15.3a (bottom), then Eq. (15.11) becomes

$$\gamma_{23}(1 + \cos\theta_0) = W_{123} = \gamma_{12} + \gamma_{23} - \gamma_{13}, \tag{15.13}$$

that is,

$$\gamma_{12} - \gamma_{23}\cos\theta_0 = \gamma_{13}, \tag{15.14}$$

$\theta_0 \simeq 90°$ ⟶ $\theta_0 = 180°$

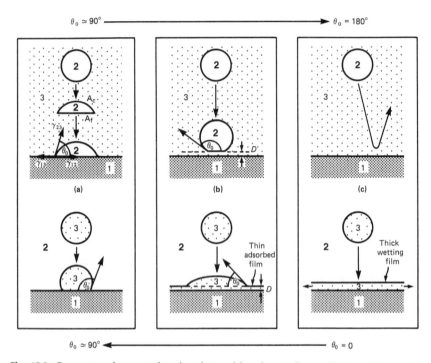

$\theta_0 \simeq 90°$ ⟵ $\theta_0 = 0$

Fig. 15.3. Contact angles can often be changed by chemically modifying surfaces or by addition of certain solute molecules into the medium that adsorb on the surfaces. For example, addition of 'surface-active' molecules such as detergents to water can cause the contact angle to increase from 0 to 180° (see Fig. 1.2d). When quartz is preheated above 300°C its hydrophilic surface silanol groups —Si(OH)—Si(OH)— give off water, leaving behind hydrophobic siloxane groups —Si—O—Si—, and the contact angle rises from 0 to about 60°. Note that the upper and lower drawings in each box are formally equivalent (with media 2 and 3 interchanged so that θ_0(bottom) = 180° − θ_0(top)).

or

$$\gamma_{23}(1 - \cos \theta_0) = W_{132}, \qquad (15.15)$$

which is the same as Eq. (15.11) with θ_0 replaced by 180° − θ_0. Thus, as might have been expected intuitively, the contact angle in Fig. 15.3a (bottom) is simply 180° − θ_0 of that in Fig. 15.3a (top).

While the above results were derived for the specific case of a spherically shaped droplet on a flat surface, the contact angle is independent of the surface geometry (Adamson, 1976, Ch. 7). Thus θ_0 is the same on a curved surface, inside a capillary or at any point on an irregularly shaped surface. Further, the contact angle θ_0 as given by the above equations is a

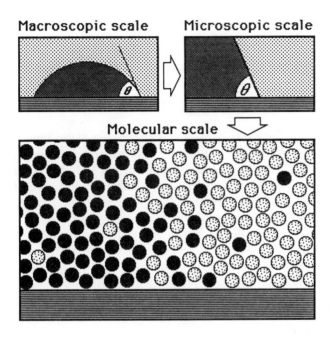

Fig. 15.4. Schematic shape of a liquid–liquid interface as it meets a solid surface seen at the macroscopic, microscopic and molecular levels. The macroscopic contact angle, θ, does not tell us anything about the shape of a surface at the molecular level.

thermodynamic, and hence a purely macroscopic, quantity—independent of the nature of the forces between the molecules so long as these are of shorter range than the dimensions of the droplet. Thus, the contact angle tells us nothing about the microscopic contact angle or the shape of the liquid profile at the point where it meets the surface (Fig. 15.4).

The above equations are more generalized forms of the famous equations of Young and Dupré derived for liquid droplets on surfaces exposed to vapour. Thus, if medium 3 in Fig. 15.3a (top) is an inert atmosphere, Eqs (15.11) and (15.12) become

$$\gamma_2(1 + \cos\theta_0) = W_{12} \qquad \text{(Young–Dupré equation)} \qquad (15.16)$$

and

$$\gamma_{12} + \gamma_2 \cos\theta_0 = \gamma_1 \qquad \text{(Young equation)} \qquad (15.17)$$

For example, for water on paraffin wax, the measured values are $\theta_0 \approx 111°$, $\gamma_1(\text{paraffin}) \approx 25 \text{ mJ m}^{-2}$, and $\gamma_2(\text{water}) = 73 \text{ mJ m}^{-2}$, from which we infer that $\gamma_{12} \approx 51 \text{ mJ m}^{-2}$ (cf. Table 15.1) and that $W_{12} \approx 47 \text{ mJ m}^{-2}$. Note that this is close to the value expected from

$$W_{12} \approx 2\sqrt{\gamma_1^d \gamma_2^d} \approx 2\sqrt{25 \times 20} \approx 45 \text{ mJ m}^{-2}.$$

There is one reported case where all four parameters of the Young equation have been independently measured (Pashley and Israelachvili, 1981; Israelachvili, 1982). This concerns a droplet of $8 \times 10^{-4} \text{ M HTAB}$ solution on a monolayer-covered surface of mica for which $\gamma_1(\text{solid}) = 27 \pm 2 \text{ mJ m}^{-2}$, $\gamma_2(\text{liquid}) = 40 \text{ mJ m}^{-2}$, $\gamma_{12}(\text{solid–liquid}) = 11 \pm 2 \text{ mJ m}^{-2}$, and $\theta_0 = 64°$, which agree with the Young equation.

In the case where medium 3 is a liquid, equilibrium can be attained at some finite distance D (Fig. 15.3b (top)) where the interaction energy W_{132} is a minimum, for example, a weak secondary minimum (Section 12.18). In such cases the contact angle is usually very low. Such phenomena occur, for example, when dissolved air bubbles or oil droplets containing lipid or surfactant monolayers (emulsion droplets) adhere weakly to each other or to a surface (Fig. 1.2d).

An analogous situation occurs when a liquid droplet attaches to a surface containing a thin physisorbed film of the same liquid (Fig. 15.3b (bottom)). Qualitatively one may say that here a small contact angle forms because the liquid rests on a surface that is of its own kind. Note that this is formally the same as that of Fig. 15.3b (top) with media 2 and 3 interchanged. Such cases occur quite often, for example, many different vapours including water and hydrocarbons adsorb as a monolayer on mica, and these liquids have a small but finite contact angle on mica of $\theta_0 < 6°$.

Finally, we may note that if the interaction between 1 and 2 across 3 is monotonically repulsive, the liquid droplet is now repelled from the surface (Fig. 15.3c (top)), and we must put $W_{132} = 0$. Such situations lead to the complete spreading of a liquid on a surface and the development of thick wetting films (Fig. 15.3c (bottom)), previously discussed in Sections 11.6 and 12.9.

15.4   HYSTERESIS IN CONTACT ANGLE AND ADHESION MEASUREMENTS

The contact angle, being a thermodynamic quantity, should be expected to be a unique value for any particular system. But it is often found that when an interface advances along a surface the 'advancing' contact angle, θ_A, is larger than the 'receding' angle, θ_R (Fig. 15.5). This is known as *contact angle*

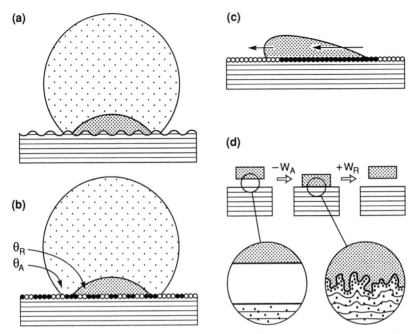

Fig. 15.5. Contact angle and adhesion hysteresis. (a) Effect of surface roughness: liquid droplets on microscopically rough surface where the microscopic contact angle is 90° in each case but the macroscopic (measured) contact angles are very different. For any particular liquid volume, these two geometries will have different Laplace pressures and one will have a lower net energy (Miller and Neogi, 1985). (b) Effect of chemical heterogeneity: droplets on a smooth but chemically heterogeneous surface where the adhesion energy W is different at different places. This gives rise to different macroscopic contact angles as in the example above. (c) and (d) Effects of molecular rearrangements on a moving droplet: the orientation of surface chemical groups often depends on the phase they are exposed to, i.e., whether liquid or vapour. If the molecular reorientation and diffusion time is too slow it results in a difference between advancing and receding adhesion energies (W_A and W_R) and thus in the advancing and receding contact angles. Other dynamic effects that can lead to hysteresis are slow adsorption and desorption of solute molecules dissolved in the liquid. Note that while the droplet is moving the temperature may not be uniform across the system. (d) When two solid surfaces come into contact, depending on the stiffness of the system, they will go through an unstable regime during which they spontaneously jump into contact from some finite separation (see Problem 10.8). A similar instability occurs on separation. The occurrence of such spontaneous jumps implies that these processes are thermodynamically irreversible, that the initial and final states are not identical, and that both the adhesion and separation processes involve dissipation of energy (in the form of heat).

hysteresis and there has been an ongoing debate as to its origins (Miller and Neogi, 1985). Clearly, the interface is not retracing its original path when it recedes, so that the process is not thermodynamically reversible. It is not immediately obvious which, if any, of the two contact angles represent the truly equilibrium value. The matter has not been easy to resolve experimentally, for example, by allowing very long equilibration times, since advancing and receding angles can be stable for very long times.

The phenomenon of contact angle hysteresis is really a manifestation of a much more important effect: the hysteresis in the *adhesion energy* of two phases, where at least one is usually a solid (Fig. 15.5d). For example, for a liquid droplet on a solid surface, it is the different values of W for the advancing and receding liquid with the surface that results in the different values for θ via the Young–Dupré equation, Eq. (15.16).

The existence of hysteresis and irreversibility usually means that a system is trapped in a metastable, non-equilibrium state. In the case of the contact angle of a liquid droplet on a solid surface this can be due to the absence of *mechanical* equilibrium (Fig. 15.5a), *chemical* equilibrium (Fig. 15.5b), or

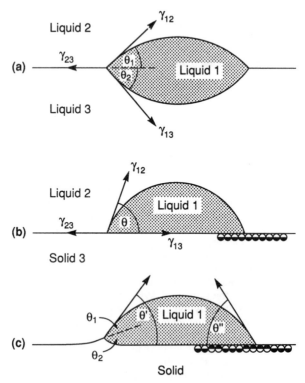

Fig. 15.6. (a) Equilibrium configuration of liquid droplet (or lens) on another liquid. (b) Non-equilibrium (but stable) configuration of liquid droplet on a solid surface. Equation 15.18 shows that θ will be greater than θ_1 but less than $(\theta_1 + \theta_2)$. (c) Microscopic and molecular-scale deformations that can occur to relax the unresolved vertical component of the interfacial tension.

a combination of these including *thermal* equilibrium (Fig. 15.5c). Let us consider these in more detail.

First, consider three liquid phases in contact with each other, as shown in Fig. (15.6a). At the three-phase boundary the three contact angles will be uniquely determined by the three interfacial tensions according to the triangle of forces rule. When this condition is satisfied true thermodynamic equilibrium will have been attained. Now consider the situation where the lower surface is a solid and where the three interfacial energies are unchanged. Clearly, the *equilibrium* geometry should also be unchanged, but in practice, if the solid is rigid and undeformable, the geometry will depend on the shape of the solid surface. Actually, a real solid *will* deform with time by creep; thus starting from Fig. 15.6b the system will slowly deform toward the true equilibrium geometry of Fig. 15.6a, but this may take billions of years.

□   ● **WORKED EXAMPLE** ●

Question:

(1) Show that $\gamma_{12} \sin \theta_1 = \gamma_{13} \sin \theta_2$ in Fig. 15.6a.
(2) If $\theta_1 = \theta_2 = 45°$ in Fig. 15.6a, what is θ in Fig. 15.6b?
(3) What balances the vertical component of the tension $\gamma_{12} \sin \theta$ in Fig. 15.6b?

Ignore gravitational effects.

Answer:

(1) At equilibrium the pressure throughout the system must be uniform. This implies that the Laplace pressures of the two interfaces are the same and that the curvature of each interface is uniform. The radii R_1 and R_2 of the upper and lower curved interfaces must therefore be related by $\gamma_{12}/R_1 = \gamma_{13}/R_2$. Since $r = R_1 \sin \theta_1 = R_2 \sin \theta_2$, we immediately obtain the desired result which shows that the resolved vertical components of the interfacial tensions are balanced, as expected.

(2) Equating vertical components we have: $\gamma_{12} \sin \theta_1 = \gamma_{13} \sin \theta_2$. Equating horizontal components we obtain: $\gamma_{12} \cos \theta_1 + \gamma_{13} \cos \theta_2 = \gamma_{23}$ and $\gamma_{12} \cos \theta + \gamma_{13} = \gamma_{23}$. Eliminating γ_{12}, γ_{13} and γ_{23} from these three equations yields:

$$\cos \theta = \cos \theta_1 - (1 - \cos \theta_1) \sin \theta_1 / \sin \theta_2. \qquad (15.18)$$

Inserting $\theta_1 = \theta_2 = 45°$ into the above we obtain $\theta = \cos^{-1}(\sqrt{2} - 1) = 65.5°$.

(3) The apparently unbalanced vertical component of the tension $\gamma_{12} \sin \theta$ must be balanced by high local stresses on the solid surface (see Problem 15.6).

This can result in elastic or even plastic deformations which cause the surface to bulge upwards (Shanahan and de Gennes, 1986) as illustrated in Fig. 15.6c, left. More importantly, it may lead to molecular rearrangements that alter the local surface energies so as to reduce these local stresses (Fig. 15.6c, right). These stress relaxation effects usually act to reduce the final contact angles θ' and θ'' below θ.   □

All the above effects can lead to hysteresis and ageing effects of contact angles and adhesion energies. Thus, values of θ and W will usually differ for advancing and receding boundaries, with W_R being generally larger than W_A, so that $\theta_A > \theta_R$. But these differences also depend on dynamic factors such as the rate at which the boundaries move.

15.5 ADHESION FORCE BETWEEN SOLID PARTICLES: THE JKR AND HERTZ THEORIES

The adhesion force of two rigid (incompressible) macroscopic spheres is simply related to their work of adhesion by

$$F = 2\pi \left(\frac{R_1 R_2}{R_1 + R_2} \right) W_{132}. \tag{15.19}$$

This general result is a direct consequence of the Derjaguin approximation, Eq. (10.18), and leads to the following special cases:

$$F = 2\pi R \gamma_{SL} \quad \text{(two identical spheres in liquid)}, \tag{15.19a}$$

$$F = 2\pi R \gamma_S \quad \text{(two identical spheres in vacuum)}, \tag{15.19b}$$

$$F = 4\pi R \gamma_S \quad \text{(sphere on flat surface in vacuum)}, \tag{15.19c}$$

$$F = 4\pi R \gamma_{SV} \quad \text{(sphere on flat surface in vapour)}. \tag{15.19d}$$

Real particles, however, are never completely rigid, and on coming into contact they deform elastically under the influence of any externally applied load as well as the attractive intersurface forces that pull the two surfaces together, this gives rise to a finite contact area even under zero external load (Fig. 15.7). One of the first attempts at a rigorous theoretical treatment of the adhesion of elastic spheres is due to Johnson, Kendal and Roberts (1971), whose theory, the 'JKR theory', forms the basis of modern theories of 'adhesion mechanics' (Pollock et al., 1978; Barquins and Maugis, 1982; Georges, 1982). In the JKR theory two spheres of radii R_1 and R_2, elastic

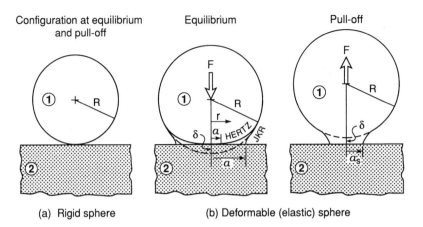

Configuration at equilibrium and pull-off   Equilibrium   Pull-off

(a) Rigid sphere   (b) Deformable (elastic) sphere

Fig. 15.7. (a) Rigid sphere on rigid surface. (b) Left: deformable (elastic) sphere on rigid surface in the absence (Hertz) and presence (JKR) of adhesion. Right: elastic adhering sphere about to separate spontaneously from adhesive contact.

moduli K, and surface energy W_{12} per unit area, will flatten when pressed together under an external load or force, F, such that their contact area will have a radius a given by

$$a^3 = \frac{R}{K}[F + 3\pi R W_{12} + \sqrt{6\pi R W_{12} F + (3\pi R W_{12})^2}] \qquad (15.20)$$

where $R = R_1 R_2/(R_1 + R_2)$. For a sphere of radius R on a flat surface of the same material (Fig. 15.7b) we may put $R_2 = \infty$, $R = R_1$ and $W_{12} = 2\gamma_{sv}$ in the above equation, so that under zero load ($F = 0$) the contact radius is finite and given by

$$a_0 = (6\pi R^2 W_{12}/K)^{1/3} = (12\pi R^2 \gamma_{sv}/K)^{1/3}. \qquad (15.21)$$

Equation (15.20) further shows that under small negative loads ($F < 0$) the solids still adhere until at some critical negative force the surfaces suddenly jump apart. For a sphere on a flat (Fig. 15.7c) this adhesion or 'pull-off' force is given by

$$F_s = -3\pi R \gamma_{sv} \qquad (15.22)$$

and separation occurs abruptly once the contact radius has fallen to

$$a_s = a_0/4^{1/3} = 0.63 a_0. \qquad (15.23)$$

Note that according to the JKR theory a finite elastic modulus, K, while having an effect on the contact area a and a_0, has no effect on the adhesion force, F_s, an interesting and unexpected result that has nevertheless been verified experimentally (see below).

The central displacement δ is given by (see Fig. 15.7b)

$$\delta = \frac{a^2}{R}\left[1 - \frac{2}{3}\left(\frac{a_0}{a}\right)^{3/2}\right]. \tag{15.24}$$

Another useful equation gives the pressure or stress distribution within the contact circle as

$$P(x) = \frac{3Ka}{2\pi R}(1 - x^2)^{1/2} - \left(\frac{3KW_{12}}{2\pi a}\right)^{1/2}(1 - x^2)^{-1/2} \tag{15.25}$$

where $x = r/a$ (see Fig. 15.7b).

When $W_{12} = 0$ we have the much simpler case of two non-adhering spheres, and the equations of the JKR theory reduce to those of the Hertz (1881) theory, viz:

Adhesion force: $F_s = 0$ (15.26)

Contact radius: $a^3 = RF/K$ (15.27)

Displacement: $\delta = a^2/R = F/Ka$ (15.28)

Pressure: $P(x) = \dfrac{3Ka(1 - x^2)^{1/2}}{2\pi R} = \dfrac{3F(1 - x^2)^{1/2}}{2\pi a^2}.$ (15.29)

The last equation shows that at the centre ($x = 0$) the pressure is $P(0) = \frac{3}{2}F/\pi a^2$, which is 1.50 times the *mean* pressure across the contact circle.

One difficulty with the JKR theory is that it predicts an infinite stress at the edge of the contact circle (at $x = 1$) where the surfaces are expected to bend infinitely sharply through 90°. This unphysical situation arises because the JKR theory is a continuum theory and implicitly assumes that the attractive forces between the two surfaces act over an infinitesimally small range. However, these infinities disappear as soon as the attractive force law between the surfaces is allowed to have a finite range, for example, by assuming a Lennard–Jones potential (Muller et al., 1980, 1983; Burgess et al., 1990, Fogden and White, 1990). But these modified JKR theories are extremely complex and can only be solved numerically with a computer. One simple conclusion, however, is that in the limit of small deformations the adhesion force changes from the 'JKR limit' of $F_s = -3\pi R\gamma$ to the so-called 'DMT limit' of $F_s = -4\pi R\gamma$ of the Derjaguin–Muller–Toporov theory (Derjaguin et al., 1975; Muller et al., 1980, 1983; Fogden and White, 1990).

Apart from its breakdown within the last few nanometres of the bifurcation boundary, most of the equations of the JKR theory and all the equations of the Hertz theory have been experimentally tested for molecularly smooth surfaces and found to apply extremely well (Johnson et al., 1971; Tabor, 1977; Israelachvili et al., 1980b; Horn et al., 1987). Thus, it has been verified that two adhering surfaces separate spontaneously once the contact radius has fallen to about 60% of the equilibrium radius a_0, in good agreement with Eq. (15.23). It has also been verified that the contact area of two surfaces as a function of applied load is excellently described by Eq. (15.27) for non-adhering surfaces and by Eq. (15.20) for adhering surfaces (Horn et al., 1987). Finally, Eq. (15.22) relating the adhesion force to the surface or interfacial energy has been found to be correct to within about 25% for a variety of surfaces in vapours or liquids and to be independent of the elastic modulus and contact area of the contacting curved surfaces (Israelachvili et al., 1980b; Shchukin et al., 1981; Shchukin, 1982).

A more serious practical limitation of the JKR and Hertz theories is that they assume perfectly smooth surfaces. Most particle surfaces are rough, and esperities as small as 1–2 nm can significantly lower their adhesion, but there is as yet no satisfactory theory for such real world situations.

Another important aspect of interparticle adhesion concerns the role of adhesion energy hysteresis. Just as in the case of a liquid droplet advancing or receding on a surface, a growing (advancing) and contracting (receding) contact area between two solid surfaces can also have different values for W or γ. This can be due to the inherent instabilities and irreversibilities associated with the bonding/debonding or loading/unloading cycles of solid surfaces (Fig. 15.5d) or because of molecular rearrangements and interdigitations occurring at the contact interface as a function of time (Landman et al., 1990). Both of these effects will give rise to an advancing adhesion energy, W_A, that is less than the receding energy, W_R, and thus to hysteresis effects and energy dissipation. The notion that even the simplest adhesion process may not always be reversible, but involves energy dissipation, has profound effects for understanding many adhesion phenomena and also provides a link between adhesion and friction. These matters, however, are outside the scope of this book.

A knowledge of interparticle adhesion forces and adhesion mechanisms in general is important for understanding many technological processes such as particle agglomeration in colloidal dispersions and during mineral separation processes, the strength of ceramics and soils, friction and lubrication, and, of course, adhesives. In Chapter 18 we shall consider the somewhat different deformations associated with the adhesion of fluid-like (rather than elastic) amphiphilic particles such as lipid bilayer vesicles.

15.6   EFFECT OF CAPILLARY CONDENSATION ON ADHESION

The mechanical and adhesive properties of many substances are very sensitive to the presence of even trace amounts of vapours in the atmosphere. For example, the adhesion of powders (Visser, 1976), the strength of quartz (Paterson and Kekulawala, 1979) and the seismic properties of rocks (Clark *et al.*, 1980) are markedly dependent on the relative humidity. All these effects are due in part to the *capillary condensation* of water around surface contact sites (e.g., in cracks and pores), which, as we shall see, can have a profound effect on the strength of adhesion joints.

Liquids that wet or have a small contact angle on surfaces will spontaneously condense from vapour into cracks and pores as bulk liquid (Fig. 15.8a,b). At equilibrium the meniscus curvature $(1/r_1 + 1/r_2)$ is related to the relative vapour pressure (relative humidity for water) p/p_{sat} by the well-known *Kelvin*

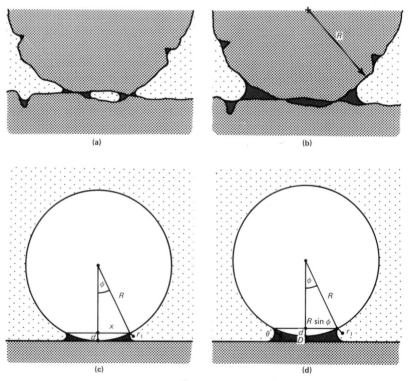

Fig. 15.8.

equation (Adamson, 1976, Ch. 2)

$$\left(\frac{1}{r_1} + \frac{1}{r_2}\right)^{-1} = r_K = \frac{\gamma V}{RT \log(p/p_s)},$$
(15.30)

where r_K is the *Kelvin radius* and V is the molar volume ($\gamma V/RT = 0.54$ nm for water at $20°C$). Thus, for a spherical concave water meniscus (putting $r_1 = r_2 = r$), we find $r = \infty$ at $p/p_{sat} = 1$, $r \approx -10$ nm at $p/p_{sat} = 0.9$, $r \approx -1.6$ nm at $p/p_{sat} = 0.5$, and $r \approx -0.5$ nm at $p/p_{sat} = 0.1$ (10% relative humidity).

What is the effect of a liquid condensate on the adhesion force between a macroscopic sphere and a surface (Fig. 15.8c)? A simple derivation is to consider the *Laplace pressure* in the liquid (Adamson, 1976, Ch. 1):

$$P = \gamma_L \left(\frac{1}{r_1} + \frac{1}{r_2}\right) \approx \frac{\gamma_L}{r_1} \qquad \text{(since } r_2 \gg r_1). $$
(15.31)

The Laplace pressure acts on an area $\pi x^2 \approx 2\pi Rd$ between the two surfaces, thus pulling them together with a force $F \approx 2\pi Rd(\gamma_L/r_1)$. For small ϕ, $d \approx 2r_1 \cos \theta$, and we obtain for the Laplace pressure contribution to the adhesion force

$$F \approx 4\pi R\gamma_L \cos \theta.$$
(15.32)

The additional force arising from the resolved surface tension around the circumference is always small compared to the Laplace pressure contribution except for $\theta \approx 90°$ when $\cos \theta \approx 0$.

An alternative derivation is to consider how the total surface free energy of the system W_{tot} changes with separation D (Fig. 15.8d). For small ϕ,

$$W_{tot} \approx 2\pi R^2 \sin^2 \phi(\gamma_{SL} - \gamma_{SV}) + \text{constant} + \text{smaller terms}$$
$$= -2\pi R^2 \phi^2 \gamma_L \cos \theta + \text{constant},$$

so that

$$F = -dW_{tot}/dD = +4\pi R^2 \phi \gamma_L \cos \theta (d\phi/dD).$$
(15.33)

Now if the liquid volume V remains constant, then since

$$V \approx \pi R^2 \sin^2 \phi(D + d) - (\pi R^3/3)(1 - \cos \phi)^2(2 + \cos \phi)$$
$$\approx \pi R^2 D\phi^2 + \pi R^3 \phi^4/4 \qquad \text{for small } \phi$$

we have $dV/dD = 0$, which gives

$$\frac{d\phi}{dD} = -\frac{1}{(R\phi + 2D/\phi)}.$$

Thus, the attractive force between the sphere and the surface due to the liquid bridge is

$$F = \frac{4\pi R\gamma_L \cos\theta}{(1 + D/d)} \tag{15.34}$$

and maximum attraction occurs at $D = 0$, where

$$F = F_{max} = 4\pi R\gamma_L \cos\theta, \tag{15.35}$$

which is the same as Eq. (15.32). More rigorous expressions, valid for large ϕ and different contact angles on each surface, are given by Orr *et al.* (1975). The case of a liquid bridge in equilibrium with vapour (where the Kelvin radius rather than the volume remains constant as D changes) is considered in Problem 15.10.

One other important parameter must be included in the above expression. This is the direct solid–solid contact adhesion force *inside* the liquid annulus, Eq. (15.19). The final result for a meniscus of radius r_1 (where $R \gg r_1$) is therefore[2]

$$F = 4\pi R(\gamma_L \cos\theta + \gamma_{SL}) = 4\pi R\gamma_{SV}. \tag{15.36}$$

In the absence of any condensing vapour (i.e., as $r_1 \to 0$) the above reduces to Eq. (15.19c): $F = 4\pi R\gamma_S$. For two spheres, R is replaced by $(1/R_1 + 1/R_2)^{-1}$ in all the above equations.

Since $\gamma_S > \gamma_{SV}$ the adhesion force should always be less in a vapour than in a vacuum. In practice, however, one often has to compare γ_{SV_1} with γ_{SV_2} depending on the relative adsorptions of different vapours from the atmosphere, and the adhesion force in air may *increase* with relative humidity if γ_{SV_1} (moist air) $> \gamma_{SV_2}$ (dry air).

Equation (15.36) shows that the adhesion force in vapour must exceed $4\pi R\gamma_L \cos\theta$. Often $\gamma_L \cos\theta$ greatly exceeds γ_{SL}, whence the adhesion force is determined solely by the surface energy of the liquid (e.g., water) and is then adequately given by Eq. (15.35).

---

[2] Fogden and White (1990) have theorized that for strong solid–solid adhesion Eq. (15.36) becomes: $F = 4\pi R(\gamma_L \cos\theta + 0.75\gamma_{SL})$, while if the meniscus radius is very small it becomes $F = 3\pi R(\gamma_L \cos\theta + \gamma_{SL})$.

Note that Eqs (15.35) and (15.36) are independent of the meniscus radius r_1 so that it is of interest to establish below what radius, or relative vapour pressure, these equations break down. McFarlane and Tabor (1950) verified that the adhesion force between glass spheres and a flat glass surface in saturated vapours of water, glycerol, decane, octane, alcohol, benzene and aniline are all given by $F = 4\pi R\gamma_L \cos\theta$ to within a few percentage points. Fisher and Israelachvili (1981) measured the adhesion forces between curved mica surfaces in various vapours such as cyclohexane and benzene and found that $F = 4\pi R\gamma_L \cos\theta$ is already valid once the relative vapour pressures exceed 0.1–0.2, corresponding to meniscus radii of only ~ 0.5 nm, i.e., about the size of the molecules. However, for water, a larger radius of ~ 2 nm appears to be needed before Eq. (15.35) is satisfied (Christenson, 1988b; Hirz et al., 1991). These results confirm the suggestion made in Section 15.2 that for molecules that interact via a simple Lennard-Jones pair potential, their bulk surface energy is already manifest at very small curvatures.

Since real particle surfaces are often rough their adhesion in vapour is not always given by Eq. (14.36). For example, the adhesion of dry sand particles is very small, and even when slightly moist the adhesion is not much different since the condensed water is only bridging small asperities (Fig. 15.8a). However, once r_K exceeds the asperity size but is still less than the particle radius R (Fig. 15.8b), the adhesion force attains its full strength of $F = 4\pi R\gamma_L \cos\theta$ (McFarlane and Tabor, 1950). Finally, when sand is completely wet the adhesion will once again be very low. It is for this reason that one can only build sandcastles with moist sand, but not with dry or completely wet sand (R. Pashley, unpublished results).

Capillary condensation also occurs when water condenses from a solvent in which it is only sparingly soluble, for example, from hydrocarbon solvents where the solubility is usually below 100 ppm. In such circumstances the presence of even 20 ppm (i.e., 20–40% of saturation) can lead to a dramatic increase in the adhesion of hydrophilic colloidal particles. Indeed, it has long been known that trace amounts of water can have a dramatic effect on colloidal stability (Bloomquist and Shutt, 1940; Parfitt and Peacock, 1978) and surfactant association (Eicke, 1980) in non-polar organic solvents; and the enhanced agglomeration of metal ores and coal particles in oils by addition of water forms the basis of several industrial separation and extraction processes (Henry et al., 1980). Christenson (1983, 1985b) found that in the presence of small amounts (< 100 ppm) of water the adhesion force between mica surfaces in benzene, octane, cyclohexane and the liquid OMCTS is given by Eq. (15.35) to within 20% with γ_L replaced by the liquid–liquid interfacial energy γ_{12}. The enhanced adhesion in such systems arises because γ_{12} is typically 35–50 mJ m^{-2} (Table 15.1)—much higher than the solid–liquid interfacial energies in the anhydrous liquids.

Finally, an interesting phenomenon occurs when two hydrophobic particles interact in water. If the contact angle exceeds 90° the above equations now predict that a *vapour* cavity should 'capillary condense' between the two surfaces, again resulting in a strongly adhesive force. This is exactly what has been observed in such systems (Christenson, 1989).

PROBLEMS AND DISCUSSION TOPICS

15.1 A liquid cylinder of radius R is connected between two flat solid walls a distance L apart as shown in Problem 1.4. Calculate the normal force between the two walls using two different methods: (i) by considering the surface energies γ_L, γ_S and γ_{SL} of the various surfaces and interfaces, and (ii) by resolving forces. You should, of course, arrive at the same result in each case. Ignore any gravitational effects.

15.2 For two surfaces interacting in a liquid their adhesion energy (i.e. their energy in final molecular contact with no liquid between them) is often well accounted for by continuum theories (e.g. the Lifshitz theory applied to the interaction across a medium) even though the form of the oscillatory force law with distance is not. Why is this? (*Hint*: split up the interaction into three separate steps.)

15.3 It is found that the surface tension of a liquid (solvent) at the liquid–air interface either (i) increases or (ii) decreases when solute molecules are dissolved in the liquid. From a consideration of the van der Waals forces between the solute molecules and the surfaces would you expect positive or negative adsorption at the liquid–air interface in the case of (i) and (ii)?

15.4 How would you expect the adhesion energy between two molecularly smooth crystalline surfaces to depend on the relative orientation of their surface crystallographic axes ('twist' angles)?

15.5 Two bodies floating on the surface of a dense liquid have contact angles as shown in Fig. 15.9. In each case, determine whether the force between the bodies due to surface tension effects is attractive or repulsive.

15.6 Estimate the local tensile pressure acting normal to a solid surface at the point where a macroscopic liquid droplet of water meets the surface? What effects could this have on the surface both at the microscopic and molecular levels?

15.7 When a liquid droplet moves along a solid surface, or a liquid filament moves forward in a capillary tube, since there is no slip at the solid–liquid interface how can this happen at the molecular level?

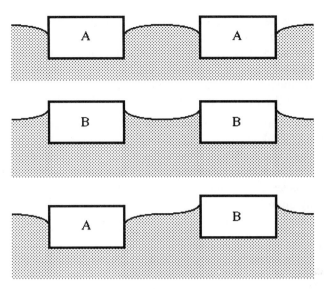

Fig. 15.9.

15.8   A liquid subtends a contact angle of θ_1 on one surface (with surface chemical groups of type 1) and an angle θ_2 on a second surface (with groups of type 2). Derive an equation giving the contact angle on a surface whose surface groups are composed of a fraction f_1 of type 1 and f_2 of type 2 (where $f_1 + f_2 = 1$). Assume additivity of interaction potentials. State what further assumptions you have made in deriving your answer.

15.9   Consider the chemical potential of single-component molecules in small droplets and in the vapour phase and derive (i) the 'Laplace equation', Eq. (15.31), for the pressure across a curved liquid–vapour interface, and (ii) the 'Kelvin equation', Eq. (15.30), for the equilibrium meniscus radius as a function of the vapour pressure. Can an atmosphere of small liquid droplets be in true equilibrium with the bulk liquid phase?

15.10   Equation (15.34) for the force between a sphere and a flat due to the Laplace pressure of the liquid bridge between them applies to the case of constant liquid volume. What equation replaces this equation when the system is at true thermodynamic equilibrium with its surrounding vapour? Obtain your answer in two ways: first by resolving forces and second by differentiating the total energy of the system with respect to separation (the two solutions should, of course, be identical). Give your answer for F as a function of D in terms of R, D, γ_L, θ and the Kelvin radius r_K. Verify that as $D \to 0$ the force is the same as for the constant volume case, Eq. (15.35).

15.11   Two molecularly smooth rigid spheres, each of radius 1 cm, are in contact in an atmosphere of water vapour at 90% relative humidity. If the surface tension of water is 73 mJ m^{-2}, what is the equilibrium adhesion force needed to pull them apart if the contact angle is 0°? Assume that the surfaces are uncharged and ignore any other forces between the surfaces across water.

Estimate how the equilibrium adhesion force will be affected if:

(i) there is also an attractive van der Waals force between the surfaces in water with a Hamaker constant of $A = 10^{-20}$ J;

(ii) there is no van der Waals force, but a strong short-range monotonically repulsive hydration force between the surfaces in water, extending about 2 nm;

(iii) the surfaces are not molecularly smooth.

(*Hint*: for any quantitative estimates of effects (ii) and (iii) first calculate the equilibrium meniscus (Kelvin) radius of the water annulus bridging the two surfaces.)

15.12   A drop of water bridges two plane parallel surfaces. The contact angle is θ. The radius of the neck, R, is much larger than the surface separation, D. Derive an equation for the force, F, between the two surfaces. Is F always attractive? Are there situations when $F = 0$?

15.13   A 2 mm thick glass disc (density: 3000 kg m^{-3}) is lying on another, larger, glass disc. Both surfaces are smooth and flat, and both are wetted by water (contact angle: 0°). When immersed in water each surface acquires a surface charge density of $\sigma = 0.2$ C m^{-2}. What is the equilibrium separation between the two surfaces when (i) the discs are totally immersed in pure water, and (ii) they are exposed to air saturated with water vapour?

15.14   For a microscopic liquid sphere would you expect its (i) density, (ii) boiling point and (iii) surface energy to be larger or smaller than the bulk material?

15.15   Owens (1970) reported the effects of immersing a surface of propylene (material A) coated with a thin film of the copolymer vinylidene chloride (material B) into water and various aqueous surfactant solutions, L. The interfacial energy of the substrate–film interface is $\gamma_{AB} = 3.5$ mJ m^{-2}, while the other surface and interfacial energies are given in Table 15.2. In which liquids would you expect the coating to spontaneously separate (debond) from the surface?

Is it surprising that the lowering of the surface tension of water by addition of surfactant is not related to the effectiveness of the surfactant solution in separating the two surfaces, i.e., to its effectiveness as a detergent?

15.16   Consider Fig. 10.6b where a deformable tape is being peeled away from a rigid surface. At first one might suppose that all that matters is that

TABLE 15.2 Experimental results on effectiveness of different liquids (L) on debonding a polymer film (B) from a surface (A)

Liquid or solution	γ_L liquid-vapour	γ_{AL} Substrate-liquid	γ_{BL} Coating-liquid	Experimental observation[1]
Pure water	72.8	27.4	13.3	NS
Sodium dodecyl sulfate	37.2	0.9	1.6	S
Sodium diisoamyl sulfosuccinate	25.6	1.2	7.9	NS
Triton X-405	42.4	2.6	0.7	S

[1] S = Spontaneous separation; NS = Did not separate.

the surface or adhesion energy (per unit area) of the two materials be high. However, it is found that easily deformable tapes can be peeled off more easily than hard tapes, and that in general the visco-elastic properties of the tape backing material is actually very important for determining the performance of an 'adhesive'. Why is this?

15.17 Two cylindrical fibres of radii R_1 and R_2 are brought into contact in an atmosphere of water vapour. The fibres have their axes at some arbitrary angle α relative to each other. Show that for fibre radii greater than about 1 μm the adhesive force between them is constant over a large range of relative humidity and is given by

$$F_{ad} = \frac{4\pi\sqrt{R_1 R_2}(\gamma_{SL} + \gamma_{LV}\cos\theta)}{\sin\alpha}. \qquad (15.37)$$

Would you expect the above equation to overestimate or underestimate the adhesion at (i) low humidities ($<50\%$), (ii) high humidities ($>99\%$), (iii) small fibre radii (<1 μm)?

PART THREE

FLUID-LIKE STRUCTURES AND SELF-ASSEMBLING SYSTEMS: MICELLES, BILAYERS AND BIOLOGICAL MEMBRANES

CHAPTER 16

THERMODYNAMIC PRINCIPLES OF SELF-ASSEMBLY

16.1 INTRODUCTION

In Part III we shall be looking at the interactions of small molecular aggregates such as micelles, bilayers, vesicles and biological membranes, which form readily in aqueous solution by the spontaneous *self-association* or *self-assembly* of certain amphiphilic molecules (see Figs 16.1 and 16.2 and Table 16.1). These structures and the systems they form—sometimes collectively referred to as *association colloids* or *complex fluids*—stand apart from the conventional colloidal particles discussed in Part II in one important respect: unlike solid particles or rigid macromolecules such as DNA, they are soft and flexible, i.e., *fluid-like*. This is because the forces that hold amphiphilic molecules together in micelles and bilayers are not due to strong covalent or ionic bonds but arise from weaker van der Waals, hydrophobic, hydrogen-bonding and screened electrostatic interactions. Thus, if the solution conditions, such as the electrolyte concentration or the pH, of an aqueous suspension of micelles or vesicles is changed, not only will this affect the interactions between the aggregates but it will also affect the intermolecular forces within each aggregate, thereby modifying the size and shape of the structures themselves. It is therefore necessary to begin by considering the factors that determine how and why certain molecules associate into various well-defined structures.

In Chapters 16 and 17 we shall be concerned with the thermodynamic and physical principles of self-assembly in general and of amphiphilic molecules such as surfactants and lipids in particular, while in the final Chapter 18 we shall investigate the various forces between amphiphilic structures and surfaces, and examine their role in the interactions of, for example, lipid vesicles and biological membranes.

Our first concern will be to formulate the basic equations of self-assembly in general statistical thermodynamic terms and then go on to investigate the relevant intermolecular interactions that determine into which structures

TABLE 16.1 Some common amphiphiles

Single-chained surfactants

Anionic	$C_{12}H_{25}$—O-SO_3^- Na^+	Sodium dodecyl sulphate (SDS or NaDS)
Anionic	$C_{18}H_{37}$—COO^- H^+	Stearic acid
Cationic	$C_{16}H_{33}$—$N^+(CH_3)_3Br^-$	Hexadecyl trimethylammonium bromide (HTAB or CTAB)
Non-ionic	$C_{12}H_{25}$—$(O-CH_2-CH_2)_5$-OH	Pentaoxyethylene dodecyl ether ($C_{12}E_5$)
Zwitterionic	Single chained lecithin (see below)	Lysolecithin

Double-chained phospholipids

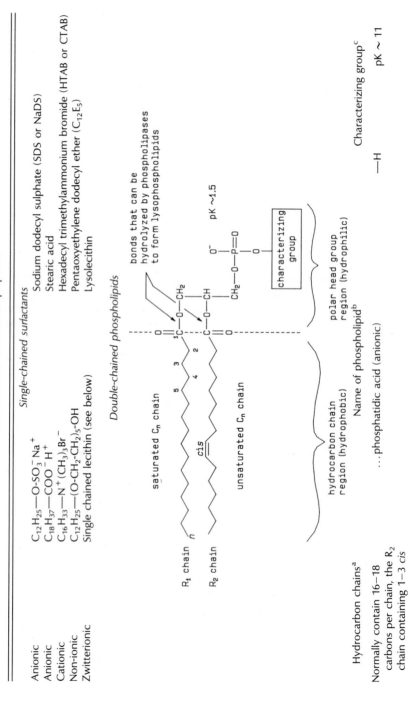

Hydrocarbon chains[a]	Name of phospholipid[b]	Characterizing group[c]
Normally contain 16–18 carbons per chain, the R_2 chain containing 1–3 *cis*	…phosphatidic acid (anionic)	—H pK ~ 11

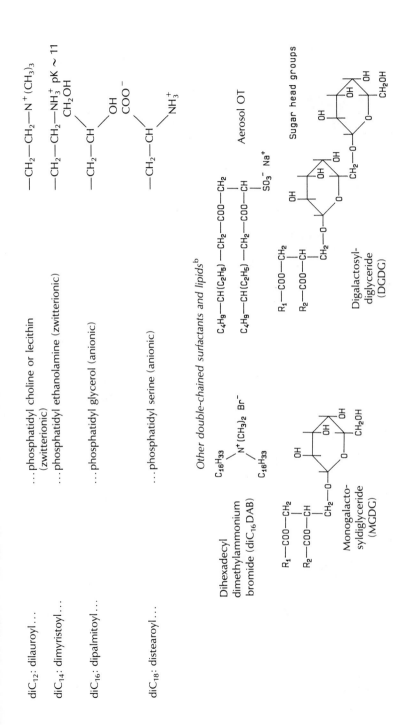

diC$_{12}$: dilauroyl...

diC$_{14}$: dimyristoyl...

diC$_{16}$: dipalmitoyl...

diC$_{18}$: distearoyl...

...phosphatidyl choline or lecithin (zwitterionic)   —CH$_2$—CH$_2$—N$^+$(CH$_3$)$_3$

...phosphatidyl ethanolamine (zwitterionic)   —CH$_2$—CH$_2$—NH$_3^+$ pK ~ 11

—CH$_2$—CH—CH$_2$OH
OH

...phosphatidyl glycerol (anionic)   —CH$_2$—CH—COO$^-$
OH

...phosphatidyl serine (anionic)   —CH$_2$—CH—NH$_3^+$

Other double-chained surfactants and lipids[b]

Dihexadecyl dimethylammonium bromide (diC$_{16}$DAB)

C$_{16}$H$_{33}$—N$^+$(CH$_3$)$_2$ Br$^-$—C$_{16}$H$_{33}$

R$_1$—COO—CH$_2$
R$_2$—COO—CH
CH$_2$—O—

Monogalacto-syldiglyceride (MGDG)

C$_4$H$_9$—CH(C$_2$H$_5$)—CH$_2$—COO—CH$_2$
C$_4$H$_9$—CH(C$_2$H$_5$)—CH$_2$—COO—CH—SO$_3^-$ Na$^+$

Aerosol OT

R$_1$—COO—CH$_2$
R$_2$—COO—CH
CH$_2$—O—

Digalactosyl-diglyceride (DGDG)

Sugar head groups

CH$_2$OH

---

[a] About 50% of biological lipids have an unsaturated chain; these increase the fluidity and hydrophilicity of bilayers.

[b] Phosphatidylcholines and phosphatidylethanolamines are the two major lipids found in animal membranes, while the galactolipids DGDG and MGDG are the major constituents of plant thylakoid membranes. Note that none of these carry a net charge at normal pH.

[c] The ionic states of the headgroups are given for aqueous dispersions at pH 7. At high pH (>11.5) phosphatidylethanolamine becomes negatively charged while at low pH (<1) it becomes positively charged.

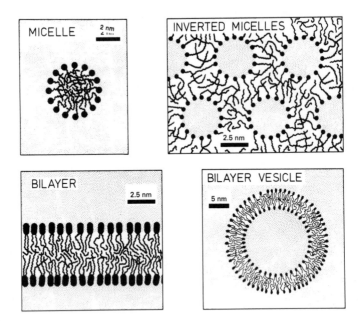

Fig. 16.1. Amphiphiles such as surfactants and lipids (see Table 16.1) can associate into a variety of structures in aqueous solutions. These can transform from one to another by changing the solution conditions such as the electrolyte or lipid concentration, pH, or temperature. In most cases the hydrocarbon chains are in the fluid state allowing for the passage of water and ions through the narrow hydrophobic regions, e.g., across bilayers. The lifetime of water molecules in lecithin vesicles is about 0.02 s, while ions can be trapped for much longer times, about 8 h for Cl^- and one month for Na^+ ions. Most single-chained surfactants form micelles, while most double-chained surfactants form bilayers, for reasons that are discussed in Chapter 17.

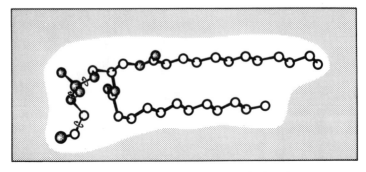

Fig. 16.2. The zwitterionic phospholipid dilauryl-phosphatidyl-ethanolamine containing two saturated hydrocarbon chains and a polar (hydrophilic) headgroup (see also Table 16.1).

different amphiphiles will assemble. We shall find that a very beautiful picture emerges that brings out the role of molecular geometry in determining the structures formed, and from which many of the physical properties of these structures can be quantitatively understood without requiring a detailed knowledge of the very complex short-range forces operating between the polar headgroups and hydrocarbon chains. (By analogy, the van der Waals equation of state contains no information on the nature and range of intermolecular forces, i.e., the force laws, and yet provides a very satisfactory description of gas–liquid phase behaviour. Indeed, when van der Waals in 1873 proposed his famous equation he knew nothing about the origin and nature of van der Waals forces.)

16.2 FUNDAMENTAL THERMODYNAMIC EQUATIONS OF SELF-ASSEMBLY

The literature on this subject is voluminous and often confusing, the most rigorous treatment being that of Hall and Pethica (1967) based on Hill's classic book on small systems thermodynamics (Hill, 1963, 1964). We shall follow the more simplified approach and notation of Tanford (1973, 1980) for micelles, which was later extended to larger lipid aggregates such as bilayers, vesicles, other micellar phases and microemulsion droplets by Nagarajan and Ruckenstein (1977, 1979), Israelachvili et al. (1976, 1977, 1980a), Wennerström and Lindman (1979), Mitchell and Ninham (1981), Szleifer et al. (1985, 1986), Blanckstein et al. (1986), Israelachvili (1987c), and Puvvada and Blanckstein (1990).

Equilibrium thermodynamics requires that in a system of molecules that form aggregated structures in solution (Fig. 16.3) the chemical potential of all identical molecules in different aggregates be the same. This may be expressed as

$$\mu = \mu_1^0 + kT \log X_1 = \mu_2^0 + \tfrac{1}{2}kT \log \tfrac{1}{2}X_2 = \mu_3^0 + \tfrac{1}{3}kT \log \tfrac{1}{3}X_3 = \ldots$$

$$\text{monomers} \qquad\qquad \text{dimers} \qquad\qquad \text{trimers}$$

or

$$\mu = \mu_N = \mu_N^0 + \frac{kT}{N} \log \left(\frac{X_N}{N} \right) = \text{constant}, \qquad N = 1, 2, 3, \ldots, \qquad (16.1)$$

where μ_N is the mean chemical potential of a molecule in an aggregate of aggregation number N, μ_N^0 the standard part of the chemical potential (the

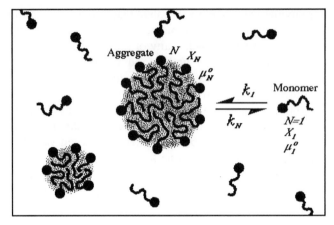

Fig. 16.3. Association of N monomers into an aggregate (e.g., a micelle). The mean lifetime of an amphiphilic molecule in a small micelle is very short, typically $10^{-5}-10^{-3}$ s.

mean interaction free energy *per molecule*) in aggregates of aggregation number N, and X_N the concentration (more strictly the activity) of molecules in aggregates of number N ($N = 1$, μ_1^0 and X_1 correspond to isolated molecules, or *monomers*, in solution). Equation (16.1) may also be derived using the familiar *law of mass action* (Alexander and Johnson, 1950) as follows: referring to Fig. 16.3 we may write

$$\text{rate of association} = k_1 X_1^N,$$

$$\text{rate of dissociation} = k_N(X_N/N),$$

where

$$K = k_1/k_N = \exp[-N(\mu_N^0 - \mu_1^0)/kT] \qquad (16.2)$$

is the ratio of the two 'reaction' rates (the equilibrium constant). These combine to give Eq. (16.1), which can also be written in the more useful (and equivalent) forms

$$X_N = N\{(X_M/M)\exp[M(\mu_M^0 - \mu_N^0)/kT]\}^{N/M} \qquad (16.3a)$$

and, putting $M = 1$,

$$X_N = N\{X_1 \exp[(\mu_1^0 - \mu_N^0)/kT]\}^N \qquad (16.3b)$$

where M is any arbitrary reference state of aggregates (or monomers) with aggregation number M (or 1). Equations (16.3) together with the conservation relation for the total solute concentration C

$$C = X_1 + X_2 + X_3 + \cdots = \sum_{N=1}^{\infty} X_N \qquad (16.4)$$

completely defines the system. Depending on how the free energies μ_1^0, μ_N^0 are defined the dimensionless concentrations C and X_N can be expressed in volume fraction or mole fraction units ($(\text{mol dm}^{-3})/55.5$ or $M/55.5$ for aqueous solutions). In particular, note that C and X_N can never exceed unity. Equation (16.2) assumes ideal mixing and is restricted to dilute systems where interaggregate interactions can be ignored. The effects of such interactions will be considered later.

□ ● WORKED EXAMPLE ●

Question: Certain types of solute molecules are found to self-assemble in solution into discrete macromolecular clusters with a fixed aggregation number N per cluster. The equilibrium between monomers (A) and aggregates (B) in the solution may be expressed in the form of a chemical reaction:

$$A + A + A + \ldots = B.$$

Let X_A and X_B be the concentrations of A and B in mole fraction units, let K be the equilibrium constant for the reaction, and let C be the *total* concentration of solute molecules in the solution.

(i) Obtain a relation between K, N, C and X_A, and show that for $K \gg 1$ and $N \gg 1$ the concentration of monomers, X_A, can never exceed $(NK)^{-1/N}$.

(ii) If $K = 10^{80}$ and $N = 20$ calculate the concentration of molecules in monomers (X_A) and in aggregates (NX_B) at $C = 2 \times 10^{-5}$, $C = 1.052 \times 10^{-4}$ and $C = 10^{-1}$. Comment on your findings. (Note that X_N in the notation of Eqs. (16.1)–(16.4) corresponds to NX_B in the present notation.)

Answer:

(i) Combining the two basic equations: $K = X_B/X_A^N$ and $C = X_A + NX_B$, we obtain $K = (C - X_A)/NX_A^N = \text{constant}$, or $X_A = [(C - X_A)/NK]^{1/N}$. Since the maximum possible value of $(C - X_A)$ is 1, we immediately find that X_A can never exceed $(NK)^{-1/N}$. For $K = 10^{80}$ and $N = 20$ this critical concentration is 0.86×10^{-4}.

(ii) Putting $K = 10^{80}$ and $N = 20$ into the above equation gives $X_A = 10^{-4}[(C - X_A)/20]^{1/20}$. Solving this for the given values of C we find:

at $C = 2 \times 10^{-5}$: $X_A = 1.99999998 \times 10^{-5}$ and $NX_B = 2 \times 10^{-13}$,

at $C = 1.052 \times 10^{-4}$: $X_A = NX_B = 0.526 \times 10^{-4}$,

at $C = 10^{-1}$: $X_A = 0.8 \times 10^{-4}$ and $NX_B = 0.9992 \times 10^{-1}$.

Thus for $C \ll 10^{-4}$ we have $X_A \approx C$ (i.e., most of the surfactant molecules remain dispersed as monomers). At $C \approx 10^{-4}$, we have $X_A \approx NX_B$ (i.e., the molecules partition equally between monomers and aggregates), while for $C \gg 10^{-4}$ we have $X_A \approx 10^{-4} \approx$ constant, and $NX_B \approx C$ (i.e., the monomer concentration remains unchanged at $\sim 10^{-4}$ as all the molecules go into aggregates). The critical concentration of $\sim 10^{-4}$, or $(NK)^{-1/N}$, is known as the *critical micelle concentration*, and is discussed further in Section 16.5.   □

Little more can be said about aggregated dispersions without specifying the form and magnitude of μ_N^0 as a function of N. This important matter will now be considered, and it is instructive to first proceed with a formal thermodynamic analysis of the equations derived so far.

16.3   CONDITIONS NECESSARY FOR THE FORMATION OF AGGREGATES

Aggregates form only when there is a difference in the cohesive energies between the molecules in the aggregated and the dispersed (monomer) states. If the molecules in different sized aggregates (including monomers) all experience the same interaction with their surroundings, the value of μ_N^0 will remain constant in different aggregates (with different N), and Eq. (16.3) becomes

$$X_N = NX_1^N \quad \text{for} \quad \mu_1^0 = \mu_2^0 = \mu_3^0 = \cdots = \mu_N^0. \tag{16.5}$$

Since $X_1 < 1$, we must have $X_N \ll X_1$ so that most of the molecules will be in the monomer state ($N = 1$). If μ_N^0 increases as N increases, Eq. (16.3) shows that the occurrence of large aggregates becomes even less probable.

The necessary condition for the formation of large stable aggregates is that $\mu_N^0 < \mu_1^0$ for some value of N, for example, when μ_N^0 progressively decreases as N increases or when μ_N^0 has a minimum value at some finite value of N. As we shall see, the exact functional variation of μ_N^0 with N also determines many of the physical properties of aggregates, such as their mean size and polydispersity. Further, since this variation may be a complex one it is clear that a number of structurally different populations may coexist

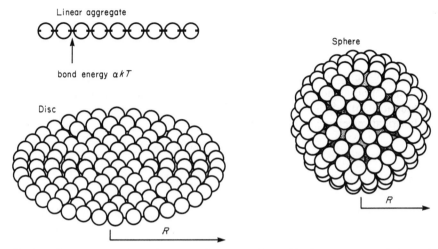

Fig. 16.4. One-, two- and three-dimensional structures formed by the association of identical monomer units in solution.

within a single phase in thermodynamic equilibrium with each other (note that X_N in Eq. (16.3) is a *distribution function* and may peak at more than one value of N).

We shall now consider the functional forms of μ_N^0 for some simple structures and by use of Eqs (16.3)–(16.4) investigate their physical properties.

16.4 VARIATION OF μ_N^0 WITH N FOR SIMPLE STRUCTURES OF DIFFERENT GEOMETRIES: RODS, DISCS AND SPHERES

One-dimensional aggregates (rods)

As mentioned above, aggregates will form if μ_N^0 decreases with N. We shall now see that in a first approximation the dependence of μ_N^0 on N is usually determined by the geometrical shape of the aggregate. Let us begin by considering a suspension of rod-like aggregates made up of linear chains of identical molecules or monomer units (strings of beads) in equilibrium with monomers in solution. Let αkT be the monomer–monomer 'bond' energy in the aggregate relative to isolated monomers in solution (Fig. 16.4). The total interaction free energy $N\mu_N^0$ of an aggregate of N monomers is therefore (remembering that the terminal monomers are unbonded)

$$N\mu_N^0 = -(N-1)\alpha kT,$$

that is,

$$\mu_N^0 = -(1 - 1/N)\alpha kT = \mu_\infty^0 + \alpha kT/N. \qquad (16.6)$$

Thus, as N increases the mean free energy μ_N^0 decreases asymptotically towards μ_∞^0, the 'bulk' energy of a molecule in an infinite aggregate. A similar expression for μ_N^0 is obtained for any type of rod-like structure (e.g., a cylindrical micelle).

Two-dimensional aggregates (discs, sheets)

Let us now look at disc-like or sheet-like aggregates (Fig. 16.4). Here the number N of molecules per disc is proportional to the area πR^2, while the number of unbonded molecules in the rim is proportional to the circumference $2\pi R$, and hence to $N^{1/2}$. The mean free energy per molecule in such an aggregate is therefore

$$\mu_N^0 = \mu_\infty^0 + \alpha kT/N^{1/2} \qquad (16.7)$$

where again α is some constant characteristic of the monomer–monomer and monomer–solvent interaction.

Three-dimensional aggregates (spheres)

Finally, let us consider spherical aggregates or small solute droplets of radius R in a solvent (Fig. 16.4). Here N is proportional to the volume $\frac{4}{3}\pi R^3$, while the number of unbonded surface molecules is proportional to the area $4\pi R^2$ and hence to $N^{2/3}$. We therefore have

$$\mu_N^0 = \mu_\infty^0 + \alpha kT/N^{1/3}. \qquad (16.8)$$

As an example, consider the association of small hydrocarbon molecules such as alkanes in water. If v is the volume per molecule, then $N = 4\pi R^3/3v$. The free energy of the sphere is given by $N\mu_\infty^0 + 4\pi R^2\gamma$, where μ_∞^0 is the bulk energy per molecule and γ the interfacial free energy per unit area (Chapter 15). Hence

$$\mu_N^0 = \mu_\infty^0 + \frac{4\pi R^2\gamma}{N} = \mu_\infty^0 + \frac{4\pi\gamma(3v/4\pi)^{2/3}}{N^{1/3}} = \mu_\infty^0 + \frac{\alpha kT}{N^{1/3}}, \qquad (16.9)$$

where

$$\alpha = \frac{4\pi\gamma(3v/4\pi)^{2/3}}{kT} = \frac{4\pi r^2 \gamma}{kT}, \tag{16.10}$$

r being the effective radius of a molecule.

We see, therefore, that for the simplest shaped structures—rods, sheets, and spheres—the interaction free energy of the molecules can be expressed as

$$\mu_N^0 = \mu_\infty^0 + \frac{\alpha kT}{N^p}, \tag{16.11}$$

where α is a positive constant dependent on the strength of the intermolecular interactions and p is a number that depends on the shape or *dimensionality* of the aggregates. As we shall see, Eq. (16.11) also applies to various micellar structures and to spherical vesicles in which the bilayers bend elastically. In particular, we note that for all these structures, μ_N^0 decreases progressively with N, which is a necessary condition for aggregate formation.

16.5 THE CRITICAL MICELLE CONCENTRATION (CMC)

Given the general functional form of μ_N^0 of Eq. (16.11) we may now ask, at what concentration will aggregates form? Incorporating Eq. (16.11) into the two fundamental equations of self-assembly, Eqs (16.3) and (16.4), leads to some very interesting conclusions. First, we note that

$$X_N = N\{X_1 \exp[(\mu_1^0 - \mu_N^0)/kT]\}^N$$
$$= N\{X_1 \exp[\alpha(1 - 1/N^p)]\}^N \approx N[X_1 e^\alpha]^N. \tag{16.12}$$

For sufficiently low monomer concentrations X_1 such that $X_1 \exp[(\mu_1^0 - \mu_N^0)/kT]$ or $X_1 e^\alpha$ is much less than unity, we have $X_1 > X_2 > X_3 > \cdots$ for all α. Thus, at low concentrations most of the molecules in the solution will be isolated monomers, i.e., $X_1 \approx C$, as shown in Fig. 16.5. However, since X_N can never exceed unity it is clear from Eq. (16.12) that once X_1 approaches $\exp[-(\mu_1^0 - \mu_N^0)/kT]$ or $e^{-\alpha}$ *it can increase no further*. The monomer concentration $(X_1)_{\text{crit}}$ at which this occurs may be called the *critical aggregation concentration* (CAC) though it is common to use the more conventional term *critical micelle concentration* (CMC) to denote the critical

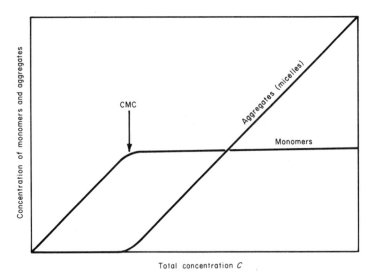

Fig. 16.5. Monomer and aggregate concentrations as a function of total concentration (schematic). Most single-chained surfactants containing 12–16 carbons per chain have their CMC in the range 10^{-2}–10^{-5} M, while the corresponding double-chained surfactants have much lower CMC values due to their greater hydrophobicity. Some important CMC values are listed in Table 16.2.

concentration of all self-assembled structures. Thus, in general

$$(X_1)_{\text{crit}} = \text{CMC} \approx \exp[-(\mu_1^0 - \mu_N^0)/kT]. \qquad (16.13)$$

If μ_N^0 is given by Eq. (16.11), we have

$$(X_1)_{\text{crit}} = \text{CMC} \approx e^{-\alpha} \qquad \text{for all } p. \qquad (16.14)$$

These two equations define the concentration at which further addition of solute molecules results in the formation of more aggregates while leaving the monomer concentration more or less unchanged at the CMC value (Fig. 16.5).

16.6   INFINITE AGGREGATES (PHASE SEPARATION) VERSUS FINITE-SIZED AGGREGATES (MICELLIZATION)

What can we say about the nature of these aggregates? This now depends very much on their shape. For simple disc-like and spherical aggregates,

Eq. (16.12) becomes

$$X_N = N[X_1 e^\alpha]^N e^{-\alpha N^{1/2}} \qquad \text{for discs } (p = \tfrac{1}{2}), \qquad (16.15)$$

$$X_N = N[X_1 e^\alpha]^N e^{-\alpha N^{2/3}} \qquad \text{for spheres } (p = \tfrac{1}{3}). \qquad (16.16)$$

Above the CMC where $X_1 e^\alpha \approx 1$, the above two equations may be approximated by $X_N \approx Ne^{-\alpha N^{1/2}}$ and $X_N \approx Ne^{-\alpha N^{2/3}}$, respectively. Now for any reasonable positive value of α, which is usually greater than 1, these equations show that apart from a few dimers, trimers, etc., *there will be very few aggregates of any appreciable size* (e.g., with $N > 5$).

Where do the molecules go above the CMC? The answer is quite simple: for discs and spheres, there is a phase transition to a separate phase, strictly to an aggregate of infinite size $(N \to \infty)$ at the CMC. Israelachvili *et al.* (1976) showed that such a transition to large macroscopic aggregates occurs whenever $p < 1$ in Eq. (16.11). This applies, quite generally, to all planar or disc-like aggregates composed of identical molecules, and it is for this reason that finite crystalline sheets, one-component lipid bilayers, and even biological membranes with exposed edges are rarely found floating about in solution. Above the CMC infinite bilayers form spontaneously from lipid monomers, although they may close up on themselves to form vesicles (discussed later).

Likewise for simple spherical structures. Here, for example, we may consider the association of oil or alkane molecules in water. On adding oil to water the molecules disperse as monomers up to the critical concentration given by Eqs (16.14) and (16.10),

$$(X_1)_{\text{crit}} \approx e^{-\alpha} \approx e^{-4\pi r^2 \gamma / kT}, \qquad (16.17)$$

above which they will separate out into a bulk oil phase, which may be considered simply as a very large spherical aggregate.

For such a system of two immiscible liquids it is clear that what we are now really talking about is the *solubility* of a solute in a solvent, where α represents the free energy of transferring a solute molecule from the solute into the solvent phase. For example, if we consider the solubility of hydrocarbons in water, we may put $\gamma \approx 50 \, \text{mJ m}^{-2}$ and $r \approx 0.2 \, \text{nm}$ for a methane molecule. Thus, the free energy of transferring a methane molecule from bulk hydrocarbon liquid into water should be approximately $4\pi r^2 \gamma \approx 2.5 \times 10^{-20} \, \text{J}$, corresponding to $\alpha \approx 6$ or about $15 \, \text{kJ mol}^{-1}$. Surprisingly, this theoretical estimate agrees well with the experimental value for the solubility or 'hydrophobic energy' of methane (see Worked Example in Section 2.3 and Problem 8.2).

The hydrophobic energy of transferring alkyl chains from water into bulk hydrocarbon (which determines their solubility) or into micelles (which determines their CMC) can be analysed in a similar fashion. Thus, for an alkane chain of radius $r \approx 0.2$ nm and an interfacial energy with water of $\gamma \approx 50$ mJ m^{-2} as above, the hydrophobic energy per unit length will be $2\pi r \gamma \approx 6 \times 10^{-11}$ J m^{-1}. Now, since the CH$_2$–CH$_2$ distance along a chain is $l = 0.126$ nm, this value corresponds to 8×10^{-21} J per CH$_2$ group added to the chain. Experimentally, one finds an increment of about 6.3×10^{-21} J (equivalent to 900 cal mol^{-1} or 3.8 kJ mol^{-1}) per CH$_2$ group added to a pure alkane chain at 25°C (Tanford, 1980). This corresponds to an increment in α of $6.3 \times 10^{-21}/kT \approx 1.5$ and thus to a lowering of the solubility of alkanes in water by e$^{-1.5} \approx 0.22$, i.e., by a factor of about four, per added CH$_2$ group (see first row of Table 16.2).

The above applies only to pure alkane chains being transferred from water into a pure bulk hydrocarbon phase. In the case of surfactant molecules being transferred into micelles or bilayers, the hydrophobic energy increment is significantly lower, ranging from 1.7 to 2.8 kJ mol^{-1} per CH$_2$ group (Table 16.2). As discussed in Section 8.7 the reduced hydrophobicity of an amphiphilic chain compared to that of a pure alkane chain is believed to be due to the proximity of the hydrophilic headgroup, and to the higher chain ordering of chains within micelles which acts to reduce the energy even more (Aniansson, 1978). The above range of values means that typical micellar CMCs fall by 0.3 to 0.5 (i.e., by a factor between 2 and 3) per CH$_2$ group added to the surfactant chain.

The important difference between alkanes and amphiphilic molecules is not so much in their solubility or CMC values but in the ability of amphiphiles to assemble into structures in which μ_N^0 reaches a minimum or constant value at some *finite* value of N. It is for this reason that the aggregates formed are not infinite ($\rightarrow$ phase separation) but of finite size ($\rightarrow$ micellization). The reasons for *why* and *how* amphiphilic molecules do this will be investigated fully in the following chapter.

16.7   Size distributions of self-assembled structures

Micelles and vesicles in equilibrium with each other in solution usually have a finite distribution of sizes about some mean value. The distribution may be narrow or broad (polydisperse), and it may be symmetrical or asymmetrical about the mean. Here we shall investigate how polydispersity comes about, starting with a consideration of aggregates for which $p = 1$ in Eq. (16.11).

TABLE 16.2 CMCs of some common surfactants and lipids showing the effects of chain length, number of chains,[a] type of headgroup, counterion, coion, salt and temperature (see Problem 16.1)

Surfactant ($R_n = C_n H_{2n+1}$)	Total number carbon atoms in chains	CMC[b] (mM)	Increment of CMC per CH$_2$ group (f)	Average energy per CH$_2$ group ($\Delta G = RT \ln f$)
Pure n-alkanes (no headgroup)	4–8	(solubility)	4.4	3.7 kJ mol^{-1} (880 cal mol^{-1})
Cationic				
Alkyl trimethylammonium bromides				
R_{10}-N(CH$_3$)$_3^+$ Br$^-$	10	66	2.1	1.8 kJ mol^{-1} (430 cal mol^{-1})
R_{12}-N(CH$_3$)$_3^+$ Br$^-$	12	15	2.1	
R_{14}-N(CH$_3$)$_3^+$ Br$^-$	14	3.5	2.0	
R_{16}-N(CH$_3$)$_3^+$ Br$^-$ (CTAB or HTAB)	16	0.9		
Alkyl trimethylammonium chlorides				
R_{10}-N(CH$_3$)$_3^+$ Cl$^-$	10	63	1.8	1.7 kJ mol^{-1} (400 cal mol^{-1})
R_{12}-N(CH$_3$)$_3^+$ Cl$^-$	12	19	2.1	
R_{14}-N(CH$_3$)$_3^+$ Cl$^-$	14	4.5	1.9	
R_{16}-N(CH$_3$)$_3^+$ Cl$^-$	16	1.3	2.0	
R_{18}-N(CH$_3$)$_3^+$ Cl$^-$	18	0.34		
Anionic				
Sodium alkyl sulphates				
R_8-SO$_4^-$ Na$^+$	8	130.	2.0	1.7 kJ mol^{-1} (410 cal mol^{-1})
R_{10}-SO$_4^-$ Na$^+$	10	33.2	2.0	
R_{12}-SO$_4^-$ Na$^+$ (SDS)	12	8.1	2.0	
R_{14}-SO$_4^-$ Na$^+$	14	2.0		

TABLE 16.2 (contd) CMCs of some common surfactants and lipids showing the effects of chain length, number of chains,[a] type of headgroup, counterion, coion, salt and temperature (see Problem 16.1)

Non-ionic				
Alkyl polyoxyethylene monoethers				
R_8-$(OCH_2CH_2)_6OH$ (C_8E_6)	8	9.8	3.3	$2.9\ kJ\ mol^{-1}$
R_{10}-$(OCH_2CH_2)_6OH$ ($C_{10}E_6$)	10	0.90	3.2	$(700\ cal\ mol^{-1})$
R_{12}-$(OCH_2CH_2)_6OH$ ($C_{12}E_6$)	12	0.087		
Zwitterionic				
Lyso-phosphatidylcholines at 25°C				
R_{10}-PC	10	7.0	3.2	$2.8\ kJ\ mol^{-1}$
R_{12}-PC	12	0.70	3.2	$(680\ cal\ mol^{-1})$
R_{14}-PC	14	0.070	3.2	
R_{16}-PC	16	0.007		
Effect of salt				
Sodium alkyl sulphates in 0.3 M NaCl at	8	67	3.1	$2.8\ kJ\ mol^{-1}$
21°C (note the lower CMCs and higher	10	6.9	3.1	$(670\ cal\ mol^{-1})$
increment factors than in salt-free water)	12	0.7		
Counterion effects				
Dodecyl sulphate (40°C) in 0.02 M of				
Cs_2SO_4	12	3.0		
K_2SO_4	12	3.5		
Na_2SO_4	12	3.8		
Li_2SO_4	12	4.0		
Co-ion effects				
Dodecyl sulphate (21°C) in 0.1 M of				
NaF	12	1.45		
NaCl	12	1.45		
NaBr	12	1.43		
NaI	12	1.38		

Temperature effects

Sodium dodecyl sulphate (SDS)			
10°C	8.7		
15°C	8.4		
20°C	8.3		
25°C	8.1 ← min	1.9	1.6 kJ mol⁻¹
30°C	8.3	1.9	(380 cal mol⁻¹)
35°C	8.4		
40°C	8.7		

Double-chained surfactants

Di-alkyl dimethylammonium chlorides

R_8R_8-N(CH$_3$)$_2^+$ Cl⁻	16	27	1.9	
$R_{10}R_{10}$-N(CH$_3$)$_2^+$ Cl⁻	20	2.0	1.9	1.6 kJ mol⁻¹ (380 cal mol⁻¹)
$R_{12}R_{12}$-N(CH$_3$)$_2^+$ Cl⁻	24	0.15		

Di-alkyl sulphates (40°C)

R_7R_6-CH-SO$_4^-$ Na⁺	14	9.70	1.5	1.2 kJ mol⁻¹ (300 cal mol⁻¹)
R_7R_7-CH-SO$_4^-$ Na⁺	15	6.65	1.6	
R_8R_7-CH-SO$_4^-$ Na⁺	16	4.25	1.8	
R_8R_8-CH-SO$_4^-$ Na⁺	17	2.35	1.6	
R_9R_9-CH-SO$_4^-$ Na⁺	19	0.94		
$R_{14}R_{14}$-CHSO$_4^-$ Na⁺	29	0.08		

Di-acyl phosphatidylcholines

R_6R_6-PC	12	1.5×10^{-2} M	2.7	2.1 kJ mol⁻¹ (500 cal mol⁻¹)
R_8R_8-PC	16	3×10^{-4} M	2.7	
$R_{10}R_{10}$-PC	20	5×10^{-6} M	2.2	
$R_{16}R_{16}$-PC (DPPC)	32	5×10^{-10} M		

[a] All values refer to saturated chains; unsaturated chains have lower interfacial energies with water and higher solubilities (Table 15.1) and are expected to have much lower CMCs than their saturated counterparts. Note that most biological lipids are unsaturated.

[b] Values in water (no added salt) at 25°C unless stated otherwise. CMC and solubility values taken from Mukerjee and Mysels (1970), Shinoda et al. (1963), Tanford (1980), Cevc and Marsh (1987), Stafford et al. (1989), and Marsh (1990).

Putting $p = 1$ in Eq. (16.12) we obtain

$$X_N = N[X_1 e^\alpha]^N e^{-\alpha}. \tag{16.18}$$

Thus, in contrast to Eqs (16.15) and (16.16) the second exponential term is now a constant rather than a rapidly decreasing function of N. Since above the CMC we have $X_1 e^\alpha \leqslant 1$, this equation shows that $X_N \propto N$ for small N, i.e., the concentration of molecules in these aggregates now *grows* in proportion to their size, and there is no phase separation. Only for very large N does the $[X_1 e^\alpha]^N$ term begin to dominate, eventually bringing X_N down to zero as N approaches infinity. The distribution is therefore highly polydisperse.

The case of $p = 1$ is in marked contrast to the case when $p < 1$ (as occurs for simple discs or spheres) where an abrupt phase transition to one infinitely sized aggregate occurs at the CMC and where the concept of a size distribution does not arise. Alternatively, for structures where $p > 1$ it can be shown that no finite or infinite sized aggregates form at any concentration so that again the concept of a size distribution does not apply (except for the few very small aggregates or clusters of molecules that are always present). Thus structures for which $p = 1$ appear to have special properties, and it is instructive to analyse this type of system in more detail.

The total concentration of molecules is given by inserting Eq. (16.18) into Eq. (16.4) as

$$\begin{aligned}
C = \sum_{N=1}^{\infty} X_N &= \sum_{N=1}^{\infty} N[X_1 e^\alpha]^N e^{-\alpha} \\
&= e^{-\alpha}[X_1 e^\alpha + 2(X_1 e^\alpha)^2 + 3(X_1 e^\alpha)^3 + \cdots] \\
&= X_1/(1 - X_1 e^\alpha)^2,
\end{aligned} \tag{16.19}$$

where we have made use of the identity

$$\sum_{N=1}^{\infty} N x^N = x/(1 - x)^2.$$

Thus,

$$X_1 = \frac{(1 + 2Ce^\alpha) - \sqrt{1 + 4Ce^\alpha}}{2Ce^{2\alpha}}. \tag{16.20}$$

Note that at low concentrations C where $Ce^\alpha \ll 1$ this gives $X_1 \approx C$, whereas at high concentrations, well above the CMC such that $Ce^\alpha \gg 1$, the above simplifies to

$$X_1 \approx (1 - 1/\sqrt{Ce^\alpha})e^{-\alpha} \leqslant e^{-\alpha}, \qquad (16.21)$$

that is, $X_1 \approx$ CMC as expected. Also above the CMC, the density distribution of *molecules* in aggregates of N molecules is given by inserting the above equation back into Eq. (16.18), yielding

$$X_N = N(1 - 1/\sqrt{Ce^\alpha})^N e^{-\alpha} \approx Ne^{-N/\sqrt{Ce^\alpha}} \qquad \text{for large } N. \qquad (16.22)$$

This function peaks at $\partial X_N/\partial N = 0$, which occurs at

$$N_{max} = M = \sqrt{Ce^\alpha}, \qquad (16.23)$$

while the *expectation value* of N, defined by $\langle N \rangle = \Sigma N X_N / \Sigma X_N = \Sigma N X_N / C$, is given by

$$\langle N \rangle = \sqrt{1 + 4Ce^\alpha}$$

$$\approx 1 \qquad \text{below the CMC,}$$

$$\approx 2\sqrt{Ce^\alpha} = 2M \qquad \text{above the CMC.} \qquad (16.24)$$

Finally, from Eq. (16.22) the density distribution of *aggregates* above the CMC is

$$X_N/N = \text{Const. } e^{-N/M} \qquad \text{for } N > M \qquad (16.25)$$

i.e., the concentration of large aggregates decays exponentially with increasing N with a characteristic decay number of M. Thus, the distribution is very broad, with the concentration of aggregates first increasing with N for small aggregates and decaying gradually to zero at large N.

The above results should apply to all dilute *one-component* aggregates for which $p = 1$ in Eq. (16.11). This includes any chain-like (polymer-like) aggregates, cylindrical micelles and fibrous structures such as microfilaments and microtubules. Later, we shall see that it also applies to spherical vesicles and microemulsion droplets whose membranes bend elastically. For all these structures, the mean aggregation number M is concentration dependent, varying with the square root of the concentration C above the CMC, and from Eq. (16.23) we note that it is also very sensitive to small changes in the interaction parameter α. Consequently, we may anticipate that the aggregation

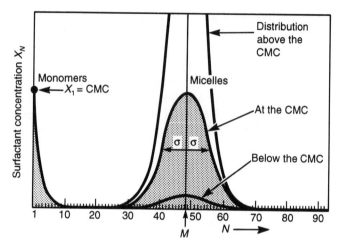

Fig. 16.6. Distribution of molecules X_N as a function of aggregation number N. Near the CMC (shaded region) we have $X_1 \approx X_M$ where the mean micellar aggregation number is M. For spherical micelles, the distribution about M is near Gaussian with standard deviation $\sigma \approx \sqrt{M}$ (see Section 17.4).

number and polydispersity of such structures in water should also be very sensitive to temperature, electrolyte concentration and pH.

Actually, there is a formal relationship between the mean aggregation number, M, the total surfactant concentration, C, and the polydispersity, σ (Fig. 16.6). This relation

$$\sigma^2 \approx \partial \log \langle N \rangle / \partial \log C \qquad (16.26)$$

is valid above the CMC (Israelachvili *et al.*, 1976) and shows that whenever the distribution is highly polydisperse, the mean aggregation number is also very sensitive to the total surfactant concentration. Conversely, monodisperse structures have aggregation numbers that do not vary with concentration.

16.8   MORE COMPLEX AMPHIPHILIC STRUCTURES

The value of p in Eq. (16.11) is constant only for aggregates composed of fairly simple molecules that self-assemble into simple geometric shapes such as spheres, discs or rods. More complex amphiphilic molecules can assemble into more complex shapes such as vesicles, interconnected rods or even three-dimensional 'periodic' structures (cf. Fig. 17.10).

Before proceeding with an analysis of the size distribution of these structures, let us first consider what causes different amphiphilic molecules to aggregate into one or another of these structures in the first place. This is determined by the types of anisotropic binding forces acting between different parts of the amphiphilic molecules. For simple (non-amphiphilic) molecules such as alkanes in water whose hydrophobic interaction with each other is largely isotropic or *non-directional*, we would expect them to coalesce and grow as small spherical droplets (for which the total surface energy is a minimum for any given N). And we have seen that for such aggregates, $p = \frac{1}{3}$ in Eq. (16.11), which results in a phase separation at the solubility limit (the effective CMC). Clearly, molecules that aggregate into linear or sheet-like structures must have asymmetric *directional* bonding. For example, cigar-shaped molecules may have their binding sites located at the ends of the molecules or axially around each molecule; the former will result in linear rod-like aggregates, the latter in sheet-like aggregates.

Further, if the molecules are also flexible, the structures they adopt will be more varied than any of the simple shapes so far considered. Thus, the energetically unfavourable regions at each end of a rod-like aggregate may be eliminated if the two ends bend and join together, resulting in a torus or 'toroidal micelle'. Similarly, the unfavourable rim energy of a disc may be eliminated by its closing up into a vesicle, which is what happens with certain classes of surfactants and lipids. In all such cases, μ_N^0 no longer decays gradually with increasing N as given by the simple equation, Eq. (16.11). Instead, it now reaches a minimum value at some *finite* value of N (say $N = M$), or it reaches a low value at $N = M$ and then remains almost constant for $N > M$ (see Fig. 17.5). Depending on the sharpness of the minimum, such a form for μ_N^0 generally results in monodisperse aggregates of mean aggregation number $N < M$, rather than infinite or polydisperse aggregates. As we shall see in the following chapter, this is what happens for certain types of vesicles and spherical micelles. Here we shall consider the consequences for the size distribution of micelles when μ_N^0 reaches a low value at some finite aggregation number M and does not continue to decrease for $N > M$.

To a high degree of accuracy the CMC may be expressed as

$$\text{CMC} \approx \exp[-(\mu_1^0 - \mu_M^0)/kT]. \tag{16.27}$$

This is the common form of the equation traditionally used in the analysis of spherical micelles. If μ_N^0 has a minimum value at $N = M$, the variation of μ_N^0 about μ_M^0 can usually be expressed in the parabolic form:

$$\mu_N^0 - \mu_M^0 = \Lambda(\Delta N)^2 \tag{16.28}$$

where $\Delta N = (N - M)$. In this case Eq. (16.3a) becomes

$$X_N = N \left\{ \frac{X_M}{M} \exp[-M\Lambda(\Delta N)^2 kT] \right\}^{N/M} \qquad (16.29)$$

and so the distribution of X_N about M will be near Gaussian (Fig. 16.6) with a standard deviation in the *aggregation number* given by

$$\sigma = \sqrt{kT/2M\Lambda}. \qquad (16.30)$$

Systems that fall into this category are spherical micelles and certain single bilayer vesicles, and we shall find that for typical values of M and Λ these can be fairly monodisperse with $\sigma/M \approx 0.1-0.3$.

16.9   EFFECTS OF INTERACTIONS BETWEEN AGGREGATES: MESOPHASES AND MULTILAYERS

So far we have ignored interaggregate interactions. These cannot be ignored at high concentrations (low water content) where, especially for surfactant and lipid dispersions, transitions to larger and more ordered *mesophase*[1] or *lyotropic liquid crystalline* structures are commonly observed. These can be ordered arrays of cylinders (*hexagonal* or *nematic* phases), stacks of bilayers (*lamellar, liposome* or *smectic* phases) or a complex three-dimensional network of interconnected surfaces (*periodic* structures forming *bicontinuous* phases). Both attractive and repulsive forces between aggregates can lead to such structural phase transitions, and it is worth considering each of these very different scenarios in turn.

First, consider the case where there are strong *repulsive* electrostatic, steric or hydration forces between the aggregates, which we shall assume are initially small spherical micelles. With increasing surfactant concentration the micelles are forced to come closer together, which is energetically unfavourable. However, if the surfactants rearrange to form an ordered array of cylinders, it is a simple matter to ascertain (see Problem 18.5) that their surfaces can now be farther apart from each other. And if they order into a stack of bilayers, their surfaces can be even further apart, all at the same surfactant

---

[1] A *mesophase* is a normal phase in the thermodynamic sense, but one that is structurally more complex than a simple liquid or solid phase. It can contain many small molecular aggregates that can be monodisperse or polydisperse, or it can have convoluted lamellar or tubular structures that link up with each other to form a repeat three-dimensional network that extends indefinitely throughout the phase. These are known as *periodic* structures, two of which are shown in Fig. 17.10.

concentration (same volume fraction). It is for this reason that many surfactant structures go from being small micelles to long cylinders to large liposomes as the surfactant content is progressively increased above about 10% by weight (Ekwall, 1975; Tiddy, 1980). Note that since these types of phase transitions arise from repulsive forces, where the aggregates are trying to get as far apart as possible within a confined volume of solution, the different phases formed fill up the whole volume of the solution.

When the structural transitions are caused by *attractive* forces the larger structures may now either separate out from, or coexist with, the smaller aggregates or monomers in solution. Such attractive forces arise between uncharged amphiphilic surfaces, e.g., those having nonionic or zwitterionic headgroups, and for charged headgroups in high salt where the electrostatic repulsion is screened. Let us again consider the transformation of small micelles or vesicles into large liposomes. Now, however, because the forces are attractive, the equilibrium separation between the bilayers in the liposomes will be at their potential-energy minimum, where the depth of the minimum is W_0 per unit area (illustrated in Fig. 18.2). While the smaller micelles are clearly favoured entropically, the liposomes could be thermodynamically more favourable if the value of W_0 is sufficiently large. The problem is to establish how these two effects compete in determining which structure is formed at the CMC and at higher concentrations.

If M is the micelle or vesicle aggregation number and $\mathbf{M}$ the liposome aggregation number ($\mathbf{M} \gg M$), then equating the chemical potentials of molecules in all the possible dispersed and aggregated states gives at equilibrium

$$\mu_1^0 + kT \log X_1 = \mu_M^0 + (kT/M) \log (X_M/M) = \mu_{\mathbf{M}}^0 + (kT/\mathbf{M}) \log (X_{\mathbf{M}}/\mathbf{M})$$

monomers   micelles/vesicles   liposomes/superaggregates

(16.31)

$$X_{\mathbf{M}}/\mathbf{M} = \{(X_M/M) \exp[M(\mu_M^0 - \mu_{\mathbf{M}}^0)/kT]\}^{\mathbf{M}/M}. \qquad (16.32)$$

The concentration at which $X_{\mathbf{M}} = X_M$ is therefore

$$(X_M)_{\mathrm{crit}} \approx M \exp[-M(\mu_M^0 - \mu_{\mathbf{M}}^0)/kT]. \qquad (16.33)$$

Thus, depending on M and the difference in the energies ($\mu_M^0 - \mu_{\mathbf{M}}^0$) per surfactant molecule in the micellar and liposome states (which includes contributions from both *intra*bilayer and *inter*bilayer interactions) the latter may form spontaneously at the CMC, or—if $(X_M)_{\mathrm{crit}}$ is greater than the CMC—at some higher concentration, while the background concentration of monomers and the smaller aggregates remains unchanged. We may conveniently term such transitions first and second CMCs. If $M \gg M$, the concentration at which large aggregates begin to form will be sharp and in all respects analogous to the first CMC. Note, too, that if we put $M = 1$ in Eq. (16.33), it reduces to Eq. (16.13) for the first CMC. Indeed, if we consider the smaller aggregates as if they were 'monomers', Eqs (16.32) and (16.33) are completely analogous to Eqs (16.3) and (16.13).

□ ● **WORKED EXAMPLE** ●

Question: In a certain system the depth of the potential energy minimum between two lipid bilayers is $W = 8 \times 10^{-2}$ mJ m^{-2}. If this is the only energy difference per molecule in a vesicle and in a liposome, estimate the 'critical liposome concentration'. Assume that the surface area occupied by each lipid molecule is 0.70 nm^2, that each vesicle contains 3000 molecules, and that the liposomes are much larger than the vesicles.

Answer: The free energy difference *per molecule* in vesicles and liposomes is

$$(\mu_M^0 - \mu_\mathbf{M}^0) \approx \tfrac{1}{2} \times 8 \times 10^{-5} \times 0.70 \times 10^{-18} = 2.8 \times 10^{-23} \text{ J},$$

or 0.0068 kT at 298 K. (*Note*: this corresponds to g_0 in Fig. 18.2.) If $M = 3000$ is the vesicle aggregation number, then from Eq. (16.33) a vesicle-to-liposome transition will occur at a lipid concentration of

$$(X_M)_{\text{crit}} = 3000 \exp(-3000 \times 0.0068) \approx 4 \times 10^{-6},$$

that is, at about 2×10^{-4} M (which may be compared with the first CMC of about 10^{-10} M typical of vesicle-forming lipids). This analysis, however, neglects any possible energy difference arising from the different bilayer curvatures in vesicles and planar bilayers which will also contribute to $(\mu_M^0 - \mu_\mathbf{M}^0)$ in Eq. (16.33). Curvature effects are discussed in Section 17.9.

□

16.10 CONCLUSION

We have gone as far as is possible with a formal analysis of the statistical thermodynamics of self-assembly, and to proceed further we must now consider the different types of interactions occurring between amphiphilic molecules in aggregates more specifically. In Chapter 17 we shall quantify these interactions and in particular investigate how the geometric shape constraints of surfactant and lipid molecules restrict their assembly into aggregates of different shapes. The important conclusion to be drawn from this chapter is that once the aggregates' shape is known their physical properties are necessarily given by the thermodynamic equations developed in this chapter, for example, a dilute dispersion of (one-component) cylindrical micelles *must* be polydisperse and their size *must* increase with concentration. Therein lies the beauty and power of thermodynamics.

PROBLEMS AND DISCUSSION TOPICS

16.1 In the so-called 'pseudo-phase' approximation of the theory of micelles, it is assumed that only two species exist in the solution: monomers and monodisperse micelles all with the same aggregation number, N. The monomers are at a concentration X_1 and the concentration of monomers in the micelles is X_N (note that the *micellar* concentration is X_N/N). For the case where $(\mu_1^0 - \mu_N^0)/kT = 10$, plot the concentration of surfactant monomers X_1 as a function of total concentration C from $C = 0$ to twice the theoretical CMC for $N = 10$, 50 and 500. For each case find the value of C at which $X_1 = X_N$, and comment on the shapes of the curves and how these values compare with the theoretical CMC as calculated using Eq. (16.13).

16.2 Ionic surfactants in water must be considered as a three-component system (unlike non-ionic surfactants which do not dissociate to give off counterions into solution). An ionic surfactant is fully dissociated in the monomer form but only a fraction f is dissociated in the micellar form. Show that above the CMC the concentration X_1 of monomers will not remain constant as shown in Fig. 16.5, but will fall according to $X_1 \approx$ constant$/C^f(1-f)^f$.

16.3 Estimate the interfacial tension γ_i of an oil–water interface at 25°C where the aqueous phase contains surfactant micelles and where the interface contains a compact (liquid-condensed) surfactant monolayer. Assume that the total surfactant concentration in the aqueous phase is $C = 10^{-2}$ M (well above the CMC), that the micelles have a mean aggregation number of $M = 100$, and that the headgroup area is $a_0 = 0.40$ nm². (*Answer:* $\gamma_i \approx (kT/Ma_0)\ln(M/C) \approx 1$ mJ m⁻².)

CHAPTER 17

AGGREGATION OF AMPHIPHILIC MOLECULES INTO MICELLES, BILAYERS, VESICLES AND BIOLOGICAL MEMBRANES

17.1 INTRODUCTION: EQUILIBRIUM CONSIDERATIONS OF AMPHIPHILIC STRUCTURES

Amphiphilic molecules such as surfactants, lipids, copolymers and proteins can associate into a variety of structures in aqueous solutions (Fig. 16.1), which can transform from one to another when the solution conditions are changed, for example, the electrolyte concentration or pH. To understand these structural aspects one requires an understanding not only of the thermodynamics of self-assembly (discussed in Chapter 16) but also of the forces between the amphiphilic molecules *within* the aggregates and how these, too, are affected by solution conditions. These two factors (thermodynamics and intra-aggregate forces), together with the strength of the inter-aggregate forces *between* aggregates in more concentrated systems, determine the equilibrium structures formed.

In this chapter we shall investigate the interaction forces between amphiphilic molecules within aggregates in more detail, and we shall see how these naturally lead to *molecular packing* considerations in determining which structures are formed. In the following chapter we consider the forces between these structures and their consequences for the equilibrium phase state of the whole system.

Before proceeding it is worth clarifying what one means by 'equilibrium structures', especially with reference to the Gibbs phase rule. Amphiphilic structures can be hard and solid-like, but they are more often soft or *fluid-like*, with the molecules in constant thermal motion within each aggregate— twisting, turning and bobbing in and out of the surface. Thus, unlike monodisperse colloidal particles, these structures have no definite size or shape, but only a *distribution* about some mean value. In Section 16.7 we saw that this distribution can sometimes be very broad.

Further, we also saw that it is possible for the equilibrium distribution to peak at more than one value of N. Thus, in principle, small aggregates such as micelles can be in thermodynamic equilibrium with large aggregates such as vesicles, all within the same one-phase system. While it may appear that a large vesicle or liposome, being a macroscopic structure, should be considered as a separate phase, this is strictly not so. The sizes of the structures play no role in the thermodynamic definition of what constitutes a single phase, which merely requires that the properties be uniform throughout the phase. Thus, in principle, structures may be very large and macroscopic and yet not constitute a separate phase if their number density in solution (or space) remains uniform throughout the whole system which, of course, must be even larger.

Genuine two-phase and three-phase systems can also occur, where monomers, micelles, vesicles or liposomes separate out into distinct phases in equilibrium with each other. However, such phase separations can take a long time to reach equilibrium, so that it is often difficult to identify experimentally the true thermodynamic state of an amphiphilic system, let alone study the structure and dynamics of the aggregates themselves.

17.2   OPTIMAL HEADGROUP AREA

The major forces that govern the self-assembly of amphiphiles into well-defined structures such as micelles and bilayers derive from the hydrophobic attraction at the hydrocarbon–water interface, which induces the molecules to associate, and the hydrophilic, ionic or steric repulsion of the headgroups, which imposes the opposite requirement that they remain in contact with water. These two interactions compete to give rise to the idea of two 'opposing forces' (Tanford, 1980) acting mainly in the interfacial region: the one tending to decrease and the other tending to increase the interfacial area a per molecule (the effective headgroup area) exposed to the aqueous phase (Fig. 17.1).

The attractive interaction arises mainly from the hydrophobic or interfacial tension forces which act at the essentially fluid hydrocarbon–water interface. This interaction may be represented by a positive interfacial free energy per unit area characteristic of the hydrocarbon–water interface of $\gamma \approx 50$ mJ m^{-2}, though as discussed in Section 16.6 there are indications that at a hydrophilic headgroup–water interface this value may be much reduced and closer to $\gamma \approx 20$ mJ m^{-2} (Parsegian, 1966; Jönsson and Wennerström, 1981; see also Table 16.2). In the case of biological lipids, about half of these usually contain one or more unsaturated C=C bonds and these have much lower interfacial

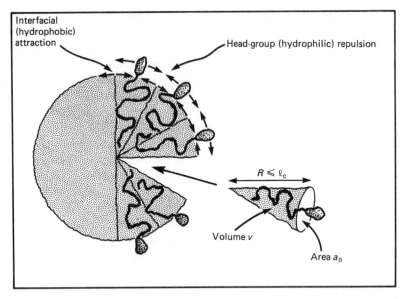

Fig. 17.1. The hydrocarbon interiors in both micelles and bilayers are normally in the fluid state (Shinitzky *et al.,* 1971; Lindblom and Wennerström, 1977). Repulsive headgroup forces and attractive hydrophobic interfacial forces determine the optimum headgroup area a_0 at which μ_N^0 is a minimum (see Fig. 17.2). The chain volume v and chain length l_c set limits on how the fluid chains can pack together, on average, inside an aggregate. Thus, the mean molecular conformation depends on a_0, v and l_c.

energies with water than the unsaturated homologues (cf. Table 15.1 where γ_{12} falls from 52 to 19 mJ m^{-2} on replacing a C—C bond by a C=C bond in octadecane).

Thus the attractive interfacial free energy contribution to μ_N^0 may be simply written as γa, where γ is expected to lie between 20 and 50 mJ m^{-2}. This simple expression is modified when account is taken of the non-purely liquid-like nature of the hydrocarbon chains, but it serves as a good first approximation.

The repulsive contributions are still too complex and difficult to formulate explicitly (Israelachvili *et al.,* 1980a; Puvvada and Blanckstein, 1990). Between mobile hydrophilic headgroups these include a steric contribution, a hydration force contribution and an electrostatic double-layer contribution if the headgroups are charged (Payens, 1955; Forsyth *et al.,* 1977). Luckily, these separate contributions do not have to be known explicitly. This is because—as in the two-dimensional van der Waals equation of state—we expect the first term in any energy expansion to be inversely proportional to the surface area occupied per headgroup a (cf. pressure $\propto 1/a^2$ in Eq. (2.29)).

The total interfacial free energy per molecule in an aggregate may therefore be written, to first order, as

$$\mu_N^0 = \gamma a + K/a, \tag{17.1}$$

where K is a constant. We shall initially assume that both these forces act in the same plane at the hydrophobic–hydrophilic interface (Fig. 17.1). The minimum energy is therefore given when $\partial \mu_N^0 / \partial a = 0$, that is,

$$\mu_N^0(\text{min}) = 2\gamma a_0, \qquad a_0 = \sqrt{K/\gamma}. \tag{17.2}$$

a_0 will be referred to as the *optimal surface area* per molecule, defined at the hydrocarbon–water interface. The interfacial energy per molecule, Eq. (17.1) may now be expressed in the more convenient form

$$\mu_N^0 = 2\gamma a_0 + \frac{\gamma}{a}(a - a_0)^2 \tag{17.3}$$

in which the unknown constant K has been eliminated, so that μ_N^0 as a function of a is now in terms of the two known or measurable parameters, γ and a_0.

We see therefore how the concept of opposing forces leads to the notion of an optimal area per headgroup at which the total interaction energy per lipid molecule is a minimum (Fig. 17.2). Moreover, for truly fluid hydrocarbon chains, the optimal area should not depend strongly on the chain length or

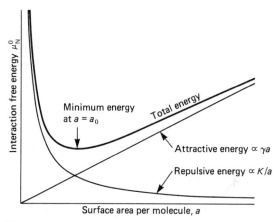

Fig. 17.2. Optimal headgroup area a_0 at which the opposing forces of headgroup repulsion and interfacial (hydrophobic) attraction are balanced.

on the number of chains, as is indeed found experimentally (Gallot and Skoulios, 1966; Reiss-Husson, 1967; Lewis and Engelman, 1983b).

The above equations, while crude, contain the essential features of interlipid interactions in micelles, bilayers and membranes. They imply that, to a first approximation, the interaction energy between lipids has a minimum at a certain headgroup area a_0, about which the energy varies parabolically (i.e., elastically). The equations ignore three second-order but nevertheless important effects, some of which are discussed later: (i) specific headgroup interactions such as ionic bridging, (ii) specific chain–chain interactions (since the hydrocarbon chains are never perfectly fluid), and (iii) the effect of surface curvature on μ_N^0.

17.3   GEOMETRIC PACKING CONSIDERATIONS

Having established the equations that adequately describe the interactions between amphiphilic molecules within an aggregate, we have yet to establish the most favoured structures. The geometry or 'packing' properties of the molecules now enter the picture. These depend on their optimal area a_0, the volume v of their hydrocarbon chain or chains, which will be assumed to be fluid and incompressible, and the maximum effective length that the chains can assume. We shall call this the *critical chain length*, l_c. This length sets a limit on how far the chains can extend; smaller extensions are allowed but further extensions are not, these being prevented by a sharp rise in the interaction energy. The critical length l_c is a semiempirical parameter, since it represents a somewhat vague cutoff distance beyond which hydrocarbon chains can no longer be considered as fluid. However, as may be expected, it is of the same order as, though somewhat less than, the fully extended molecular length of the chains l_{max} (Tanford, 1973, 1980; Israelachvili *et al.*, 1976, 1977). According to Tanford, for a saturated hydrocarbon chain with n carbon atoms:

$$l_c \leqslant l_{max} \approx (0.154 + 0.1265n)\,\text{nm} \tag{17.4}$$

and

$$v \approx (27.4 + 26.9n) \times 10^{-3}\,\text{nm}^3. \tag{17.5}$$

Note that for large n, $v/l_c \approx 0.21\,\text{nm}^2 \approx$ constant, which is close to the minimum cross-sectional area that a hydrocarbon chain can have.

Once the optimal surface area a_0, hydrocarbon chain volume v and critical length l_c are specified for a given molecule—all these being measurable or estimable—one may ascertain which structures the molecules can pack into consistent with these geometric constraints. It turns out that these can be satisfied by a great variety of different structures. However, since μ_N^0 will be roughly the same for all these structures (since a_0 is the same) entropy will favour the structure with the smallest aggregation number, say, at $N = M$, and this structure is unique! Larger structures will be entropically unfavoured, while smaller structures, where packing constraints force the surface area a to increase above a_0, will be energetically unfavoured.

It has been shown previously (Israelachvili *et al.*, 1976) that for lipids of optimal area a_0, hydrocarbon volume v and critical chain length l_c, the value of the dimensionless *packing parameter* or *shape factor*, v/a_0l_c, will determine whether they will form spherical micelles ($v/a_0l_c < \frac{1}{3}$), non-spherical micelles ($\frac{1}{3} < v/a_0l_c < \frac{1}{2}$), vesicles or bilayers ($\frac{1}{2} < v/a_0l_c < 1$), or 'inverted' structures ($v/a_0l_c > 1$). Each of these structures corresponds to the minimum-sized aggregate in which all the lipids have minimum free energy. These structures will now be described in turn.

17.4 SPHERICAL MICELLES

For molecules to assemble into spherical micelles, their optimal surface area a_0 must be sufficiently large and their hydrocarbon volume v sufficiently small that the radius of the micelle R will not exceed the critical chain length l_c. From simple geometry we have, for a spherical micelle of radius R and mean aggregation number M (Fig. 17.1),

$$M = 4\pi R^2/a_0 = 4\pi R^3/3v, \tag{17.6}$$

namely,

$$R = 3v/a_0, \tag{17.7}$$

so that only for

$$\frac{v}{a_0l_c} < \frac{1}{3} \tag{17.8}$$

will the amphiphiles be able to pack into a spherical micelle with their headgroup areas equal to a_0 and with the micelle radius R not exceeding l_c.

An example of such micelle-forming amphiphiles is the 12-carbon chain

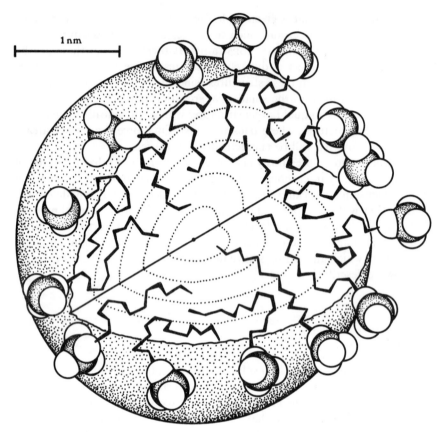

Fig. 17.3. A sodium dodecylsulphate (SDS) micelle drawn to scale. The micelle contains 60 sodium dodecylsulphate molecules. The hydrocarbon chains pack at liquid hydrocarbon density in the core where they are almost as disordered as in the bulk liquid state. Each of the five spherical shells contains approximately the correct number of chain segments to ensure even chain packing density throughout. Note that all segments of the chain spend an appreciable proportion of time near the micelle surface. Thus, even though the core is almost completely devoid of water each segment samples the hydrophilic environment. Drawing based on calculations by Gruen (1981) and Gruen and de Lacey (1984).

sodium dodecyl sulphate surfactant (SDS) in water (Fig. 17.3) where experimentally $M \approx 74$ (Cabane *et al.*, 1985). Putting $n = 12$ into Eq. (17.5) gives $v = 0.3502 \, \text{nm}^3$. Equation (17.6) then gives $a_0 \approx 0.57 \, \text{nm}^2$, and Eq. (17.7) gives for the optimal micelle radius: $R \approx 1.84 \, \text{nm}$. Now for a 12-carbon chain, Eq. (17.4) gives $l_c \approx 1.67 \, \text{nm}$, which is 0.17 nm short of the required (optimal) radius. Thus, for SDS micelles in water, $v/a_0 l_c \approx 0.37$, which means that they just cannot pack into spheres and so must be slightly non-spherical.

● WORKED EXAMPLE ●

Question: Below what aggregation number will SDS micelles in water be spherical and how could this be achieved in practice.

Answer: After some thought or algebra, using Eqs (17.6)–(17.8), it becomes clear that the answer is simply

$$M = \frac{4\pi l_c^3}{3v} = \frac{4\pi[(0.154 + 0.1265 \times 12)10^{-9}]^3}{3[(27.4 + 26.9 \times 12)10^{-30}]} = 56.$$

We have seen that experimentally, $M = 74$ and $v/a_0 l_c \approx 0.37$ for SDS in water. Since v and l_c are fixed, the only way to reduce $v/a_0 l_c$ to the required value of 0.33 is to raise a_0 by about 10%. In practice, this could be achieved by raising the pH of the solution. This would increase the degree of ionization of the negatively charged headgroups which increases the repulsion between them (see Problem 12.7), resulting in an increase in a_0. The required value of a_0 is given by Eq. (17.8) as $3v/l_c = 0.63$ nm^2. If the surfactant were cationic, we would decrease the pH. ☐

The mean size of spherical micelles is relatively insensitive to the surfactant concentration above the CMC, and the micelles are fairly monodisperse. The standard deviation σ in the aggregation number about the mean (at $N \approx M$ where $a = a_0$) may be obtained by first noting that μ_N^0 of Eq. (17.3) can be written as

$$\mu_N^0 = \mu_M^0 + \frac{\gamma}{a}(a - a_0)^2. \tag{17.9}$$

Now for a spherical micelle, we have $N = 4\pi R^2/a = 4\pi R^3/3v = 36\pi v^2/a^3$. Equation (17.9) may therefore be expressed as

$$\mu_N^0 - \mu_M^0 = \Lambda(N - M)^2, \qquad \text{where } \Lambda = \gamma a_0/9M^2. \tag{17.10}$$

Substituting this value for Λ into Eq. (16.30) gives

$$\sigma = \sqrt{(9kT/2\gamma a_0)M}. \tag{17.11}$$

Typically, for γ lying in the range 20–50 mJ m^{-2} and $a_0 \approx 0.60$ nm^2, we therefore expect

$$\sigma \approx \sqrt{M}, \tag{17.12}$$

for example, for $M \approx 60$, $\sigma \approx 8$. Aniansson *et al.* (1976) found that for a variety of sodium alkyl sulphate micelles, the value of $\sigma/\sqrt{M}$ varies between 1 and 2. The distribution or spread about M is thus fairly narrow but by no means sharp, as illustrated in Fig. 16.6.

17.5   NON-SPHERICAL AND CYLINDRICAL MICELLES

Most lipids that form spherical micelles have charged headgroups since this leads to large headgroup areas a_0. Addition of salt partially screens the electrostatic inter-headgroup repulsion and thereby reduces a_0. Those lipids that possess smaller headgroup areas such that $\frac{1}{3} < v/a_0 l_c < \frac{1}{2}$ cannot pack into spherical micelles but can form cylindrical (rod-like) micelles. Falling into this category are single-chained lipids possessing charged headgroups in high salt (e.g., SDS, CTAB) or those possessing uncharged, non-ionic or zwitterionic headgroups (e.g., $C_{12}E_5$, lysolecithin).

As discussed in Section 16.7 rod-like aggregates must have very unusual properties: they are large and polydisperse, and their mean aggregation number is very sensitive to the total lipid concentration C. According to Eq. (16.24) above the CMC, their mean aggregation number should increase proportionally to $\sqrt{C}$, which is indeed found to be the case experimentally (Mazer *et al.*, 1976; Missel *et al.*, 1980).

It is important to note that the unusual properties of cylindrical micelles are due entirely to 'end effects': at each end the lipids are forced to pack into hemispherical caps with a headgroup area a determined by $v/a l_c = \frac{1}{3}$, so that $a > a_0$ since $v/a_0 l_c > \frac{1}{3}$. The unfavourable energy of these end lipids determines the magnitude of the interaction parameter α in Eq. (16.24), which increases as a_0 decreases. Since $\langle N \rangle = 2\sqrt{C e^\alpha}$ it is not surprising that the growth of cylindrical micelles is very sensitive to changes in temperature, chain length, and—for ionic lipids—to ionic strength (Missel *et al.*, 1980, 1983; Malliaris *et al.*, 1985; Lin *et al.*, 1990; Wennerström and Lindman, 1979). For example, their sensitivity to increasing ionic strength arises from its effect on decreasing a_0, which increases α. As might be expected the mean aggregation number increases dramatically with increased ionic strength. Thus, the aggregation number of SDS micelles in 0.6 M NaCl is ~ 1000 compared to ~ 60 in water. The unfavourable end energy of cylindrical micelles may be eliminated if the two ends join, thereby forming a toroidal micelle (the two-dimensional analogue of a vesicle). However, toroidal micelles have so far not been observed.

17.6 BILAYERS

Lipids that form bilayers are those that cannot pack into micellar structures due to their small headgroup area a_0 or because their hydrocarbon chains are too bulky to fit into such small aggregates while maintaining the surface area at its optimal value. For bilayer-forming lipids, the value of $v/a_0 l_c$ must lie close to 1, and this requires that for the same headgroup area a_0 and chain length l_c, their hydrocarbon volume v must be about twice that of micelle-forming lipids (for which $v/a_0 l_c$ is in the range $\frac{1}{3}$ to $\frac{1}{2}$). Therefore, lipids with *two* chains are likely to form bilayers, and indeed most of them do. For example, single-chained lysolecithins form small but non-spherical micelles while lecithins with two alkyl chains, such as di-C_{14} and di-C_{16} lecithins, form bilayers (Fig. 17.4). Similarly, CTAB forms micelles while the double-chained analogue forms bilayers.

The doubling of the chains also affects other aggregate properties, both static and dynamic. First, it increases the hydrophobicity of the lipids, which in turn drastically lowers their CMC: compare the CMCs of common micelle-forming lipids (10^{-2}–10^{-5} M) with those of bilayer-forming lipids (10^{-6}–10^{-10} M). We shall return to comment on the biological significance of such low CMCs later. Second, it increases the lifetimes or 'residence times', τ_R, of the molecules within aggregates. Self-assembled structures are usually highly dynamic, with the molecules in constant thermal motion within the aggregates as well as exchanging with monomers in the bulk solution (Pfeiffer et al., 1989). This diffusive exchange may be understood as an activation process, whereby a certain activation energy ΔE has to be surmounted before a molecule can escape from the micelle or bilayer into the bulk solution. Consider the molecules within a bilayer to be jumping about with some characteristic 'collision time', τ_0. Since the probability of a molecule leaving the bilayer each time it hits the interface is $e^{-\Delta E/kT}$, the mean lifetime of a molecule is therefore $\tau_0/e^{-\Delta E/kT}$. After a little thought it becomes clear that ΔE is essentially the same as $(\mu_1^0 - \mu_N^0)$ in Eq. (16.13), which allows us finally to express the residence time in the simple form

$$\tau_R = \tau_0/e^{-\Delta E/kT} \approx 55\tau_0/\text{CMC}, \qquad (17.13)$$

where the CMC is in units of M. Typical motional correlation times for amphiphiles in micelles and bilayers are in the range $\tau_0 = 10^{-9}$–10^{-7} s. Thus for micelles and bilayers we find

$$\tau_R \text{ (micelles)} \sim 55 \times 10^{-9}/10^{-3} \sim 10^{-4} \text{ s},$$

$$\tau_R \text{ (bilayers)} \sim 55 \times 10^{-7}/10^{-10} \sim 10^{+4} \text{ s}.$$

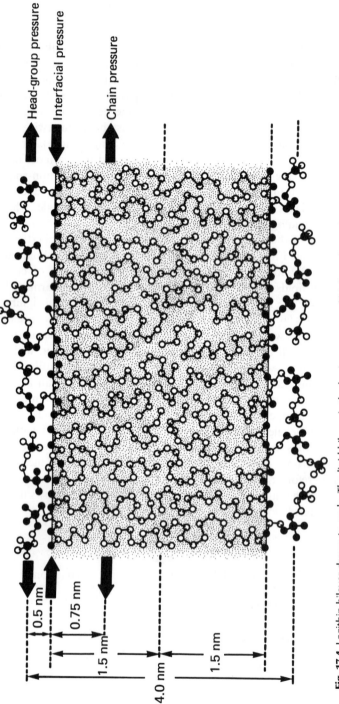

Fig. 17.4. Lecithin bilayer drawn to scale. The lipid bilayer is the basic structure of biological membranes, and most membrane lipids contain two hydrocarbon chains. The lipids diffuse rapidly in the plane of the bilayer, covering a distance of about 1 μm in 1 s. They also cross the bilayer from one side to the other ('flip-flop'), as well as exchange with lipids in the solution, but at much slower rates, of the order of hours (cf. the lifetime of single-chained surfactants in micelles of 10^{-5} to 10^{-3} s). (From Israelachvili et al., 1980a.)

These values are typical of those measured (Wennerström and Lindman, 1979; Wimley and Thompson, 1990). Furthermore, from the measured CMCs of lipids (Table 16.2) Eq. (17.13) suggests that exchange rates should fall by a factor of about 4–10 per two CH_2 groups added to the chains, in good agreement with measured values of about 8 for phospholipids (Homan and Pownall, 1988). Note, however, that residence times depend on the individual molecules and not on the structures. Thus, when a double-chained lipid bilayer hosts a single-chained surfactant molecule, the residence time of the guest lipid will be much shorter than that of the host lipids.

Another dynamic effect in bilayers is transbilayer lipid exchange, or the 'flip-flop' of molecules from one side to the other. This may also be viewed as a diffusive exchange process, except that the energy ΔE in Eq. (17.13) now refers to the energy needed to put the hydrophilic headgroup into the hydrophobic region of the bilayer. However, other mechanisms for flip-flop have also been proposed such as the diffusion of molecules around the walls of transient pores (see Fig. 17.9 and Problem 17.9). Flip-flop energies are generally higher than those for lipid exchange so that flip-flop times are longer and typically range from minutes to days (10^2–10^5 s) and, unlike interbilayer exchange, depend more on the headgroup than on the chains (Homan and Pownall, 1988; Wimley and Thompson, 1990).

When a bilayer is stretched from its equilibrium state it expands elastically. For fluid bilayers the area compressibility modulus k_a may be readily estimated from Eq. (17.3) since by definition

$$\text{elastic energy} = \tfrac{1}{2}k_a(a - a_0)^2/a, \qquad (17.14)$$

which gives

$$k_a \approx 2\gamma \text{ per monolayer}, \qquad (17.15a)$$

$$\approx 4\gamma \text{ per bilayer}. \qquad (17.15b)$$

Assuming $\gamma \approx 20$–50 mJ m^{-2}, we therefore expect $k_a \approx 4\gamma \approx 80$–$200$ mJ m^{-2} for bilayers. This estimate compares well with measured values on fluid lipid bilayers and free biological cell membranes which range from 100–230 mJ m^{-2} (Kwok and Evans, 1981; Evans and Rawicz, 1990; Marsh, 1990). Bilayers also have an elastic bending or curvature modulus, k_b, which affects the energy of curved bilayers and vesicles (Section 17.9). Both the area and bending moduli play an important role in determining the thermal fluctuation forces between bilayers, discussed in Section 18.6.

So far, it has been implicitly assumed that the hydrocarbon chains in micelles and bilayers are in the fluid state. At room temperature this is the case for most micelle-forming single-chained surfactants, as well as for

TABLE 17.1 Chain melting (phase transition) temperatures, T_c, of some common double-chained lipid bilayers in water (at pH 7) in order of increasing T_c

Lipid (giving number of carbons per chain)	Headgroup type[a] and chain melting temperature,[b] T_c (°C)				Melting point of n-alkane with same number of carbon atoms
	PC	PG⁻	PS⁻	PE	
Saturated					
Dilauroyl (12)	−2	0	13	30	−9.6
Dimyristoyl (14)	23	24	36	49	5.9
Dipalmitoyl (16)	41	41	52	64	18.2
Distearoyl (18)	55	55	68	74	28.2
Unsaturated (cis)					
Dioleoyl (18)	−22	−18	−7	−16	−30

[a] PC: phosphatidylcholine (zwitterionic); PG⁻: phosphatidylglycerol (negatively charged); PS⁻: phosphatidylserine (negatively charged); PE: phosphatidylethanolamine (zwitterionic).
[b] Compiled from Cevc and Marsh (1987) and Marsh (1990).

bilayer-forming double-chained lipids having less than 14 carbons per chain. Unsaturated and branched chained lipids remain in the fluid state down to much lower temperatures. Some typical lipid chain melting temperatures, T_c, are given in Table 17.1. We see that these are somewhat above the melting points of the corresponding n-alkane, i.e., without the headgroup. At temperatures below T_c, bilayers cease to be fluid-like: the elastic moduli increase, lateral diffusion and flip-flop rates fall, etc. However, bilayers below T_c do not always freeze into crystalline solids, but often retain some of their fluid-like properties, for example, the headgroups may still be highly mobile even though the chains are not. Such bilayers are usually referred to as being in the *gel* state. For an excellent book covering all aspects of the physical properties of lipid bilayers see *Phospholipid Bilayers* by Cevc and Marsh (1987), and the *CRC Handbook of Lipid Bilayers* (Marsh, 1990) for factual details.

17.7 VESICLES

Under certain conditions it becomes more favourable for closed spherical bilayers (vesicles) to form rather than infinite planar bilayers. This arises since in a closed bilayer the energetically unfavourable edges are eliminated at a finite, rather than infinite, aggregation number, which is also entropically favoured. Thus, so long as the lipids in a curved bilayer can maintain their areas at their optimal value, vesicles should be the preferred structures. What then determines the radii of vesicles? First, let us note that if $v/a_0 l_c = 1$, only planar bilayers will form. For a bilayer to curve, the lipids in the outer monolayer must be able to pack, on average, into truncated cones. This requires that $v/a_0 l_c < 1$. Simple geometric considerations show (Israelachvili et al., 1976) that for $\frac{1}{2} < v/a_0 l_c < 1$, the radius of the smallest vesicle that may be formed without forcing the headgroup area a in the outer monolayer to exceed a_0 is given by

$$R_c \approx l_c \left[\frac{3 + \sqrt{3(4v/a_0 l_c - 1)}}{6(1 - v/a_0 l_c)} \right] \approx \frac{l_c}{(1 - v/a_0 l_c)} \qquad (17.16)$$

which is the *critical radius* below which a bilayer cannot curve without. introducing unfavourable packing strains on the lipids. So long as a vesicle's radius does not fall below R_c the lipids in both the inner and outer monolayers can pack with their surface areas at the optimal value a_0 and with the two hydrocarbon chain regions not exceeding l_c. For $R < R_c$, a must exceed a_0 in the outer monolayer and such vesicles are energetically unfavoured, while for $R > R_c$ the vesicles are entropically unfavoured. For a vesicle of radius R_c and bilayer hydrocarbon thickness $t \approx 2v/a_0$, the aggregation number is

$$N \approx 4\pi[R_c^2 + (R_c - t)^2]/a_0. \qquad (17.17)$$

As an example, we may apply the above equations to the much-studied egg lecithin vesicles for which $a_0 \approx 0.717 \, \text{nm}^2$ and $v \approx 1.063 \, \text{nm}^3$. Taking $l_c \approx 1.75 \, \text{nm}$ (i.e., $v/a_0 l_c \approx 0.85$), Eqs (17.16) and (17.17) then yield $R_c \approx 11 \, \text{nm}$, $N \approx 3000$, $t \approx 3.0 \, \text{nm}$, and an outside-to-inside lipid ratio of $R_c^2/(R_c - t)^2 \approx 1.9$, all in good agreement with measured values.

Many people believe that vesicles represent the prototypes of early living cells, and it has now been demonstrated that certain biological lipids as well as synthetic surfactants and mixtures can spontaneously self-assemble into stable monodisperse vesicles (Brunner et al., 1976; Gains and Hauser, 1983; Hauser et al., 1983, 1990; Hauser, 1984; Talmon et al., 1983; Kaler et al., 1989; Madani and Kaler, 1990).

Finally, for lipids with very small optimal headgroup areas (e.g., $a_0 < 0.42 \, \text{nm}^2$ for double-chained lipids) or with bulky polyunsaturated chains (large v, small l_c), their value of $v/a_0 l_c$ will exceed unity. According to Eq. (17.16), when $v/a_0 l_c > 1$, R_c becomes negative. What this means is that such lipids form 'inverted' micellar structures or precipitate out of solution (e.g., MGDG, unsaturated phosphatidylethanolamines, negatively charged lipids in the presence of Ca^{2+} ions, cholesterol).

17.8   FACTORS AFFECTING CHANGES FROM ONE STRUCTURE TO ANOTHER

We have seen that the geometric packing properties of different lipids may be conveniently expressed in terms of the packing parameter $v/a_0 l_c$ characteristic for each lipid in a given solution environment, the value of which determines the type of aggregate formed. Table 17.2 illustrates the structures formed by some common lipids, and how these can be modified by their ionic environment, temperature, chain unsaturation, etc., as will now be summarized.

(i) *Factors affecting headgroup area.* Lipids with smaller headgroup areas (high $v/a_0 l_c$) form larger vesicles, less-curved bilayers, or inverted micellar phases. For anionic headgroups, this can be brought about by increasing the salt concentration, particularly Ca^{2+}, or lowering the pH. This also has the effect of straightening (condensing) the chains.

(ii) *Factors affecting chain packing.* Introducing chain branching and unsaturation, particularly of *cis* double bonds, reduces l_c and thus increases $v/a_0 l_c$. Similar effects occur when the effective volume, v, of the chains is increased due to the penetration of organic molecules such as low MW alkanes into the chain regions.

Both the above effects lead to larger vesicles and ultimately to inverted structures. In the case of microemulsions (surfactant/water/oil mixtures), they lead to larger 'oil-in-water' droplets and ultimately to inverted 'water-in-oil' droplets.

(iii) *Effects of temperature T.* Changing the temperature can alter both a_0 and l_c so that these effects are more subtle and generally less well understood. For double-chained lipids in the fluid state ($T > T_c$), increasing T increases the hydrocarbon chain motion involving *trans-gauche* isomerization and thereby reduces their limiting length l_c. If the headgroup area a_0 does not change significantly, as occurs for charged headgroups, the net effect is an increase in $v/a_0 l_c$. However, changing T can also change a_0. For example, the areas of polyoxyethylene headgroups *decrease* with increasing T due to their increased hydrophobicity (Puvvada and Blanckstein, 1990), and this acts to enhance the increase in $v/a_0 l_c$. But the areas of more hydrophilic headgroups usually *increase* with T due to the increased steric repulsion between them, and this acts to decrease $v/a_0 l_c$. Thus, with increasing temperature non-ionic spherical micelles grow in size and become more cylindrical (Kato and Seimiya, 1986; Herrington and Sahi, 1988), while charged micelles shrink (Missel *et al.*, 1980). Zwitterionic micelles appear to behave somewhere in between, and their aggregation number hardly changes with temperature (Malliaris *et al.*, 1985).

(iv) *Lipid mixtures.* When an aggregate is composed of a lipid mixture, so long as the different molecules mix ideally and do not phase-separate, the

TABLE 17.2 Mean (dynamic) packing shapes of lipids and the structures they form

Lipid	Critical packing parameter $v/a_0 l_c$	Critical packing shape	Structures formed
Single-chained lipids (surfactants) with large head-group areas: *SDS in low salt*	< 1/3	Cone	Spherical micelles
Single-chained lipids with small head-group areas: *SDS and CTAB in high salt, nonionic lipids*	1/3-1/2	Truncated cone	Cylindrical micelles
Double-chained lipids with large head-group areas, fluid chains: *Phosphatidyl choline (lecithin), phosphatidyl serine, phosphatidyl glycerol, phosphatidyl inositol, phosphatidic acid, sphingomyelin, DGDG[a], dihexadecyl phosphate, dialkyl dimethyl ammonium salts*	1/2-1	Truncated cone	Flexible bilayers, vesicles
Double-chained lipids with small head-group areas, anionic lipids in high salt, saturated frozen chains: *phosphatidyl ethanolamine, phosphatidyl serine + Ca^{2+}*	~1	Cylinder	Planar bilayers
Double-chained lipids with small head-group areas, nonionic lipids, poly *(cis)* unsaturated chains, high *T*: *unsat. phosphatidyl ethanolamine, cardiolipin + Ca^{2+} phosphatidic acid + Ca^{2+} cholesterol, MGDG[b]*	> 1	Inverted truncated cone or wedge	Inverted micelles

[a] DGDG, digalactosyl diglyceride, diglucosyl diglyceride.
[b] MGDG, monogalactosyl diglyceride, monoglucosyl diglyceride.

aggregate properties may be treated, in a first approximation, in terms of some mean packing parameter intermediate between those of the individual components (Carnie et al., 1979). The sizes of vesicles may thus be conveniently increased or decreased by adding an appropriate amount of another component whose packing parameter is larger or smaller than that of the host lipid. In the case of microemulsion droplets, their sizes are often modulated in this way by adding a 'cosurfactant'. In other cases, totally new structures may be obtained by a suitable choice of lipid additive. For example, micelle-forming lysolecithin ($v/a_0 l_c < 0.5$) and non-aggregate forming cholesterol ($v/a_0 l_c > 1.0$) mix in certain proportions to form bilayer vesicles ($0.5 < v/a_0 l_c < 1.0$).

The above effects of headgroup size, ionic strength, unsaturation, and temperature are illustrated in Table 17.2 for dilute lipid structures in aqueous solutions. It is worth noting that micelles, bilayers and other self-assembling structures also form in other hydrogen-bonding liquids, such as ethylene glycol, formamide and hydrazine ($N_2 H_4$), but their properties have not been as extensively studied. The main requirement for the solvent appears to be that it have a high interfacial energy with hydrocarbons (see Table 15.1).

17.9   CURVATURE ELASTICITY OF BILAYERS AND MEMBRANES

The above treatment offers a rough and ready recipe for analysing the packing properties of lipid structures, but it is nevertheless incomplete. The only restriction we have so far considered is that the fluid chains cannot extend away from the headgroup further than a certain distance l_c. This leads to an increased energy only for radii below R_c, but for $R > R_c$ there is no curvature dependence because the lipid molecules can now arrange themselves with their areas at the optimal value. However, due to our simplifying assumptions we have missed two other contributions to the energy, which will now be investigated.

(i) *Headgroup repulsion.* The attractive and repulsive interfacial forces that determine a_0 and μ_N^0 in Eq. (17.3) were both assumed to act in the same plane, at the hydrocarbon–water interface, at which the surface area per molecule has been defined. This is likely to be true for the attractive interfacial tension force but not for the headgroup repulsive forces, which are likely to be centred at some finite distance D above the interface as shown in Figs 17.1 and 17.4.

(ii) *Chain repulsion.* It was assumed that hydrocarbon chains are entirely fluid and do not oppose any distortion until they become forced to extend beyond l_c. In other words, it was assumed that there is no curvature dependence of μ_N^0 due to the non-fluidity of chains. Gruen and de Lacey (1984), inspired by the earlier theory of Marcelja (1974), analysed the energetics of chain packing in both micelles and bilayers and found that the assumption of fluidity in bilayers is much less valid than it is in micelles. They concluded that $l_c \approx l_{max}$ in spherical and cylindrical micelles, but that $l_c \approx 0.7l_{max}$ in bilayers (cf. the assumed value of $l_c \approx 1.75$ nm $\approx 0.75l_{max}$ for egg lecithin vesicles in Section 17.7). Further, the restriction of chain freedom in bilayers gives rise to an additional lateral chain pressure acting inside the hydrocarbon region, i.e., at some negative distance $-D$ from the hydrocarbon–water interface (Fig. 17.4).

It can be shown that both the above effects result in an additional curvature dependence of μ_N^0 on R. For bilayers this has the form

$$\Delta E = -2\gamma tD/R^2 = \tfrac{1}{2}k_b/R^2 \text{ per unit area,} \tag{17.18}$$

where t is the bilayer thickness and D the distance out from the hydrocarbon–water interface where the repulsive forces are centred (Figs 17.1 and 17.4). By expressing the curvature or bending energy as $\tfrac{1}{2}k_b/R^2$ we adopt a convention where k_b is, by definition, the *bending modulus* of an elastic sheet. Using typical bilayer parameters such as $\gamma = 20$ mJ m^{-2}, $t = 3.0$ nm and $D = \pm 0.3$ nm, we obtain $k_b = \pm 4\gamma tD \approx \pm 7 \times 10^{-20}$ J. This is within the range of (positive) values measured for fluid bilayers: $(2–20) \times 10^{-20}$ J (Evans and Rawicz, 1990; Marsh, 1990; Abillon and Perez, 1990). However, surfactant *monolayers*, as occur at oil–water interfaces, appear to have much lower bending moduli, of the order of 5×10^{-21} J (Safinya et al., 1986).

Note that depending on where the repulsive and attractive forces are located, k_b can be positive or negative. In case (i) above, when the headgroup repulsion dominates, D is positive so that k_b is *negative*. Thus, headgroup repulsion *favours* bending right from the start. In case (ii) where the chain repulsion dominates, D is negative and k_b is therefore positive. Thus, chain repulsion acts to oppose bending. Let us consider these two cases in more detail (see also Israelachvili et al., 1976, 1980a, 1987c).

Positive curvature modulus $(k_b > 0)$

The interaction energy per vesicle of large radius R $(R > R_c)$ and aggregation number N now becomes

$$N\mu_N^0 = N\mu_\infty^0 + (\tfrac{1}{2}k_b/R^2)4\pi R^2 = N\mu_\infty^0 + 2\pi k_b \tag{17.19}$$

and thus

$$\mu_N^0 = \mu_\infty^0 + 2\pi k_b/N \qquad (17.20)$$

which is the same as Eq. (16.6) with $\alpha = 2\pi k_b/kT$. Thus, if C is the total lipid concentration (in mol dm^{-3}/55.5), the mean aggregation number M of elastic bilayer vesicles will be given by Eq. (16.23) as

$$M = \sqrt{Ce^\alpha} = e^{\pi k_b/kT}\sqrt{C} \qquad (17.21)$$

and the number density of vesicles will decay exponentially with their aggregation number, N, according to Eq. (16.25):

$$\text{vesicle distribution} \quad X_N/N = \text{Const. } e^{-N/M}. \qquad (17.22)$$

Thus, for typical dilute lipid concentrations of $C = 10^{-4}$ mol dm^{-3} and below, we see that if $k_b > 2 \times 10^{-20}$ J, vesicles will be large ($M > 10\,000$) and polydisperse, and their mean size will increase with the total lipid concentration (according to $R \propto C^{1/4}$). But if $k_b < 2 \times 10^{-20}$ J, the bending modulus is too low to have any significant effect, and vesicle populations will now be small and monodisperse, with a mean radius $R \approx R_c$ determined by the critical packing parameter (Fig. 17.5).

Eventually, at high enough concentrations, large vesicles must always transform into extended bilayers, either because of space requirements or because of the attractive forces between them, as discussed in Section 16.8.

Negative curvature modulus ($k_b < 0$)

A negative elasticity favours bending of a bilayer right from the start and leads to smaller vesicles than expected from simple packing considerations. The effect is enhanced for large repulsive headgroups, as well as for shorter hydrocarbon chains. Thus, below a certain chain length no stable vesicles should form; instead cylindrical or spherical micelles become the preferred structures as occurs for double-chained lecithins with 12 or fewer carbons per chain.

The above analysis should apply to any spherical structure in solution bounded by an elastic bilayer, monolayer or membrane, for example, microemulsion droplets. Figure 17.5 shows how the molecular interaction energy and size distribution of a vesicle population depends on the relative magnitudes of the bending modulus and molecular packing.

The simple 'packing model' developed so far takes us about as far as one

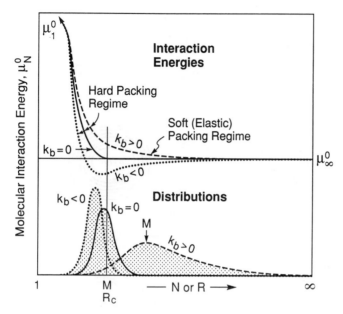

Fig. 17.5. Schematic variation of interaction energy per molecule in a vesicle with aggregation number N. Solid lines: no curvature elasticity ($k_b = 0$). Dashed lines: positive curvature elasticity ($k_b > 0$). Dotted lines: negative curvature elasticity ($k_b < 0$). The corresponding distributions of vesicle sizes (X_N/N) is also shown. Note that for large positive k_b the distribution is determined mainly by k_b (soft packing regime), while for small or negative k_b it is determined by $v/a_0 l_c$ (hard packing regime).

can go without resorting to much more sophisticated theoretical methods in which the various intermolecular and interaggregate interactions are treated to a full statistical thermodynamic analysis or a computer simulation. These have recently been extended from looking at small micelles (Jönsson and Wennerström, 1981; Gruen, 1985; Szleifer *et al.*, 1985, 1986; Jokela *et al.*, 1987; Blanckstein *et al.*, 1986) to looking at bilayers (Leermakers and Scheutjens, 1988; Egberts and Berendsen, 1988; Cevc and Marsh, 1987; De Loof *et al.*, 1991) including interbilayer interactions (Granfeldt and Miklavic, 1991) and even more complex self-assembling structures and biological membranes (Pastor, 1990).

17.10   BIOLOGICAL MEMBRANES

Membranes are the most common cellular structures in both animals and plants (Fig. 17.6) where they are involved in almost all aspects of cellular

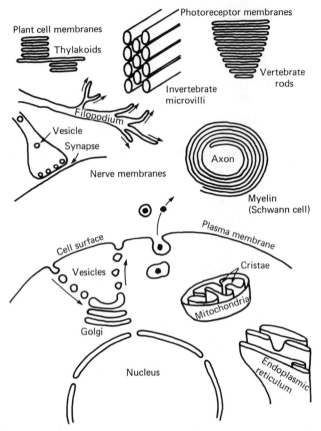

Fig. 17.6. Cellular membranes are thin sheets of lipids and proteins. Most biological membranes offer little resistance to bending.

activity ranging from simple mechanical functions such as motility, food entrapment and transport to highly specific biochemical processes such as energy transduction, immunological recognition, nerve conduction and biosynthesis (for general books on biomembranes see Bittar, 1980; Harrison and Lunt, 1980; Jain, 1988). Biological membranes are very complex and varied. They commonly contain 50 or more different proteins and a host of phospholipids and glycolipids with various headgroups, numbers of chains, chain lengths and degrees of unsaturation, as well as steroids (e.g., cholesterol) and other amphiphilic molecules. Yet in spite of their complexity there are many aspects of membrane structure that may be qualitatively understood in terms of the concepts we have already outlined, and it is best to start with a consideration of membrane lipids.

17.11   MEMBRANE LIPIDS

Most biological membrane lipids are *double-chained* phospholipids or glycolipids, with *16 to 18* carbons per chain, one of which is *unsaturated* or *branched* (see Table 16.1). These properties are not accidental but carefully designed by nature to ensure

(i) that biological lipids will self-assemble into thin bilayer membranes that can compartmentalize different regions within a cell as well as protect the inside of the cell from the outside;

(ii) that because of their extremely low CMC the membranes remain intact even when the bathing medium is grossly depleted of lipids; and

(iii) that because of the unsaturation or branching the membranes are in the fluid state at physiological temperatures.

For example, as shown in Table 17.1, the chain melting temperatures T_c of saturated di-C_{18} phospholipids are well above room temperature, whereas the unsaturated lipids have their T_c below $0°C$. As a consequence of this fluidity most biological membranes can easily deform and bend until limited by packing constraints. Fluid membranes also allow various solute molecules to pass through them and protein molecules to diffuse along them.

A further important aspect of lipid chain fluidity is that different lipid types can pack together, that is, mutually accommodate each other as well as other molecules, while remaining within a planar or curved bilayer configuration. In addition, the curvature can be regulated by altering the ratio of its constituent lipids. For example, addition of cone-shaped lipids such as lysolecithin (Table 17.2) to lecithin vesicles results in smaller vesicles since a mixture of such lipids can pack into more highly curved bilayers. Actually, only small amounts of single-chained lysolipids can be incorporated into bilayers or biological membranes before they break up (become solubilized) into small vesicles or micelles.

On the other hand, addition of wedge-shaped lipids such as phosphatidyl-ethanolamine and cholesterol increases the radius of bilayers, straightens the hydrocarbon chains and in general reduces the fluidity (thereby causing the 'hardening') of membranes. Again, only a limited amount of cholesterol and phosphatidylethanolamine can be incorporated into bilayers before the bilayer structure is destroyed, at the point where the fluid lipids can no longer accommodate these wedge-shaped lipids. It is probably for this reason that natural biological membranes never contain high amounts of both cholesterol and phosphatidylethanolamine (the two most common wedge-shaped lipids in animal membranes). As might be expected, cholesterol and lysolecithin— neither of which can form a stable bilayer by itself—can combine in certain

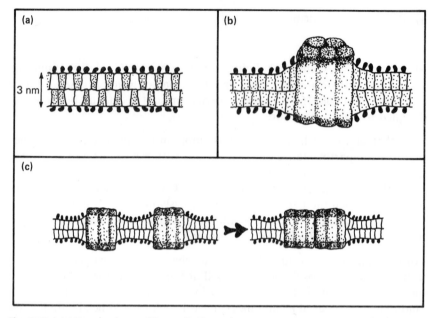

Fig. 17.7. (a) Mixture of two different lipids packing together within a planar membrane. The shaded lipids are cone-shaped ($v/a_0l_c < 1$); the white lipids are wedge-shaped ($v/a_0l_c > 1$). (b) Packing constraints induced in the hydrocarbon chain regions of lipids around a protein molecule, which may be relaxed when proteins aggregate, as shown in (c).

proportions to form bilayers. These and other physical properties of *mixed* lipid bilayers, vesicles, and biological membranes have been discussed by Carnie *et al.* (1979), Murphy (1982), and Kaler *et al.* (1989).

It is instructive to note that the most common lipids in animal cells are the two phospholipids phosphatidylcholine (PC) and phosphatidylethanol-amine (PE), while in plant cells they are the glycolipids digalactosyl diglyceride (DGDG) and monogalactosyl diglyceride (MGDG). In each case the first lipid packs as a truncated cone ($v/a_0l_c < 1$) while the second packs as a wedge ($v/a_0l_c > 1$). Thus, depending on the ratio of these lipid types in a bilayer, they can pack together into planar bilayers (Fig. 17.7a) or into bilayers of varying curvature and flexibility, a facility that is made use of by the lipid synthesis machinery of cells (Section 17.12). It is also worth noting that each of these four lipids has an uncharged headgroup. Their headgroup interactions are therefore due entirely to steric hydration forces that are fairly insensitive to changes in the ionic environment of the cytoplasm. The invariant packing properties of PC, PE, DGDG and MGDG make these lipids the ideal structural building blocks for stable membrane organization.

17.12   MEMBRANE PROTEINS AND MEMBRANE STRUCTURE

Membrane proteins are long-chained polypeptide polymers consisting of a long string of amino acid residues. The particular sequence of residues determines the *primary structure* of a protein, and the total molecular weight can reach 500 000. Compared to membrane lipids, membrane proteins are structurally rigid. The chains fold into cylindrical α-helical segments or β-pleated sheets (the *secondary structure* of proteins), which then self-organize into a globule (the *tertiary structure*). Soluble proteins have a totally hydrophilic surface, whereas membrane-associated proteins are usually amphiphilic with their surface exposing both hydrophobic and hydrophilic regions (Capaldi, 1982). If the geometry is right, such proteins can be incorporated into a lipid bilayer where the hydrophobic region spans the bilayer and where the hydrophilic residues are exposed to the aqueous phase on one or both sides of the membrane (Fig. 17.7b).

When proteins are incorporated into a lipid bilayer in the fluid state they usually induce stresses on the lipids in their vicinity. Such perturbed lipids are known as *boundary lipids*. Stresses arise if the hydrophobic regions of the protein and bilayer have different lengths, for then the lipids become stretched or compressed in order to accommodate the protein. This results in a shift of the lipid headgroup area from the optimal value. Such packing stresses around a protein usually involve more than one lipid layer, as illustrated in Fig. 17.7b. Additional lipid–protein interactions include any specific electrostatic or hydrogen-bonding interactions between the hydrophilic headgroups and the exposed amino acid residues of the proteins. Such lipid–protein interactions often result in a preferential clustering of specific lipids around a protein—the favoured lipids being those that can be packed most easily around the protein or specifically interact with it. For a review of lipid–protein interactions in membranes, see Benga and Holmes (1984).

We now turn to consider the forces between proteins within a fluid membrane. In principle, this two-dimensional system can be treated in the same way as a three-dimensional system, and we saw in Chapters 2 and 6 that a 2-D van der Waals equation of state can be used to analyse phase transitions in monolayers and membranes. In biomembranes, however, the forces between the proteins are more complex than the simple attractive van der Waals-dispersion interaction discussed earlier. Both attractive and repulsive forces can now occur between various membrane components. Repulsive forces arise between proteins if they have a strong affinity for the lipids, for example, via ionic bonds. Attractive forces arise if the proteins can bind to each other, e.g., via Ca^{2+} or molecular bridges, or if the protein–lipid packing mismatch is too great. In the latter case, the stresses on the lipids

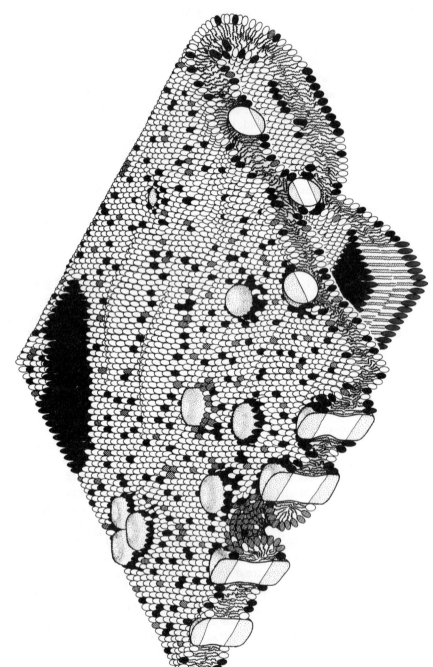

Fig. 17.8. Schematic figure of a biological membrane. The lipid contents of different membranes vary from as little as 25% up to 80% by

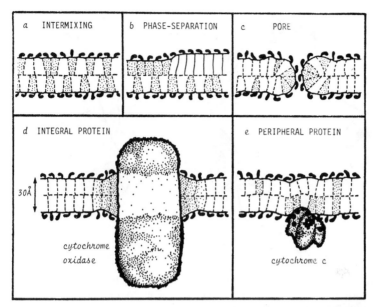

Fig. 17.9. Mean packing conformations of mixed lipid and lipid-protein membranes, showing how local packing stresses may cause clustering of specific lipids and/or non-bilayer shapes (shaded lipid regions). All the figures have been drawn to scale. Note the relatively large size of the cytochrome oxidase protein molecule which protrudes greatly from the bilayer. (From Israelachvili *et al.,* 1980a.)

may be relieved if the proteins aggregate, as illustrated in Fig. 17.7c. As examples of this phenomenon, it has been found that the state of aggregation of rhodopsin and bacteriorhodopsin incorporated into phospholipid bilayers can be altered by varying the lipid chain length (Chen and Hubbell, 1973; Lewis and Engelman, 1983a), and the organization of photosynthetic membrane proteins has also been found to depend on lipid composition (Siegel *et al.,* 1981).

A biological membrane is a dynamic structure. Both the lipids and the proteins move about rapidly in the plane of the membrane (Singer and Nicholson, 1972). However, heterogeneous domains and local clustering of lipids and proteins also occur (Figs 17.8 and 17.9), and these are important for the normal functioning of the membrane and its components. How does a cell maintain and regulate the structural integrity of its membranes? The main criterion for membrane stability appears to be that the heterogeneous lipid mixture should be able to self-assemble into bilayers, even though individual species, e.g., cholesterol, may not (Rilfors *et al.,* 1984). Thus, many organisms change their lipid composition in response to a change in ambient temperature, for example, synthesizing more *unsaturated* lipids at *lower*

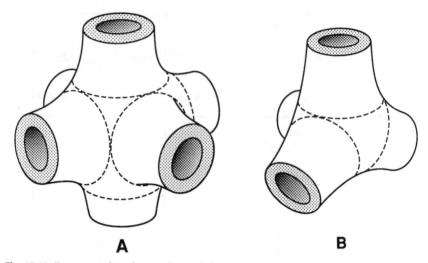

A   **B**

Fig. 17.10. Two examples of 'periodic' and 'bicontinuous' membrane structures in water, where a single lipid bilayer or biomembrane folds its way through the whole of space, at the same time separating two aqueous compartments from each other. Such membranous structures occur naturally in leaves.

temperatures. This ensures the stability of the bilayer structure since the higher packing parameter (higher $v/a_0 l_c$) of unsaturated lipids offsets the lowering of $v/a_0 l_c$ at lower temperatures (cf. Table 17.2). Likewise, Wieslander *et al.* (1980) showed that the *in vivo* response of the bacterium *Acholeplasma laidlawii* to external stimuli, such as temperature changes and incorporation of fatty acids and cholesterol, is such that the cell synthesizes just those lipids that will compensate for these stimuli so as to maintain packing compatibility within the membrane, for example, synthesizing more DGDG ($v/a_0 l_c < 1$) and less MGDG ($v/a_0 l_c > 1$) after cholesterol (for which $v/a_0 l_c > 1$) is introduced.

These examples illustrate the manner in which cells control their membrane structure. They do not do this by a crude mechanical pushing and pulling of components. The cell synthesizes the appropriate lipids and proteins and then leaves them to do their job.

PROBLEMS AND DISCUSSION TOPICS

17.1 With the help of Table 16.2 and other literature sources, give possible explanations for the following:

(i) The CMCs of surfactants are higher than the solubilities of the corresponding alkanes with the same number of carbon atoms.

(ii) The CMC increment per CH_2 is greater for single-chained than for double-chained surfactants.

(iii) The smaller the bare ion size of the counterion the larger is its effect on lowering the CMC.

(iv) Co-ions have almost no effect on the CMC.

(v) The CMC increment per CH_2 group is greater for non-ionic than for ionic and zwitterionic surfactants.

(vi) Addition of salt to an ionic micelle decreases the CMC and increases the CMC increment per CH_2 group to the value for non-ionic micelles.

(vii) The CMC often decreases then increases with temperature.

(viii) Chain *branching* does not have as strong an effect on the CMC as chain *unsaturation*, which significantly lowers the CMC.

(*Hint*: think of how electrostatic, steric-hydration and interchain interactions affect the headgroup area and packing parameter, $v/a_0 l_c$, and how this in turn affects micelle size, μ_N^0 and the CMC.)

17.2 You are God. You are dissatisfied with earth and have decided to raise its average temperature by 20°C. How would you modify the membrane lipids to make them more suitable as membrane structure regulators in their new environment? Consider such properties as chain length, degree of branching and unsaturation, type of headgroup, etc.

17.3 It is found that amphiphiles often associate into small micelles in hydrophobic liquids such as decane. Consider how intermolecular forces and molecular geometry (packing) contribute to this phenomenon. Make a sketch, similar to Fig. 17.1, showing how such micelles might look. How would you expect the size and structure of such micelles to be affected by the presence of small amounts of water in the liquids?

17.4 Two small spherical oil droplets of radius R in water coalesce into one larger droplet. If γ is the interfacial free energy of the interface, obtain an expression for the net change in surface energy during this process.

If each of the original droplets also carries a net charge Q evenly spread on its surface, what is the additional electrostatic contribution to the total energy change?

If $\gamma = 50 \, \text{mJ m}^{-2}$ and $Q = 100e$ calculate the critical radius above which the total energy change is negative (i.e. when the coalescence becomes energetically favourable). Assume that the dispersion is very dilute and that the solvent is pure water with no added electrolyte.

17.5 Describe and contrast the different mechanisms of the adsorption from solution of *surfactants* and of *polymers*.

17.6 What types of polymers and in what solvents would you expect polymers to self-assemble into aggregates, and how would these differ from surfactant aggregates such as micelles and bilayers.

17.7 Derive Eq. (17.16).

17.8 Derive Eq. (17.18).

17.9 Amphiphilic surfaces do not always have simple spherical, cylindrical or planar geometries. They can be highly convoluted, where at any point the curvature is theoretically defined by $(1/r_1 + 1/r_2)$. This can lead to interesting situations where what appears as a highly curved surface may actually have no curvature at all. The following two examples nicely illustrate this:

(i) Show that the membranous structures in Fig. 17.9 need not necessarily have a net curvature at any point on their surfaces, i.e., that they are everywhere theoretically *flat*.

(ii) Figures 17.8 and 17.9 show membranes containing holes. Such holes can be short-lived transient 'pores' involving only lipid molecules, or longer-lived 'channels' usually associated with a protein molecule. Transient lipid pores can provide a low-energy path for ions as well as lipids to traverse across membranes. The energetics of a lipid pore requires a careful analysis of how the molecules pack along the pore walls, whose geometry is like that of the inner surface of a doughnut or torus. Analyse and discuss whether cone-shaped or wedge-shaped lipids are more suitable for pore formation.

CHAPTER 18

THE INTERACTIONS BETWEEN LIPID BILAYERS AND BIOLOGICAL MEMBRANES

18.1 INTRODUCTION

In Chapters 11–14 we investigated the four main types of forces acting between surfaces in liquids: van der Waals, electrostatic, solvation (hydration) and steric forces. For a typical colloidal system of rigid particles in water, it is rare for more than two of these forces to be dominating the interaction at any one time. In contrast to this, the forces between the highly mobile amphiphilic surfaces of fluid bilayers and biological membranes can have all four operating simultaneously, as well as other—more specific—types of interactions. In addition, when two fluid-like structures such as two vesicles come together, the forces between them often cause them to change their shape. In this final chapter we shall first review these forces, starting with the two DLVO forces, and then see how they act together in regulating the morphologies and properties of complex fluid microstructures.

18.2 ATTRACTIVE VAN DER WAALS FORCES

Between two planar surfaces the van der Waals interaction energy is given by

$$W(D) = -A/12\pi D^2 \qquad \text{per unit area of surface,} \qquad (18.1)$$

where A is the non-retarded *Hamaker constant* (Chapter 11). The van der Waals forces between amphiphilic structures are generally small for the following three reasons. First, between hydrocarbon phases across water, A is relatively small, lying in the range $4-7 \times 10^{-21}$ J (Section 11.5). Biological membranes may be expected to have higher values due to the presence of proteins whose refractive index is generally higher than that of fluid hydrocarbon chains: 1.55–1.60 compared to 1.42–1.48.

Second, Eq. (18.1), when applied to bilayers or membranes, is only strictly valid at separations D below about 3 nm. Above 3 nm the Hamaker constant is no longer 'constant' but diminishes progressively as D increases due to retardation effects and to the finite thickness of bilayers (cf. Section 11.12 and Problem 11.12), and by 10 nm the force is usually much less than half the value given by the non-retarded equation, Eq. (18.1).

Third, in the presence of electrolyte there is an additional reduction of A at finite separations due to the ionic *screening* of the zero-frequency contribution $A_{v=0}$, which is now given by (Mahanty and Ninham, 1976, p. 202):

$$A_{v=0}(D) = A_{v=0}(0)2\kappa De^{-2\kappa D}, \qquad \text{for } \kappa D \gg 1, \qquad (18.2)$$

where κ is the Debye length and where for hydrocarbons across water $A_{v=0}(0) = 3 \times 10^{-21}$ J (Section 11.5). Equation (18.2) is accurate to within 15% for $\kappa D > 2$. Thus, in 0.15 M NaCl solution, where $\kappa^{-1} = 0.8$ nm, the zero-frequency term, which accounts for about half of the nonretarded force, has essentially disappeared by $D = 1.5$ nm.

Marra (1985, 1986a) measured the van der Waals force law between two bilayer surfaces composed of the uncharged sugar-headgroup lipid, MGDG. In distilled water there was an attractive force at surface separations from $D = 4$ nm down to the adhesive well at $D = 1$ nm. Both the measured force-law and adhesion energy were consistent with a van der Waals-type interaction with a nonretarded Hamaker constant of $A = 7 \times 10^{-21}$ J. However, in 0.15 M NaCl (physiological saline conditions) the Hamaker constant over the same range had fallen to $A = 3.5 \times 10^{-21}$ J, as expected due to screening of the zero-frequency contribution.

In conclusion, the van der Waals force between bilayers and membranes, especially in high salt, is fairly weak and has an effective range of at most 15 nm, beyond which it is too weak to be of any major significance.

18.3   ELECTROSTATIC (DOUBLE-LAYER) FORCES

The double-layer repulsion, unlike the van der Waals attraction, is much more sensitive to the type and concentration of electrolyte present, the pH, and the surface charge density or potential. An approximate expression for the double-layer interaction energy between two surfaces is given by Eq. (12.47), while exact solution are plotted in Figs 12.10 and 12.11.

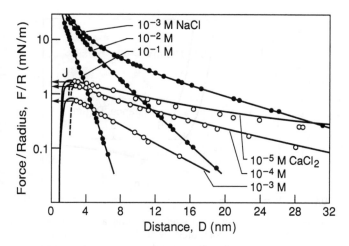

Fig. 18.1. Measured repulsive forces between two DSPG bilayers in various aqueous electrolyte solutions at 22°C. The solid lines are the theoretically predicted DLVO force profiles assuming a Hamaker constant of $A = 6 \times 10^{-21}$ J. In NaCl solutions the double-layer repulsion corresponds to two fully charged surfaces. In $CaCl_2$ solutions the surface charge has been reduced due to ion binding, and from the force maxima at J the two surfaces jump into strong adhesive bilayer–bilayer contact at $D = 0$–0.4 nm. Note that in $CaCl_2$ solutions at small separations the measured forces (dashed lines) are more attractive than expected from the DLVO theory (solid lines), an effect that is believed to be due to attractive *ion correlation* forces. (Adapted from Marra, 1986b.)

As described in Chapter 12, the interplay between attractive van der Waals forces and repulsive double-layer forces forms the basis of the so-called DLVO theory of colloid science. Both of these forces have now been directly measured between charged anionic or cationic surfactant and lipid bilayers in aqueous solutions of both monovalent 1:1 and divalant 2:1 electrolytes (Pashley and Israelachvili, 1981; Marra and Israelachvili, 1985; Marra, 1985, 1986a, b, c; Pashley et al., 1986; Pashley and Ninham, 1987; Claesson et al., 1989; see also Fig. 12.5).

Figure 18.1 shows the measured forces between negatively charged bilayers of DSPG in dilute electrolyte solutions of NaCl and $CaCl_2$. The effectiveness of divalent cations in reducing the double-layer repulsion, even at very low concentrations, is quite remarkable, and appears to parallel their effectiveness as adhesogens and fusogens of surfactant bilayers and biological membranes. Indeed, in the presence of divalent counterions, measured short-range forces are often significantly more attractive than can be accounted for by van der Waals forces (see jumps J in Fig. 18.1; Marra, 1986b, c; Khan et al., 1985), an effect that has been attributed to the additional attractive van der Waals-type *ion-correlation* force between the highly polarizable divalent ions

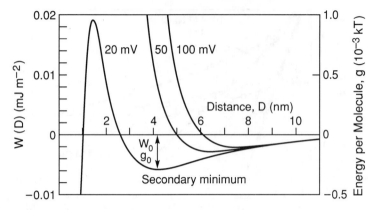

Fig. 18.2. Theoretical DLVO interaction energy per unit area between two amphiphilic surfaces in 0.15 M NaCl (kindly computed by R. Pashley). The non-retarded Hamaker constant is assumed to be $A = 6 \times 10^{-21}$ J and the surfaces are assumed to interact at constant surface potential ψ_0. The three curves at $\psi_0 = 100$, 50 and 20 mV correspond to surface charge densities of 1e per 1, 3 and 8.5 nm^2, respectively. The energy per molecule $g = \frac{1}{2} W a_0$ is calculated for a headgroup area of $a_0 = 0.4$ nm^2. Note that in the innermost secondary minimum the 'adhesion energy' of $W_0 \approx 6 \times 10^{-3}$ mJ m^{-2} corresponds to an energy of about $g_0 = 3 \times 10^{-4} kT$ per molecule.

within the diffuse double layers (Guldbrand *et al.*, 1984; Kjellander and Marcelja, 1984, 1985c, 1986a, b).

At higher monovalent electrolyte concentrations the double-layer repulsion generally diminishes (Chapter 12), but because of the relatively weak van der Waals attraction between lipid bilayers it usually remains strong enough to keep the surfaces apart, even for low surface potentials and in high salt (so long as it remains monovalent). This is illustrated in Fig. 18.2 for two bilayers in physiological saline solution ($\sim$0.15 M NaCl) where a weak secondary minimum occurs at 4–6 nm for typical values of ψ_0 and A.

The situation in practice is often much more complex. Below 3 nm repulsive hydration and steric forces may dominate the interaction depending on the hydrophilicity and thermal motion of the surface groups. On the other hand, attractive ion-correlation and hydrophobic forces may dominate depending on the stresses on the bilayers and the presence of divalent ions. These additional forces will be discussed below. Note, too, that the surfaces of real biological membranes are not smooth but locally rough due to protruding proteins (Fig. 17.9), which further complicates the application of the DLVO theory at small distances.

18.4 HYDRATION FORCES

As we saw in Chapter 13, repulsive hydration forces arise whenever water molecules bind strongly to hydrophilic surface groups because of the energy needed to dehydrate these groups as two surfaces approach each other. Between two solid crystalline surfaces the hydration force is usually oscillatory. The oscillations have a periodicity of the diameter of the water molecule, about 0.25 nm, and reflect the ordering of water molecules into semidiscrete layers between the smooth, rigid surfaces. Between bilayer surfaces no such ordering into well-defined layers is possible because (i) the headgroups are rough on the scale of a water molecule, and (ii) the surfaces are usually thermally mobile giving rise to a steric repulsion. Consequently, any oscillatory force becomes smeared out and one is left with only the monotonic component (Caffrey and Bilderback, 1983). It is probably for this reason that only smoothly decaying steric-hydration forces have ever been measured between amphiphilic surfaces in both water and nonaqueous liquids.

Monotonically repulsive hydration forces were first proposed to arise between amphiphilic surfaces by Langmuir (1938) and were first measured between surfactant monolayers across soap films by Clunie et al. (1967), as shown in Fig. 13.8A. These forces have now been measured between various amphiphilic colloidal surfaces and bilayers using either hydrostatic or osmotic pressure techniques (Homola and Robertson, 1976; LeNeveu et al., 1976; Parsegian et al., 1979; McIntosh and Simon, 1986) or the Surface Forces Apparatus technique (Marra and Israelachvili, 1985; Marra, 1985, 1986a; Claesson et al., 1989). The agreement between the various techniques is surprisingly good (Horn et al., 1988b). However, there are uncertainties in how to define the separation between these essentially fluid-like interfaces, but these have to some extent been resolved by directly measuring the electron density distribution across the aqueous phase (McIntosh and Simon, 1986).

The range of the steric-hydration forces so far measured between various surfactant and lipid bilayers is about 1–3 nm, below which they rise steeply and roughly exponentially, with decay lengths ranging from 0.08 to 0.64 nm (Marsh, 1989; Rand and Parsegian, 1989). Steric-hydration forces often dominate over DLVO forces at small separations, preventing the coalescence in a primary minimum of bilayers, vesicles and biological membranes. In particular, these forces are responsible for the lack of strong adhesion or aggregation of bilayers and vesicles composed of uncharged lipids such as lecithin and those possessing polyoxyethylene and sugar headgroups.

Unfortunately, because of the very short range of these forces between amphiphilic surfaces (see Fig. 13.8B), it is usually not possible to disentangle the contributions arising from pure hydration (solvation) effects and thermal fluctuation (steric) forces. As described in Chapter 14, the overlapping of the mobile headgroups protruding out from two approaching bilayers produces a variety of repulsive steric or osmotic-type forces between them. These forces are also roughly exponentially repulsive (cf. Eqs (14.3), (14.6) and (14.12)) and, as discussed in Section 18.6, their magnitude and range are consistent with the measured values. The finite sizes of headgroups introduce non-ideal terms to these forces which act to enhance the repulsion (just as the finite size of molecules enhance the pressure of a van der Waals gas). How, then, can one distinguish between steric repulsion and hydration repulsion? The role of hydration in headgroup interactions can be considered either as (i) simply increasing the effective size of the headgroup even further, as in the case of the hydrated sizes of ions (Section 4.5); or (ii) as having a much larger effect, one that effectively introduces a separate 'hydration' force contribution to the other forces. This question remains controversial. In this section we shall investigate effects that can be unambiguously attributed to hydration. In the following sections we shall explore the steric forces.

Some lipid headgroups are intrinsically hydrophilic, for example, those containing $N(CH_3)_3^+$, ethylene oxide or sugar groups, while others become hydrophilic on binding hydrated ions, especially Na^+, Li^+, or Mg^{2+}. The first class of lipids have *intrinsic* hydration forces, the second have hydration forces that can be *regulated* by ion binding or 'ion exchange' (Chapter 12). We now consider some examples and consequences of regulated hydration forces.

Figure 18.3 shows the measured forces between bilayers of a double-chained anionic surfactant as a function of salt concentration. For NaCl concentrations below 10^{-2} M the forces are well-described by the DLVO theory: the double-layer repulsion dominates at separations greater than 2 nm, and the van der Waals attraction dominates at smaller separations. But at concentrations above 10^{-2} M NaCl and high pH a strongly repulsive hydration force comes in at separations below 2 nm that dominates over the van der Waals attraction and prevents the surfaces from coming into adhesive contact as before. This behaviour is believed to arise from the binding of hydrated Na^+ ions to the surfaces under conditions of high salt and high pH, and is qualitatively very similar to that occurring between colloidal particle surfaces, as described in Section 13.5.

There are many other, less direct but more relevant, examples where repulsive hydration forces appear to be induced by ion binding or exchange. Thus, negatively charged phospholipid vesicles aggregate or fuse in dilute $CaCl_2$ solutions, as expected from the DLVO theory, but not in concentrated NaCl solutions, where the DLVO theory predicts strong irreversible adhesion in molecular contact (Day et al., 1980). Also, phosphatidylserine vesicles fuse in dilute (approximately millimolar) $CaCl_2$ solutions but not in $MgCl_2$

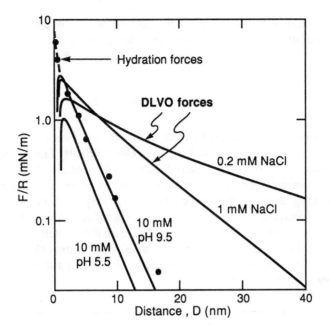

Fig. 18.3. Forces between layers of di-hexadecyl phosphate in aqueous NaCl solutions. The solid lines are the theoretical DLVO forces which agree with the measured forces except in 10^{-2} M NaCl at pH 9.5, where the extra repulsion at $D < 2$ nm (dashed line) cannot be explained by the DLVO theory. (From Claesson *et al.*, 1989.)

solutions (Wilschut *et al.*, 1981), most likely because the more hydrated Mg^{2+} ions prevent the surfaces from coming into sufficiently close contact to fuse (McIver, 1979).

As a further example of cation-regulated hydration forces, Princen *et al.* (1980) studied the interactions between oil-in-water emulsion droplets, whose surfaces were covered by an alkyl sulphate monolayer. They found that the adhesion between the droplets in concentrated salt solutions decreases with the increasing hydration of the cation bound to the surfaces, namely KCl (maximum adhesion) > NaCl > LiCl (almost no adhesion). Other examples of the role of hydration in bilayer and membrane interactions were discussed by McIver (1979).

Non-ionic and zwitterionic bilayers usually display intrinsic hydration forces that are not affected much by solution conditions. This is because there is little ion binding to such surfaces. For example, uncharged phosphatidylcholine (PC) and phosphatidylethanolamine (PE) bilayers are both known to adsorb Ca^{2+} and Mg^{2+} ions. This gives rise to a double-layer repulsion between the now positively charged surfaces, but the short-range steric-hydration repulsion is not affected and is simply additive with the (variable) double-layer force (Marra and Israelachvili, 1985).

18.5 LIMITATIONS OF THE HYDRATION MODEL

The origin of the monotonically repulsive hydration forces seen in so many amphiphilic systems remains unclear. Marcelja's theory (Marcelja and Radic, 1976; Gruen and Marcelja, 1983) based on the polarization (alignment) induced by surfaces or surface groups on the water molecules adjacent to them has long been the most appealing. This theory predicts an exponential repulsion of unspecified decay length, though a value of about 0.25 nm—the size of a water molecule and close to some measured decay lengths—appeared reasonable. However, molecular dynamics simulations (Kjellander and Marcelja, 1985a, b) failed to reproduce this force and the matter remains theoretically unresolved.

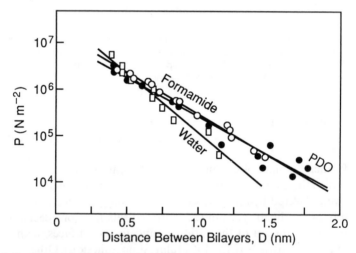

Fig. 18.4. Pressure between egg lecithin bilayers in water, formamide and 1,3-propanediol (PDO: $HO—CH_2—CH_2—CH_2—OH$). The forces may be approximated by exponential functions such as $P = P_0 e^{-D/\lambda}$. The trends are for P_0 to decrease as λ increases, and for λ to increase with decreasing interfacial tension γ_i of the hydrocarbon–solvent interface: viz. 53 mJ m^{-2} with water, 32 mJ m^{-2} with formamide, and 21 mJ m^{-2} with PDO (Table 15.1). These trends are as expected if the forces are due to steric-protrusion effects, described by Eq. (18.5). For solvents having much lower interfacial tensions with hydrocarbons, e.g., 1,2-propanediol, the aggregates become progressively more diffuse and the pressure approaches the osmotic limit. (Reproduced from McIntosh et al., 1989a, with permission.)

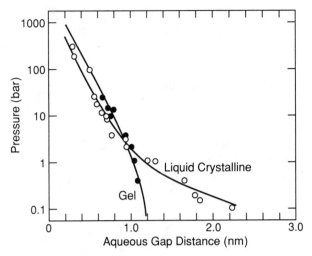

Fig. 18.5. Repulsive forces in water between uncharged monoelaidin bilayers in the *gel state* and monocaprylin bilayers in the *liquid-crystalline state*, as measured by the Osmotic Pressure technique (McIntosh *et al.,* 1989b). At short distances, below about 1 nm, the forces between both bilayers are roughly exponential with a decay length of ~0.13 nm. However, between the liquid-crystalline bilayers the tail end of the interaction decays as $1/D^3$, and appears to be well described by Eq. (18.3) for the undulation force between bilayers having a bending modulus of $k_b = 3.2 \times 10^{-20}$ J. (Reproduced from McIntosh *et al.,* 1989b, with permission.)

Experimentally, too, the picture has become more complex and increasingly less consistent with a purely hydration origin for these forces. First, recent measurements on different lipids show that the characteristic decay lengths of these forces span a range from 0.08 to 0.64 nm (Lyle and Tiddy, 1986; Marsh, 1989; Rand and Parsegian, 1989), an eight-fold variation that can no longer be associated with the size of the water molecule or some other characteristic property of water. Second, as shown in Fig. 18.4, very similar exponentially repulsive forces also exist between bilayers in a variety of non-aqueous solvents (Persson and Bergenståhl, 1985; McIntosh *et al.,* 1989a). This shows that these forces are not unique to water.

Third, the forces between *free bilayers*, as measured using osmotic pressure techniques, are generally of longer range than those between *supported bilayers*, as measured using force balance techniques (Horn *et al.,* 1988b; Pezron *et al.,* 1990). This suggests that some thermal fluctuation forces are operating between the free bilayers (cf. Fig. 14.8) which become suppressed when the bilayers are supported on a rigid surface.

Finally, measurements of the short-range repulsion between bilayers generally increases with temperature, and many surfactant bilayers do not swell at all when in the *solid-crystalline state*, but do swell as soon as the

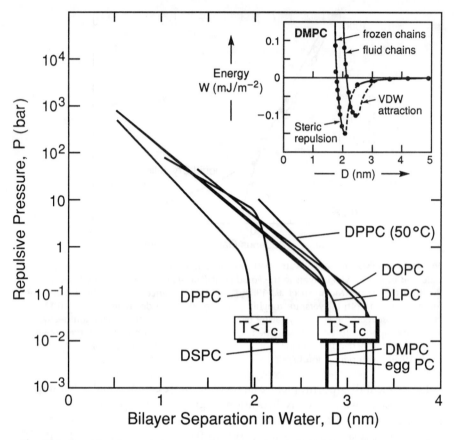

Fig. 18.6. Repulsive forces between various free phospholipid bilayers in water in the *liquid-crystalline* and *fluid* states, i.e., below and above the chain melting temperature, T_c, as measured by the Osmotic Pressure technique. *Inset*: forces between supported DMPC bilayers below and above T_c ($\sim$24°C) as measured by the Surface Forces Apparatus technique. (Adapted from Israelachvili, 1985; Marra and Israelachvili, 1986.)

temperature is raised to the *gel* state (Tiddy, 1980). On further increasing the temperature from the *gel* to the *liquid-crystalline* state there is a further increase in the magnitude and range of the repulsion (Fig. 18.5). Above the chain melting temperature, T_c, the repulsion between the now *fluid* bilayers increases once again (Fig. 18.6). Within the hydration model, this trend would suggest that the water structure is increasing with increasing temperature. This is very unlikely: with increasing temperature, as the amphiphilic molecules become more disordered, one would expect the same to occur for the water molecules in the gap between them.

All the above observations suggest that these forces have an entropic origin. This was first proposed by Helfrich in 1978 whose theory of *undulation* forces (reviewed by Servuss and Helfrich, 1987, 1989) revealed the first of a number of steric forces between fluid-like amphiphilic surfaces. We now turn to consider these forces.

18.6   STERIC FORCES

The idea that repulsive steric forces are important in the interactions between amphiphilic surfaces has grown steadily over the last few years. With the apparent failure of models based on purely electrostatic or solvation interactions to explain these forces, attention has recently focused on the role of thermal fluctuation interactions. These were discussed in some detail in Chapter 14, but mainly in relation to the interactions between polymer-coated surfaces. Here we shall attempt to make a quantitative assessment of their possible role in the interactions between bilayer and other amphiphilic surfaces.

Four different types of repulsive steric forces can arise between bilayers interacting in any liquid i.e., independent of the solvent. These are the *undulation* force (Fig. 14.7), the *peristaltic* force (Fig. 14.8), the *protrusion* force (Fig. 14.8) and the *steric overlap* force between surface-anchored polymer 'brushes' or headgroups (Fig. 14.1). Below we summarize the approximate equations appropriate for each of these forces (per unit area) as previously derived in Chapter 14:

Undulation:
$$F \approx \frac{(kT)^2}{2k_b D^3}, \tag{18.3}$$

Peristaltic:
$$F \approx \frac{(kT)^2}{5k_a D^5}, \tag{18.4}$$

Protrusion:
$$F \approx \frac{2.7\Gamma kT\, e^{-D/\lambda}}{\lambda}, \qquad \text{where } \lambda \approx kT/\alpha_p, \tag{18.5}$$

Headgroup overlap:
$$F \approx 100\Gamma^{3/2}kT\, e^{-D/\lambda}, \qquad \text{where } \lambda \approx L/\pi, \tag{18.6}$$

where, for the latter interaction, L is the effective thickness of the fluctuating headgroup region and $2L$ is the range of the interaction.

Of the four steric forces listed above, only the undulation force is expected to have an indefinitely long range; the other three forces will decay rapidly to zero beyond a certain distance roughly equal to the lengths of the lipid molecules. Safinya *et al.* (1986) carried out an x-ray study of the repulsive forces between two oil–water interfaces each bounded by a sodium dodecyl sulphate (SDS) monolayer for which the bending modulus is small ($k_b \approx 5 \times 10^{-21}$ J). The undulation force was therefore large and accurately measurable, and the results confirmed the inverse third distance dependence of the Helfrich equation, Eq. (18.3). McIntosh *et al.* (1989b), and Abillon and Perez (1990), also measured a $1/D^3$ tail in the interaction between bilayers whose bending moduli are about ten times higher (see Fig. 18.5).

At smaller separations, below about 2 nm, the protrusion and headgroup overlap forces are expected to dominate the undulate repulsion. Figure 18.7 shows theoretical plots of the above four steric interactions, computed using values appropriate for double-chained lipid bilayers in the fluid state. The attractive van der Waals interaction given by Eq. (18.1) is also shown for comparison.

The computed curves of Fig. 18.7, taken together, appear to account satisfactorily for the magnitude and range of the force profiles measured between the much-studied egg lecithin bilayers. However, given the many contributions to the total interaction and the experimental uncertainties of defining the zero of separation, a detailed comparison between theory and experiment is not yet possible (though initial computer simulations by Granfeldt and Miklavic, 1991, do indicate the importance of protrusion and headgroup overlap forces in bilayer interactions). Some expected semiquantitative trends, however, may be established based on the protrusion force, as follows.

Protrusion energies α_P can be estimated either from interfacial energy values γ_i or from the incremental changes in the CMC values per added CH_2 group (Israelachvili and Wennerström, 1990). An energy increment per added CH_2 group of f corresponds to a protrusion energy per unit length of $\alpha_P = kT \ln f / 0.1265 \times 10^{-9} = n3.25 \times 10^{-11} \ln f$ J m^{-1} where n is the number of chains and 0.1265 nm is the CH_2–CH_2 distance along a single chain. From Table 16.2 the range of values measured for f therefore correspond to protrusion energies in the range $\alpha_P = (1.5 - 4.0) \times 10^{-11}$ J m^{-1} for single saturated chains, and $\alpha_P = (3.0 - 5.6) \times 10^{-11}$ J m^{-1} for double chains. These in turn correspond to protrusion decay lengths in the range $\lambda \approx kT/\alpha_p \approx 0.07$–$0.27$ nm. This essentially covers most of the measured range. Lower λ values should apply to saturated, double-chained surfactants and lipids; higher values to single-chained surfactants (lysolipids) especially with unsaturated chains (since these are more hydrophilic than saturated chains, as shown in Table 15.1). Thus, we should expect somewhat longer decay lengths for unsaturated lipids, as observed: compare

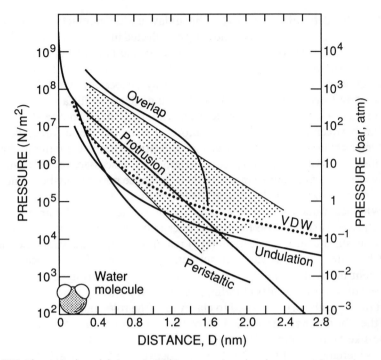

Fig. 18.7. Theoretical *undulation, peristaltic, protrusion, headgroup overlap* and *van der Waals* forces between two bilayers in the fluid state in water at $T = 25°C$. The following values were used in the computations, these being typical for phospholipid bilayers in the fluid state. For the van der Waals force: Hamaker constant, $A = 5 \times 10^{-21}$ J. For the undulation force: bending modulus, $k_b = 10^{-19}$ J. For the peristaltic force: area expansion modulus, $k_a = 150$ mJ m^{-2}. For the protrusion force: surface density of protruding groups or chains, $\Gamma = 2 \times 10^{18}$ m^{-2}; protrusion energy, $\alpha_P = 2.5 \times 10^{-11}$ J m^{-1} (corresponding to $\lambda \approx 0.17$ nm). For the headgroup overlap force: mean separation between headgroups, $s = 0.8$ nm; layer thickness of hydrated headgroup region, $L = 0.8$ nm (corresponding to $\lambda \approx 0.25$ nm). Note that all the forces plotted are repulsive except for the van der Waals force, which is attractive, but which has been plotted together with the other forces for convenience of comparison. The shaded region gives the limits of the repulsive forces measured between egg lecithin bilayers (McIntosh and Simon, 1986; Rand and Parsegian, 1989; McIntosh et al., 1989a). Pure saturated-chain bilayers generally exhibit shorter range repulsions.

$\lambda \approx 0.14$–0.26 nm for fluid DMPC and DPPC with $\lambda \approx 0.17$–0.34 nm for egg PC and unsaturated fluids DSPCs (McIntosh and Simon, 1986; Marsh, 1989). We should also expect lysolipids to have longer decay lengths, also as observed: the highest λ values so far reported are $\lambda \approx 0.44$–0.64 nm for lysoPS and lysoPC (Marsh, 1989).

Steric protrusion forces are also expected to arise in non-aqueous liquids, and to be determined more by the unfavourable interaction energies of hydrocarbon chains with these liquids (as reflected in their interfacial energies, γ_i, and CMC values) than in any property of the liquids alone. This trend, too, is borne out by experiments (Fig. 18.4).

It appears, therefore, that the previously suspected 'hydration' forces, intimately associated with water structuring effects at surfaces, may be no more than a collection of complex steric interactions. If so, the short-range forces between bilayers are really more akin to the steric forces between polymer-covered surfaces than to any hydration effects. Genuine hydration interactions will, of course, be operating, as discussed in Section 18.4. However, in a first approximation, their influence may simply be to increase the effective size of the 'hydrated' headgroups.

Whether water structuring at amphiphilic surfaces provides an additional long-range force between them as it does between solid, crystalline surfaces such as silicate surfaces (Chapter 13) remains an open question. Certainly, the hydrated state of water at surfactant–water interfaces appears to be different but with no indication of any long-range ordering beyond the first 'layer' of water molecules (Lindman et al., 1987). Clearly more experimental and theoretical work is needed to clarify fully the roles and interdependencies of the short-range steric and solvation/hydration interactions between fluid-like amphiphilic interfaces in water.

A different type of steric force occurs between the hydrocarbon chains of surfactant monolayers interacting across hydrocarbon liquids. Such situations arise between water-in-oil microemulsion droplets as well as when liquid alkanes are present in the central regions of bilayers. Gruen and Haydon (1981) studied this type of system, both experimentally and theoretically, and concluded that low molecular weight alkanes penetrate into the hydrocarbon chains, leading to an effective entropic repulsion between the opposing monolayers. This causes them to swell. Higher molecular weight alkanes penetrate progressively less and the interaction eventually becomes attractive. The van der Waals force between hydrocarbon chains across hydrocarbon liquids is, of course, small and plays only a minor role in such systems.

18.7   HYDROPHOBIC FORCES

As we saw in Section 13.6, the attractive hydrophobic interaction between hydrocarbon molecules or surfaces in water is of surprisingly long range and much stronger than the van der Waals attraction at small separations. In the case of free (unstressed) bilayers, the hydrophilic headgroups 'shield' the underlying hydrocarbon groups from the aqueous phase, which effectively

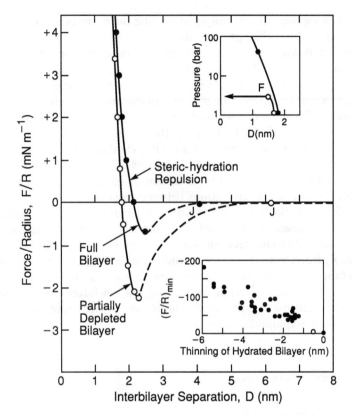

Fig. 18.8. Forces between supported DLPC and DMPC layers in water as a function of decreasing bilayer thickness (equivalent to increasing the headgroup area above the optimal value). Headgroup areas of supported monolayers and bilayers can be controlled (stressed) in an artificial way by either depositing less lipid on the surfaces than needed for full coverage, or by starving the solution of molecules which depletes the amount adsorbed (Helm *et al.,* 1989). *Bottom inset*: increasing adhesion force, $(F/R)_{min}$, measured between progressively thinner bilayers. *Top inset*: spontaneous fusion, F, occurring between two thinned bilayers once they come within about 1 nm of each other.

masks the hydrophobic interaction between them. However, when bilayers are subjected to a stretching force, they expand laterally, and the increased hydrophobic area exposed to the aqueous phase now allows the hydrophobic interaction to emerge.

Figure 18.8 shows the measured forces between lecithin bilayers supported on mica surfaces as a function of their headgroup area. A long-range, strongly attractive hydrophobic force emerges as more hydrocarbon groups become exposed to the aqueous phase. It is remarkable that the attraction could be enhanced so significantly by simply increasing the mean headgroup area by a few per cent above the optimal value.

In solution, bilayers and membranes can be stressed by applying an electric field across them (see Problem 18.6), by the osmotic swelling of cells or vesicles (Servuss and Helfrich, 1987, 1989), or by introducing local stresses on the lipids via ion binding or packing mismatches with other membrane components. The enhanced hydrophobic attraction brought about by such stresses (Bailey et al., 1990) may be the primary cause leading to the fusion of bilayers and membranes (Papahadjopoulos et al., 1977; Chernomordik et al., 1986; Helm and Israelachvili, 1991).

Hydrophobic forces can also be enhanced by increasing the temperature. Such effects occur in the interactions between certain nonionic micelles and bilayers, for example, those composed of surfactants with polyoxyethylene headgroups (Claesson et al., 1986b). However, it is still not clear whether the increased adhesion between such aggregates arises solely from an increased hydrophobic attraction, or whether some of it also comes from the reduced range of the steric repulsion between the less hydrated headgroups.

Thus, as in the case of repulsive steric-hydration forces, we still do not really understand the separate contributions of the steric and hydrophobic forces between amphiphilic surfaces.

18.8   SPECIFIC INTERACTIONS

Here we shall look at non-covalent forces that give rise to very strong adhesion, or binding, between two surfaces or molecular groups. Such interactions are often referred to as being 'specific' even though, as we shall see, it is difficult to define a specific interaction unambiguously.

Electrostatic

Some electrostatic interactions are unusually strong: for example, the adhesion of phosphatidylserine bilayers in the presence of Ca^{2+} ions. Also, certain divalent ions can trigger conformational or chemical changes in proteins and membranes while others do not. These, too, are normally referred to as being specific. However, since the normal (non-specific) Coulombic interaction between ionic bonds can be very strong (Section 3.3), it is not obvious that this should be called a specific interaction. Nevertheless, it *is* worth noting that some ions can give rise to extremely strong binding, even in water.

Bridging

In Chapter 14 we noted that polymer bridges between two surfaces can cause an effective 'bridging' attraction between them (Fig. 14.1). Bridging forces also arise in biological systems: for example, polymer bridges have been found to connect myelin membranes where their function is presumably to prevent these membranes from moving too far from each other (Rand *et al.*, 1979).

'Lock-and-key' or ligand–receptor interactions

It has long been known that many biological interactions such as those involved in immunological recognition (ligand–receptor interactions) and

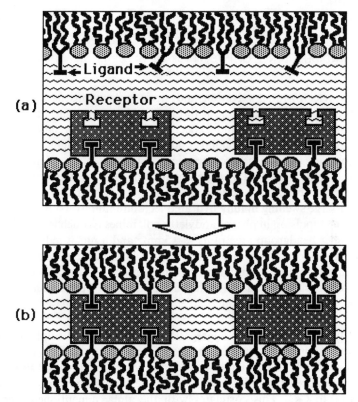

Fig. 18.9. Example of 'lock and key' binding mechanism developed by nature to obtain strong and highly specific non-covalent bonds between biomolecules such as ligands and receptors, antibodies and antigens. Shown here is a schematic of the binding between biotin (a ligand) and streptavidin (a protein receptor), whose binding energy of 88 kJ mol^{-1} (about $35kT$ per bond) is one of the highest known for a ligand–receptor interaction.

cell–cell contacts are *totally* specific for one, and only one, molecule. Early this century, biologists such as Paul Erlich proposed that certain biomolecules may have a perfect geometrical fit which allows them to bind together extremely strongly via a purely mechanical mechanism that is essentially a molecular 'lock and key'. Figure 18.9 shows a schematic illustration of one such mechanism, that between biotin and avidin, whose binding energy is about 35 kT per bond, but where no covalent bond is formed (Wilchek and Bayer, 1990). It is clear that nature has here developed an extremely efficient binding mechanism whereby non-covalent adhesive junctions having the effective strength of covalent bonds can be switched on, or 'locked', quickly and with minimal expenditure of energy. One presumes that these types of bonds can be 'unlocked' equally easily, e.g., by a change in the pH or solution conditions.

18.9 Interdependence of intermembrane and intramembrane forces

Since the electrostatic, solvation and steric interactions *between* different bilayers are essentially the same as those acting between adjacent headgroups *within* one bilayer, we should expect to find many correlations between interbilayer forces and intrabilayer forces. Thus, we might expect larger headgroup areas a_0 to be accompanied by larger repulsive forces between bilayers. This is borne out by experiments. For example, the large hydration of the lecithin headgroup results in a large surface area of $a_0 \approx 0.7$ nm^2 as well as a large swelling in fully hydrated lecithin multilayers. By contrast the headgroup repulsion in phosphatidylethanolamines is much less, which leads to a smaller headgroup area of $a_0 \approx 0.5$ nm^2 and a much reduced swelling.

Analogous correlations occur for charged lipid bilayers where in general a decrease in pH or addition of divalent cations reduces the electrostatic headgroup repulsion and hence the surface area per lipid and also leads to reduced bilayer swelling in multilayer phases and to increased adhesion of vesicles.

As a further example, the aggregation number of alkyl sulphate micelles increases as the electrolyte is changed from LiCl $\rightarrow$ NaCl $\rightarrow$ KCl $\rightarrow$ CsCl (Missel et al., 1982). This implies a reduced headgroup area a_0, which arises from the reduced hydration repulsion of the bound counterions as we go from Li$^+$ to Cs$^+$. This correlates with the already mentioned increasing adhesion between oil-in-water emulsion droplets, stabilized by alkyl sulphate monolayers, as monitored by the contact angles between them (Princen et al., 1980).

In the area of biological membrane interactions, many studies of membrane and vesicle fusion have found that proteins redistribute during the fusion process, and even in the absence of fusion both protein and lipid redistributions have been reported to occur at membrane–membrane contact sites. For example, the inner surface of the outer mitochondrial membrane (see Fig. 17.6) has a non-random distribution of anionic sites, with a lower

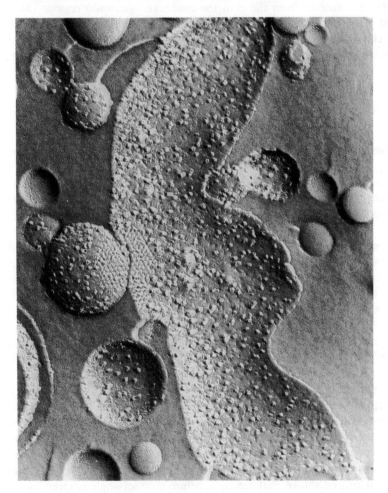

Fig. 18.10. Freeze-fracture electron micrograph of reconstituted plant thylakoid membranes containing plant lipids and the two major thylakoid proteins known as photosystem I (mean diameter 9 nm) and the light-harvesting complex of photosystem II (LHC-II of mean diameter 7 nm). In the presence of 5 mM Mg^{2+} the LHC-II aggregate into two-dimensional crystalline patches. These act as adhesion sites between membranes, as seen here and also in intact thylakoids. The adhering vesicle in the figure has a diameter of 220 nm. (Micrograph: I. J. Ryrie.)

density of sites in those regions facing the inner membrane. However, the distribution of anionic sites on the totally exposed outer surface is random (Hackenbrock and Miller, 1975). In plant thylakoid membranes (Fig. 17.6) the random distribution of proteins characteristic of the isolated membranes becomes non-random on stacking: the stacked (grana) regions contain different proteins from those in the unstacked (stroma) regions (Staehelin and Arntzen, 1979; Anderson and Andersson, 1982; Ryrie, 1983). Figure 18.10 illustrates this phenomenon, showing how two 'reconstituted membranes' have adhered at the sites where specific proteins have aggregated.

18.10   ADHESION

When a liquid droplet settles on a surface, or when two soap bubbles adhere, they distort into truncated spheres. For a drop on a surface (Fig. 15.3) the contact angle θ is given by the Young–Dupré equation: $\gamma(1 + \cos \theta) = W$, where γ is the liquid surface tension, and W is the adhesion energy (work of adhesion) per unit area of the solid–liquid interface. Similar deformations occur when two vesicles adhere (Fig. 18.11), but the equations describing these deformations are different. This is because the energy of a pure liquid surface is characterized by γ whereas that of an elastic bilayer is described by its elastic area and bending moduli, k_a and k_b. Thus, for a liquid surface, a change in its surface area by ΔA is accompanied by a surface free energy change of $\gamma \Delta A$, while for an elastic membrane this becomes replaced by $\frac{1}{2}k_a(\Delta A)^2/A_0$, where A_0 is the initial, unstressed, area.

Unlike the case of a liquid of constant volume adhering to a surface, there is no unique equation that describes the equilibrium state of a vesicle interacting with a flat surface or with another vesicle. The reason for this has to do with the high permeability of vesicles to water, ions and other solute molecules, which ensures that the vesicle volume will change as soon as it becomes deformed. Two scenarios will be considered: in the first, two initially spherical and unstressed vesicles adhere and rapidly deform until their adhesion and elastic energies balance. If the volume of the vesicles remains constant, i.e., if no water leaves the now compressed vesicle interiors, then it can be shown that the equilibrium contact angle is given by

$$2k_a(1 - \cos \theta) = W \bigg/ \left\{ \frac{(3 - \cos \theta)}{[2(1 + \cos \theta)^{1/2}(2 - \cos \theta)]^{2/3}} - 1 \right\} \qquad (18.7)$$

where θ is here defined as the contact angle measured *outside* the vesicles (as in Fig. 18.11), and W is the adhesion energy per unit area of the two

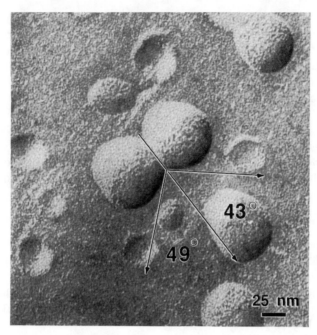

Fig. 18.11. Adhesion of two lecithin vesicles imaged by transmission electron microscopy (TEM). $\rho = 20$ mM, $\Delta\rho \approx 10$ mM, $\sigma = 0.52$ mN m^{-1}, $W = 0.30$ mJ m^{-2}. (Micrograph: J. A. N. Zasadzinski.)

surfaces as before. It may be further established that for two identical vesicles of initial radius R,

(i) the total adhesion energy at contact equilibrium is proportional to R^2 rather than to R as for rigid (undeformable) spheres;
(ii) the adhesion force is $2\pi RW$, just as for rigid spheres;
(iii) for typical values of $k_a \approx 100$ mJ m^{-2}, the contact angle θ will exceed $45°$ once the work of adhesion W exceeds 1 mJ m^{-2}. When this occurs the total surface area of each vesicle will be stretched by more than 2% of its initial (unstressed) value. Since most lipid bilayers and biological membranes cannot be stretched beyond 2–4% without rupturing (Wobschall, 1971; Evans and Skalak, 1980), this could lead to rupture, followed by fusion of the burst vesicles.

In the second scenario, we consider two adhering vesicles of initial radius R as above, but where water can now permeate out of the vesicles so as to relax the lateral tension on the stretched bilayers. If there is only pure water both inside and outside the vesicles, then the expulsion of water would continue until the two vesicles have collapsed. But if the solution is rich in salt or some non-permeating solute the water will stop its outward diffusion once the increased osmotic pressure, $\Delta P = \Delta\rho kT$, inside the vesicles balances the lateral tension σ of the bilayers, i.e., when $\Delta P(\pi R^2) = (2\pi R)\sigma$. At equilibrium, σ is also related to the work of adhesion, W, by the usual expression $W = 2\sigma(1 - \cos\theta)$. Thus we may write:

$$W = 2\sigma(1 - \cos\theta) = \Delta\rho(1 - \cos\theta)RkT. \tag{18.8}$$

The increased solute concentration inside the vesicles, $\Delta\rho$, is directly related to the volume change of the vesicles as defined by θ according to

$$\Delta\rho = \rho\left\{\frac{(3 - \cos\theta)^{3/2}}{2(1 + \cos\theta)^{1/2}(2 - \cos\theta)} - 1\right\} \tag{18.9}$$

where ρ is the initial concentration. Thus, from a knowledge of ρ, R and θ one may deduce σ and W (Bailey et al., 1990). Since water permeates fairly rapidly across lipid bilayers and biological membranes, this type of equilibrium is more likely than that described by Eq. (18.7), and is the type shown in Fig. 18.11.

While the above equations may be formally correct the situation is far more complex in practice. First, a complete treatment must also include the bending energy change during vesicle deformations. Second, we have already seen that by stretching bilayers their adhesion energy, W, increases due to an increased hydrophobic attraction. Thus, in general, we cannot assume that W is independent of θ or σ for bilayers. Third, any changes in the solution conditions that alter W may also alter other bilayer properties such as the elastic moduli. Thus, in general, we cannot assume that k_a is constant either. In certain cases a number of additional processes and relaxations may also be involved during vesicle adhesion. These include flip-flop, a lateral redistribution of lipids and proteins, phase separation (in multicomponent systems), lipid exchange with other bilayers, and slow diffusion of ions and other solute molecules across the vesicle walls.

18.11 FUSION

When the adhesive forces between two vesicles are sufficiently strong, they will undergo a complete transformation by fusing into larger or completely different structures. However, compared to all the feverish goings on in a cell, the fusion of biological membranes *in vivo* is not a regular event. On the other hand, there are specialized membranes and vesicles whose function it is to fuse at some well-chosen time and place, notably during synaptic nerve transmission, exocytosis (vesicle incorporation) and pinocytosis (vesicle shedding), as illustrated in Fig. 17.6. Vesicle fusion occurs very rapidly and is usually over within 0.1–1 ms. Consequently, it has not yet been possible to 'trap' the intermediate stages of fusion, for example, by rapidly freezing or fixing the fusing membranes prior to imaging them with an electron microscope. More recent interest in vesicles as potential drug carriers has also stimulated research into the way vesicles adhere and fuse with each other and with planar membranes (Chernomordik et al., 1986; Sowers, 1987).

Pure lipid vesicles (e.g., phosphatidylcholines) in water are often stable for months but fuse rapidly (within minutes) into larger vesicles when the temperature is lowered much below the chain melting temperature, T_c. This is because the curved bilayers of small vesicles become highly stressed below T_c as the chains now attempt to line up and pack into less curved bilayers. The stressed vesicles easily rupture during collisions, and then fuse with each other to form larger, less curved, vesicles. It is important to appreciate that this type of fusion process is driven, not by a change in the *intervesicle* forces (see below), but by a change in the *intravesicle* or internal packing forces that have induced fragility in the vesicles.

As we established in the previous section, one should also be able to induce vesicles to fuse by increasing their adhesion energy beyond a certain threshold value. To some extent this is borne out by experiments: for example, negatively charged lipid vesicles of phosphatidylserine, phosphatidic acid and phosphatidylglycerol aggregate then fuse in the presence of 1–10 mM $CaCl_2$. Such concentrations of $CaCl_2$ certainly reduce the electrostatic repulsion and greatly enhance the short-range attraction between negatively charged bilayers (Fig. 18.1). But here, too, the adsorption of Ca^{2+} ions to the bilayer surfaces also changes the packing stresses of the lipids in much the same way as lowering the temperature below T_c, an effect that also facilitates fusion. Calcium is usually involved in cellular fusion processes, but the fusion of biological vesicles or membranes, which contain calcium-specific proteins, is often triggered by much lower calcium concentrations, usually 1–10 μM Ca^{2+}.

The roles of different forces in leading to adhesion and fusion are still far from clear. There are examples of strong adhesion without fusion and of fusion with no or little adhesion. The two most important forces that lead to fusion appear to be ion binding and the hydrophobic interaction. The first probably acts by destabilizing membranes, the second by pulling the hydrocarbon interiors of bilayers together (Papahadjopoulos et al., 1977; Chernomordik et al., 1986; Helm and Israelachvili, 1991).

The precise molecular rearrangements accompanying fusion processes are still unknown, and it is also not known whether there is one general fusion

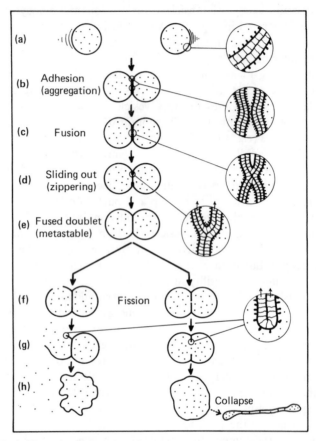

Fig. 18.12. Probable molecular events taking place during the adhesion, fusion and fission of bilayers and membranes. (a) *Approach* of two vesicles. (b) *Adhesion* (aggregation): bilayers in tension with consequent exposure of more hydrophobic surface regions to each other. Rupture or local breakthrough may now occur, leading to... (c–e) *Monolayer fusion* or *hemi-fusion*: via the rapid sliding out of the outer monolayer lipids (zippering fusion). The final stage (e) can be a long-lived metastable state. Note that so far the vesicles' contents remain 'internalized'. (f–h) *Full fusion, bilayer fusion* or *fission* (right side): here the final vesicle (h) has all the original outer monolayer lipids on the outside and all the inner monolayer lipids on the inside. The vesicles' contents remain 'internalized'.

Full fusion (left side). Alternative fusion pathway. Not all the vesicles' contents are 'internalized'. Here both the outer and inner monolayer lipids of one of the vesicles go to the outer monolayer of the final vesicle, resulting in high stresses in the membrane that would have to be alleviated somehow (e.g., via flip-flop). This fission process is therefore energetically unfavourable compared to the one on the right.

mechanism or whether many different pathways exist, each specific to a particular system. One of the problems with fusion has always been to explain how the short-range repulsion is overcome by membranes to enable them to fuse. Horn (1984) and Helm *et al.* (1989) investigated the molecular rearrangements associated with the fusion of supported bilayers. They found that the fusion mechanism was very similar for a range of different lipid and surfactant bilayers. In this mechanism the bilayers do not have to 'overcome' some repulsive, e.g. hydration or steric, force barrier before they fuse. Instead, once two bilayer surfaces are within about 1 nm of each other (Fig. 18.8) local deformations and molecular rearrangements occur which expose hydrophobic groups that strongly attract each other across the aqueous gap. In this way the short-range repulsive forces are 'bypassed' via local 'breakthrough' mechanisms (Fig. 18.12). It was also found that fusion via this type of mechanism could be significantly enhanced by increasing the hydrophobic attraction between the bilayers simply by stretching them, an effect that may explain the enhanced adhesion and fusion of membranes subjected to osmotic or electric field gradients (Zimmerberg *et al.*, 1980; Zimmermann and Scheurich, 1981; Akabas *et al.*, 1984; Fisher and Parker, 1984; Young *et al.*, 1984; Bailey *et al.*, 1990) since these stresses increase the hydrophobicity of membranes (see Problem 18.6).

In conclusion, vesicle and membrane fusion may result from a number of factors. First, vesicles may fuse because of the increased adhesion force between them, brought about by some ionic or osmotic change. Second, membranes may become internally stressed or destabilized by changes in the ionic environment, temperature or packing stresses. If such changes favour larger vesicles, they may lead to rupture followed by the association of the ruptured vesicles. In such cases fusion would not be due to intervesicle forces *per se*. It is clear that more specific binding mechanisms are involved in the fusion of biological membranes, though there are indications (Horn, 1984; Chernomordik *et al.*, 1986) that the fusion processes of both lipid vesicles and biological membranes proceed via the same intermediate stages as shown in Fig. 18.12.

PROBLEMS AND DISCUSSION TOPICS

18.1 (i) List four types of repulsive forces and four types of attractive forces between amphiphilic surfaces, and describe realistic situations where each would dominate the interaction over a certain distance regime.

(ii) Will increasing the repulsion between self-assembled structures increase or decrease their mean size?

18.2   In Fig. 18.2, if the solution concentration were increased to 1 M NaCl show that the secondary minimum at $\psi_0 = 50 \, \text{mV}$ moves in from $D = 6 \, \text{nm}$ to about $D = 2 \, \text{nm}$ and its depth increases about tenfold.

18.3   Some lipid headgroups can be quite large, being composed of a number of segments. They can also be highly mobile. If we model such headgroups as free polymer molecules pinned at one end of the surface, then the repulsive force between two such surfaces will have the form given by Eq (14.3) or (14.5). Consider the interaction of two non-ionic bilayers of $C_{12}E_6$ across water where each E_6 headgroup occupies an area of $0.5 \, \text{nm}^2$ (i.e., $s = 0.7 \, \text{nm}$), and consists of $n = 6$ freely jointed segments spaced $l = 0.2 \, \text{nm}$ apart. Ignoring all other forces other than the attractive van der Waals force between the hydrocarbon chains across water, decide which equations are appropriate for this system and then calculate the equilibrium separation between two such bilayers (i) in pure water, and (ii) in 0.15M NaCl solution, at 25°C.

18.4   Consider a lipid bilayer in the fluid state in water where the hydrophilic headgroups interact with each other via a repulsive steric-hydration force whose interaction pair potential is given by $w(r) = +(C/r)e^{-r/\lambda}$, where r is the distance between headgroups (each of hard-core diameter σ), and where C and λ are constants. In addition, there is an attractive force between the lipids described by the normal expression for the interfacial energy of a hydrocarbon–water interface, viz. γa, where γ is the interfacial energy per unit area, and a is the mean area occupied per molecule (Section 17.2). Using an analysis similar to that used to derive the two-dimensional van der Waals equation of state, derive an expression for the 'optimal' area a_0 (the equilibrium headgroup area) occupied by each lipid in the bilayer in terms of C, σ, λ and γ.

Next, consider two opposing bilayers at a distance D apart in water whose headgroups occupy the same optimal area a_0. If each headgroup is also assumed to interact with all the headgroups in the opposite bilayer via the same repulsive potential function as given by the above equation, show that the repulsive 'hydration' pressure $P(D)$ between two such bilayers is a pure exponentially decaying function of their separation D, and obtain $P(D)$ in terms of D, γ, σ and λ.

18.5   The phase diagrams of many amphiphile–water systems, which include both surfactant and amphiphilic polymer systems, display a common feature, where on increasing the surfactant concentration above 20–50% there is a series of structural transitions having the following characteristic sequence: spherical micelles → aligned cylindrical micelles → oriented planar bilayers (lamellar phase) → 'inverted' cylindrical structures. The transitions between these one-phase systems are usually separated by narrow two-phase regions.

Explain this phenomenon. If the various surfactant aggregates repel each other with a strongly repulsive steric-hydration force coming in sharply at a surface separation of 1.5 nm, and if the fully extended length of the hydrocarbon chain in each type of aggregate is also 1.5 nm, estimate the volume fractions of surfactant at which the transitions from spheres to cylinders and cylinders to lamellae will occur.

18.6 (i) Calculate the lateral tension (σ, in units of $mN\ m^{-1}$) that must be applied to a bilayer for its mean thickness to decrease by 2%.

(ii) Calculate the electric potential difference (in volts) that must be applied across the bilayer in water for its thickness to change by 2%. Will the thickness increase or decrease?

What possible effects could these stresses have on the forces between two adjacent bilayers? (Assume an initial bilayer thickness of 3.0 nm, an elastic modulus of $k_a = 150\ mJ\ m^{-2}$, and a dielectric constant of the hydrocarbon core of $\varepsilon = 2.2$.)

18.7 Describe three realistic ways in which the adhesion of biomembranes can be enhanced by increasing the strength of an attractive force between them, and three by decreasing the strength of a repulsive force.

18.8 Are the following statements true when applied to the interaction in a liquid or vapour of any two unconstrained but similar electroneutral particles or surfaces?

(i) The total purely electrostatic contribution to the interaction is always attractive.

(ii) Apart from the hard-core repulsion at molecular contact, all the repulsive contributions have an entropic origin.

REFERENCES

Abillon, O. and Perez, E. (1990). *J. Phys. (France)* **51**, 2543–2556.

Abraham, F. F. (1978). *J. Chem. Phys.* **68**, 3713–3716.

Adam, N. K. and Stevenson, D. G. (1953). *Endeavour* **XII** (45), 25, 32.

Adamson, A. W. (1976). *Physical Chemistry of Surfaces*, 3rd ed., Wiley, New York and London.

Akabas, M. H., Cohen, F. S. and Finkelstein, A. (1984). *J. Cell Biol.* **98**, 1063–1071.

Alder, B. J., Hoover, H. G. and Young, D. A. (1968). *J. Chem. Phys.* **49**, 3688–3696.

Alexander, A. E. and Johnson, P. (1950). *Colloid Science*, ch. XXIV, Oxford University Press (Clarendon), London and New York.

Alexander, S. J. (1977). *Physique* **38**, 983–987.

Allen, M. P. and Tildesley, D. J. (1987). *Computer Simulation of Liquids*, Clarendon Press, Oxford, and Oxford University Press, New York.

Almog, Y. and Klein, J. (1985). *J. Colloid Interface Sci.* **106**, 33–44.

Amis, E. S. (1975). In *Solutions and Solubilities* (M. R. J. Dack, ed.), Part 1, pp. 105–193, Wiley (Interscience), New York and London.

Anderson, J. M. and Andersson, B. (1982). *Trends Biochem. Sci.* **7**, 288–292.

Aniansson, E. A. G. (1978). *J. Phys. Chem.*, **82**, 2805–2808.

Aniansson, E. A. G., Wall, S. N., Almgren, M., Hoffmann, H., Kielmann, I., Ulbricht, W., Zana, R., Lang, J. and Tondre, C. (1976). *J. Phys. Chem.* **80**, 905–922.

Asakura, S. and Oosawa, F. (1954). *J. Chem. Phys.* **22**, 1255–1256.

Asakura, S. and Oosawa, F. (1958). *J. Polymer Sci.* **33**, 183–192.

Attard, P. and Batchelor, M. T. (1988). *Chem. Phys Lett.* **149**, 206–211.

Aveyard, R. and Saleem, S. M. (1976). *J. Chem. Soc. Faraday Trans. I* **72**, 1609–1617.

Bailey, A. I., Price, A. G. and Kay, S. M. (1970). *Spec. Discuss. Faraday Soc.* **I**, 118–127.

Bailey, S. M., Chiruvolu, S., Israelachvili, J. N. and Zasadzinski, J. A. N. (1990) *Langmuir* **6**, 1326–1329.

Banerjee, A., Ferrante, J. and Smith, J. R. (1991) In *Fundamentals of Adhesion* (L. H. Lee, ed.), pp. 325–348, Plenum, New York.

Barquins, M. and Maugis, D. (1982). *J. Méc. théor. appl.* **1**, 331–357.

Ben Naim, A., Wilf, J. and Yaacobi, M. (1973). *J. Phys. Chem.* **77**, 95–102.

Benga, G. and Holmes, R. P. (1984). *Prog. Biophys. Molec. Biol.* **43**, 195–257.

Bittar, E. E. (1980). *Membrane Structure and Function*, Wiley-Interscience, New York.

Blake, T. D. (1975). *J. Chem. Soc. Faraday Trans. I* **71**, 192–208.

Blanckstein, D., Thurston, G. M. and Benedek, G. B. (1986). *J. Chem. Phys.* **85**, 7268–7288.

Bloomquist, C. R. and Shutt, R. S. (1940). *Ind. Eng. Chem.* **32**(6), 827–831.

Bockris, J. O'M. and Reddy, A. K. N. (1970). *Modern Electrochemistry*, vol. 1, ch. 2, Plenum, New York.

Bondi, A. (1968). *Physical Properties of Molecular Crystals, Liquids, and Glasses*, Wiley, New York and London.

Bradley, R. S. (1932). *Phil. Mag.* **13**, 853–862.

Brunner, J., Skrabal, P. and Hauser, H. (1976). *Biochim. Biophys. Acta* **455**, 322–331.

Buffat, Ph. and Borel, J-P. (1976). *Phys. Rev. A* **13**, 2287–2298.

Buffey, I. P., Brown, W. B. and Gebbie, H. A. (1990). *J. Chem. Soc. Faraday Trans.* **86**, 2357–2360.

Burgess, A. K., Hughes, B. D. and White, L. R. (1990)–unpublished results.

Cabane, B., Duplessix, R. and Zemb, T. (1985). *J. Physique* **46**, 2161–2178.

Caffrey, M. and Bilderback, D. H. (1983). *Nuclear Inst. Methods* **208**, 495–510.

Capaldi, R. A. (1982). *Trends. Biochem. Sci.* **7**, 292–295.

Carnie, S. L. and Chan, D. Y. C. (1982). *Adv. Colloid Interface Sci.* **16**, 81–100.

Carnie, S., Israelachvili, J. N. and Pailthorpe, B. A. (1979). *Biochim. Biophys. Acta* **554**, 340–357.

Castellan, G. W. (1972). *Physical Chemistry*, 2nd ed., Addison–Wesley, Reading, Massachusetts.

Cevc, G. and Marsh, D. (1987). *Phospholipid Bilayers*, Wiley, New York.

Chan, D. Y. C. and Horn, R. G. (1985). *J. Chem. Phys.* **83**, 5311–5324.

Chan, D. Y. C. and Mitchell, D. J. (1983). *J. Colloid Interface Sci.* **95**, 193–197.

Chan, D. Y. C., Healy, T. W. and White, L. R. (1976). *J. Chem. Soc. Faraday Trans. 1* **72**, 2844–2865.

Chan, D. Y. C., Mitchell, D. J., Ninham, B. W. and Pailthrope, B. A. (1980a). *J. Chem. Soc. Faraday Trans. 2* **76**, 776–784.

Chan, D. Y. C., Pashley, R. M. and White, L. R. (1980b). *J. Colloid Interface Sci.* **77**, 283–285.

Chandler, D. (1987). *Introduction to Modern Statistical Mechanics*, ch. 6, Oxford University Press, Oxford.

Chen, Y. S. and Hubbell, W. L. (1973). *Exp. Eye Res.* **17**, 517–532.

Chernomordik, L. V., Melikyan, G. B. and Chizmadzhev, Y. A. (1987). *Biochim. Biophys. Acta* **906**, 309–352.

Christenson, H. (1983). *J. Chem. Phys.* **78**, 6906–6913. Also PhD Thesis, Australian National University, Canberra, Australia.

Christenson, H. K. (1985a). *Chem. Phys. Lett.* **118**, 455–458.

Christenson, H. K. (1985b). *J. Colloid Interface Sci.* **105**, 234–249.

Christenson, H. K. (1986). *J. Phys. Chem.* **90**, 4–6.

Christenson, H. K. (1988a). *J. Disp. Sci. Technol.* **9**, 171–206.

Christenson, H. K. (1988b). *J. Colloid Interface Sci.* **121**, 170–178.

Christenson, H. K. and Claesson, P. M. (1988). *Science* **239**, 390–392.

Christenson, H. K. and Horn, R. G. (1983). *Chem. Phys. Lett.* **98**, 45–48.

Christenson, H. K. and Horn, R. G. (1985). *Chemica Scripta* **25**, 37–41.

Christenson, H. K. and Israelachvili, J. N. (1984). *J. Chem. Phys.* **80**, 4566–4567.

Christenson, H. K., Claesson, P. M., Berg, J. and Herder, P. C. (1989). *J. Phys. Chem.* **93**, 1472–1478.

Christenson, H. K., Fang, J., Ninham, B. W. and Parker, J. L. (1990). *J. Phys. Chem.* **94**, 8004–8006.

Christenson, H. K., Gruen, D. W. R., Horn, R. G. and Israelachvili, J. N. (1987). *J. Chem. Phys.* **87**, 1834–1841.

Christou, N. I., Whitehouse, J. S., Nicholson, D. and Parsonage, N. G. (1981). *Symp. Faraday Soc.* **16**, 139–149.

Ciccotti, G., Frenkel, D. and McDonald, I. R. (1987). *Simulation of Liquids and Solids*, North Holland, Amsterdam.

Claesson, P. M. and Christenson, H. K. (1988). *J. Phys. Chem.* **92**, 1650–1655.

Claesson, P., Carmona-Ribeiro, A. M. and Kurihara, K. (1989). *J. Phys. Chem.* **93**, 917–922.

Claesson, P., Horn, R. G. and Pashley, R. M. (1984). *J. Colloid Interface Sci.* **100**, 250–263.

Claesson, P. M., Blom, C. E., Herder, P. C. and Ninham, B. W. (1986a). *J. Colloid Interface Sci.* **114**, 234–242.

Claesson, P. M., Kjellander, R., Stenius, P. and Christenson, H. K. (1986b). *J. Chem. Soc. Faraday Trans. I* **82**, 2735–2746.

Clark, V. A., Tittmann, B. R. and Spencer, T. W. (1980). *J. Geophys. Res.* **85**, 5190–5198.

Clunie, J. S., Goodman, J. F. and Symons, P. C. (1967). *Nature* **216**, 1203–1204.

Coakley, C. J. and Tabor, D. (1978). *J. Phys. D* **11**, L77–L82.

Cohen-Stuart, M. A., Cosgrove, T. and Vincent, B. (1986). *Adv. Colloid Interface Sci.* **24**, 143–239.

Cosgrove, T. (1990). *J. Chem. Soc. Faraday Trans.* **86**, 1323–1332.

Cotton, F. A. and Wilkinson, G. (1980). *Advanced Inorganic Chemistry*, 4th ed., p. 1188, Wiley, New York.

Coulson, C. A. (1961). *Valence*, 2nd ed., ch XIII, Oxford University Press, London and New York.

Cowley, A. C., Fuller, N. L., Rand, R. P. and Parsegian, V. A. (1978). *Biochemistry* **17**, 3163–3168.

Croxton, C. A. (1975). *Introduction to Liquid State Physics*, Wiley, New York.

Dang, L. X., Rice, J. E. and Kollman, P. A. (1990). *J. Chem. Phys.* **93**, 7528–7529.

Dasent, W. E. (1970). *Inorganic Energetics*, Penguin Books, Harmondsworth.

Dash, J. G. (1989). *Contemp. Phys.* **30**, 89–100.

Dashevsky, V. G. and Sarkisov, G. N. (1974). *Mol. Phys.*, **27**, 1285–1290.

Davies, M. (1965). *Some Electrical and Optical Aspects of Molecular Behaviour*, Pergamon, Oxford.

Day, E. P., Kwok, A. Y. W., Hark, S. K., Ho, J. T., Vail, W. J., Bens, J. and Nir, S. (1980). *Proc. Natl. Acad. Sci. USA* **77**, 4026–4029.

de Boer, J. H. (1936). *Trans. Faraday Soc.* **32**, 10–38.

de Gennes, P. G. (1979). *Scaling Concepts in Polymer Physics*, Cornell University Press, Ithaca, NY. (2nd printing, 1985.)

de Gennes, P. G. (1981). *Macromolecules* **14**, 1637–1644.

de Gennes, P. G. (1982). *Macromolecules* **15**, 492–500.

de Gennes, P. G. (1985). *C. R. Acad. Sci. (Paris)* **300**, 839–843.

de Gennes, P. G. (1987a). *Adv. Colloid Interface Sci.* **27**, 189–209.

de Gennes, P. G. (1987b). *C. R. Acad. Sci. (Paris)* **305**, 1181–1184.

De Loof, H., Harvey, S. C., Segrest, J. P. and Pastor, R. W. (1991). *Biochemistry* **30**, 2099–2113.

Del Pennino, U., Mazzega, E., Valeri, S., Alietti, A., Brigatti, M. F. and Poppi, L. (1981). *J. Colloid Interface Sci.* **84**, 301–309.

Denbigh, K. G. (1940). *Trans. Faraday Soc.* **36**, 936–948.

Derjaguin, B. V. (1934). *Kolloid Zeits.* **69**, 155–164.

Derjaguin, B. V. and Abrikossova, I. I. (1954). *Discuss. Faraday Soc.* **18**, 24–41.

Derjaguin, B. V. and Churaev, N. V. (1974). *J. Colloid Interface Sci.* **49**, 249–255.

Derjaguin, B. V. and Kusakov, M. M. (1939). *Acta Physiocochim. URSS* **10**, 25–44, 153–174.

Derjaguin, B. V. and Landau, L. (1941). *Acta Physicochim. URSS* **14**, 633–662.

Derjaguin, B. V. and Titijevskaia, A. S. (1954). *Discuss. Faraday Soc.* **18**, 24–41.

Derjaguin, B. V., Abrikossova, I. I. and Lifshitz, E. M. (1956). *Quart. Rev. Chem. Soc.* **10**, 295–329.

Derjaguin, B. V., Muller, V. M. and Toporov, Y. P. (1975). *J. Colloid Interface Sci.* **53**, 314–326.

Derjaguin, B. V., Rabinovich, Y. I. and Churaev, N. V. (1978). *Nature* **272**, 313–318.

Diederichs, K., Welte, W. and Kreutz, W. (1985). *Biochim. Biophys. Acta* **809**, 107–116.

Dolan, A. K. and Edwards, S. F. (1974). *Proc. R. Soc. Lond.* **A337**, 509–516.

Donners, W. A. B., Rijnbout, J. B. and Vrij, J. (1977). *J. Colloid Interface Sci.* **61**, 249–260.

Ducker, W. A., Senden, T. J. and Pashley, R. M. (1991). *Nature* (in press).

Dymond, J. H. (1981). *J. Phys. Chem.* **85**, 3291–3294.

Dzyaloshinskii, I. E., Lifshitz, E. M. and Pitaevskii, L. P. (1961). *Adv. Phys.* **10**, 165–209.

Egberts, E. and Berendsen, H. J. C. (1988). *J. Chem. Phys.* **89**, 3718–3726.

Eicke, H. F. (1980). *Top. Curr. Chem.* **87**, 85–145.

Ekwall, P. (1975). *Adv. Liq. Cryst.* **1**, 1–142.

Elimelech, M. (1990). *J. Chem. Soc. Faraday Trans.* **86**, 1623–1624.

Ertl, H. and Dullien, F. A. L. (1973). *AIChE J.* **19**, 1215–1223.

Evans, D. F., Tominaga, T. and Davis, H. T. (1981). *J. Chem. Phys.* **74**, 1298–1305.

Evans, E. and Needham, D. (1988). *Macromolecules* **21**, 1822–1831.

Evans, E. and Rawicz, W. (1990). *Phys. Rev. Lett.* **64**, 2094–2097.

Evans, E. A. and Skalak, R. (1980). *Mechanics and Thermodynamics of Biomembranes*, CRC Press, Boca Raton, FL.

Evans, R. and Parry, A. O. (1990). *J. Phys.: Condens. Matter* **2**, SA15–SA32.

Ferrante, J. and Smith, J. R. (1985). *Phys. Rev. B*, **31**, 3427–3434.

Fischer, G. and Schmid, F. X. (1990). *Biochemistry* **29**, 2205–2212.

Fisher, L. R. and Israelachvili, J. N. (1981). *Colloids Surf.* **3**, 303–319.

Fisher, L. R. and Parker, N. S. (1984). *Biophys. J.* **46**, 253–258.

Fleer, G. J. (1988). *Surfactants Sci. Ser.* **27**, 105–158.

Flory, P. J. (1953). *Principles of Polymer Chemistry*, Cornell University Press, Ithaca, NY.

Flory, P. J. (1969). *Statistical Mechanics of Chain Molecules*, J. Wiley, New York.

Fogden, A. and White, L. R. (1990). *J. Colloid Interface Sci.* **138**, 414–430.

Forsyth, P. A., Marcelja, S., Mitchell, D. J. and Ninham, B. W. (1977). *Biochim. Biophys. Acta* **469**, 335–344.

Forsyth, P. A., Marcelja, S., Mitchell, D. J. and Ninham, B. W. (1978). *Adv. Colloid Interface Sci.* **9**, 37–60.

Fowkes, F. F. (1964). *Ind. Eng. Chem.* **56**(12), 40–52.

Fowler, P. W., Lazzeretti, P. and Zanas, W. (1989). *Mol. Phys.* **4**, 853–865.

Franks, F. (1972–1982). *Water: a Comprehensive Treatise* (F. Franks, ed.), vols 1–7, Plenum, New York and London.

Frens, G. and Overbeek, J. Th. G. (1972). *J. Colloid Interface Sci.* **38**, 376–387.

Gains, N. and Hauser, H. (1983). *Biochim. Biophys. Acta* **731**, 31–39.

Gallot, B. and Skoulios, A. (1966). *Kolloid Z. U. Z. Polymere* **208**, 37–43.

Gee, M. L. and Israelachvili, J. N. (1990). *J. Chem. Soc. Faraday Trans.* **86**, 4049–4058.

Gee, M. L., Healy, T. W. and White, L. R. (1989). *J. Colloid Interface Sci.* **131**, 18–23.

Gee, M. L., Healy, T. W. and White, L. R. (1990). *J. Colloid Interface Sci.* **140**, 450–465.

Gee, M. L., Tong, P., Israelachvili, J. N. and Witten, T. A. (1990) *J. Chem. Phys.* **93**, 6057–6064.

Georges, J. M. (ed.) (1982). *Microscopic Aspects of Adhesion and Lubrication*, Elsevier, Amsterdam.

Gingell, D. and Parsegian, V. A. (1972). *J. Theor. Biol*, **36**, 41–52.

Glaser, M. A. and Clarke, N. A. (1990). *Phys. Rev. A* **41**, 4585–4588.

Good, R. J. and Elbing, E. (1970). *Ind. Eng. Chem.* **62**(3), 54–78.

Grahame, D. C. (1953). *J. Chem. Phys.* **21**, 1054–1060.

Granfeldt, M. K. and Miklavic, S. J. (1991). *J. Phys. Chem.* **95**, 6351–6360.

Gregory, J. (1970). *Adv. Colloid Interface Sci.* **2**, 396–417.

Gregory, J. (1973). *J. Chem. Soc. Faraday Trans.* 2 **69**, 1723–1728.

Gregory, J. (1975). *J. Colloid Interface Sci.* **51**, 44–51.

Gruen, D. W. R. (1981). *J. Colloid Interface Sci.* **84**, 281–283.

Gruen, D. W. R. (1985). *J. Phys. Chem.* **89**, 146–152, 153–163.

Gruen, D. W. R. and de Lacey, E. H. B. (1984). In *Surfactants in Solution* (K. L. Mittal and B. Lindman, eds), vol. 1, pp. 279–306, Plenum, New York.

Gruen, D. W. R. and Haydon, D. A. (1981). *Biophys. J.* **33**, 167–188.

Gruen, D. W. R. and Marcelja, S. (1983). *J. Chem. Soc. Faraday Trans.* 2 **79**, 225–242.

Guggenheim, E. A. (1949). *Thermodynamics*, North Holland, Amsterdam.

Guldbrand, L., Jönsson, B., Wennerström, H. and Linse, P. (1984). *J. Chem. Phys.* **80**, 2221–2228.

Hackenbrock, C. R. and Miller, K. J. (1975). *J. Cell Biol.* **65**, 615–630.

Hadziioannou, G., Patel, S., Granick, S. and Tirrell, M. (1986). *J. Am. Chem. Soc.* **108**, 2869–2876.

Hall, D. G. and Pethica, B. A. (1967). In *Nonionic Surfactants* (M. J. Schick, ed.), ch. 16, Marcel Dekker, New York.

Hamaker, H. C. (1937). *Physica* **4**, 1058–1072.

Hamnerius, Y., Lundstrom, I., Paulsson, L. E., Fontell, K. and Wennerström, H. (1978). *Chem. Phys. Lipids* **22**, 135–140.

Hansma, P. K., Elings, V. B., Marti, O. and Bracker, C. E. (1988). *Science*, **242**, 209–216.

Harrison, R. and Lunt, G. G. (1980). *Biological Membranes*, 2nd ed., Halsted Press, Wiley, New York.

Hasted, J. B. (1973). *Aqueous Dielectrics*, ch. 4, Chapman and Hall, London.

Hauser, H. (1984). *Biochim. Biophys. Acta* **772**, 37–50.

Hauser, H., Gains, N. and Müller, M. (1983). *Biochemistry* **22**, 4775–4781.

Hauser, H., Mantsch, H. H. and Casal, H. L. (1990). *Biochemistry* **29**, 2321–2329.

Healy, T. W. and White, L. R. (1978). *Adv. Colloid Interface Sci.* **9**, 303–345.

Healy, T. W., Chan, D. and White, L. R. (1980). *Pure Appl. Chem.* **52**, 207–219.

Healy, T. W., Homola, A., James, R. O. and Hunter, R. J. (1978). *Faraday Discuss. Chem. Soc.* **65**, 156–163.

Helfrich, W. (1978). *Z. Naturforsch.* **33a**, 305–315.

Helm, C. A. and Israelachvili, J. N. (1991). *Methods in Enzymology.* (In press.)

Helm, C. A., Israelachvili, J. N. and McGuiggan, P. M. (1989). *Science* **246**, 919–922.

Henderson, J. R. (1986). *Mol. Phys.* **59**, 89–96.

Henderson, D. and Lozada-Cassou, M. (1986). *J. Colloid Interface Sci.* **114**, 180–183.

Henry, J. D. Jr., Prudich, M. E. and Chak, L. (1980). *Colloids Surf.* **1**, 335–348.

Herrington, T. M. and Sahi, S. S. (1988). *J. Colloid Interface Sci.* **121**, 107–120.

Hertz, H. (1881). *J. Reine Angew. Math.* **92**, 156–171. (See also in *Miscellaneous Papers*, Macmillan, London, 1896, p. 146.)

Hertz, H. G. (1973). In *Water: a Comprehensive Treatise* (F. Franks, ed.), vol. 3, ch. 7, Plenum, New York and London.

Hesse, M. B. (1961). *Forces and Fields*, Thomas Nelson, London.

Hesselink, F. Th. (1971). *J. Phys. Chem.* **75**, 65–71.

Hesselink, F. Th., Vrij, A. and Overbeek, J. Th. G. (1971). *J. Phys. Chem.* **75**, 2094–2103.

Hiemenz, P. C. (1977). *Principles of Colloid and Surface Chemistry*, Dekker, New York and Basel.

Hill, T. L. (1963, 1964). *Thermodynamics of Small Systems*, Benjamin, New York.

Hirschfelder, J. O., Curtiss, C. F. and Bird, R. B. (1954). *Molecular Theory of Gases and Liquids*, Wiley, New York; Chapman and Hall, London.

Hirz, S. J., Homola, A. M., Hadziioannou, G. and Frank, C. W. (1991). *Langmuir*, (in press).

Hobbs, P. V. (1974). *Ice Physics*, ch. 2, Clarendon Press, Oxford.

Hogg, R., Healy, T. W. and Fuerstenau, D. W. (1966). *Trans. Faraday Soc.* **62**, 1638–1651.

Hollins, G. T. (1964). *Proc. Phys. Soc.* **84**, 1001–1016.

Homan, R. and Pownall, H. J. (1988). *Biochim. Biophys. Acta* **938**, 155–166.

Homola, A. and Robertson, A. A. (1976). *J. Colloid Interface Sci.* **54**, 286–498.

Honig, E. P. and Mul, P. M. (1971). *J. Colloid Interface Sci.* **36**, 258–272.

Horn, R. G. (1984). *Biochim Biophys. Acta* **778**, 224–228.

Horn, R. G. (1990). *J. Am. Ceram. Soc.* **73**(5), 1117–1135.

Horn, R. G. and Israelachvili, J. N. (1981a). *J. Physique* **42**, 39–52.

Horn, R. G. and Israelachvili, J. N. (1981b). *J. Chem. Phys.* **75**, 1400–1411.

Horn, R. G. and Israelachvili, J. N. (1988). *Macromolecules* **21**, 2836–2841.

Horn, R. G., Clarke, D. R. and Clarkson, M. T. (1988a). *J. Materials Res.* **3**, 413–416.

Horn, R. G., Israelachvili, J. N. and Pribac, F. (1987). *J. Colloid Interface Sci.* **115**, 480–482.

Horn, R. G., Smith, D. T. and Haller, W. (1989a). *Chem. Phys. Lett.* **162**, 404–408.

Horn, R. G., Hirz, S. J., Hadziioannou, G. H., Frank, C. W. and Catala, J. M. (1989b). *J. Chem. Phys.* **90**, 6767–6774.

Horn, R. G., Israelachvili, J. N., Marra, J., Parsegian, V. A. and Rand, R. P. (1988b). *Biophys. J.* **54**, 1185–1187.

Hough, D. B. and White, L. R. (1980). *Adv. Colloid Interface Sci.* **14**, 3–41.

Hu, Y. W., Van Alsten, J. and Granick, S. (1989). *Langmuir* **5**, 270–272.

Hunter, R. J. (1989). *Foundations of Colloid Science*, vol. 1, Clarendon Press, Oxford.

Ingersent, K., Klein, J. and Pincus, P. (1986). *Macromolecules* **19**, 1374–1381.

Ingersent, K., Klein, J. and Pincus, P. (1990). *Macromolecules* **23**, 548–560.

Israelachvili, J. N. (1972). *Proc. R. Soc. Lond. A* **331**, 39–55.

Israelachvili, J. N. (1973). *J. Chem. Soc. Faraday Trans. 2* **69**, 1729–1738.

Israelachvili, J. N. (1974). *Q. Rev. Biophys.* **6**, 341–387.

Israelachvili, J. N. (1982). *Adv. Colloid Interface Sci.* **16**, 31–47.

Israelachvili, J. N. (1985). *Chemica Scripta* **25**, 7–14.

Israelachvili, J. N. (1986). *J. Colloid Interface Sci.* **110**, 263–271.

Israelachvili, J. N. (1987a). In *Physics of Complex and Supermolecular Fluids* (S. A. Safran and N. A. Clark, eds) pp. 101–114, Wiley, New York.

Israelachvili, J. N. (1987b). *Accounts Chem. Res.* **20**, 415–421.

Israelachvili, J. N. (1987c). In *Surfactants in Solution*, vol. 4 (K. L. Mittal and P. Bothorel, eds) pp. 3–33, Plenum, New York.

Israelachvili, J. N. (1989). *Chemtracts–Analy. Phys. Chem.* **1**, 1–12.

Israelachvili, J. N. and Adams, G. E. (1978). *J. Chem. Soc. Faraday Trans. I* **74**,

975–1001.

Israelachvili, J. N. and Kott, S. J. (1988). *J. Chem. Phys.* **88**, 7162–7166.

Israelachvili, J. N. and McGuiggan, P. M. (1990). *J. Mater. Res.* **5**(10), 2223–2231.

Israelachvili, J. N. and Ninham, B. W. (1977). *J. Colloid Interface Sci.* **58**, 14–25.

Israelachvili, J. N. and Pashley, R. M. (1982a). In *Biophysics of Water* (F. Franks, ed.), pp. 183–194, Wiley, New York.

Israelachvili, J. N. and Pashley, R. M. (1982b). *Nature* **300**, 341–342.

Israelachvili, J. N. and Pashley, R. M. (1983). *Nature* **306**, 249–250.

Israelachvili, J. N. and Tabor, D. (1972). *Proc. R. Soc. Lond. A* **331**, 19–38.

Israelachvili, J. N. and Tabor, D. (1973). *Prog. Surf. Membr. Sci.* **7**, 1–55.

Israelachvili, J. N., Homola, A. M. and McGuiggan, P. M. (1988). *Science* **240**, 189–191.

Israelachvili, J. N. and Wennerström, H. (1990). *Langmuir* **6**, 873–876.

Israelachvili, J. N., Kott, S. J. and Fetters, L. (1989). *J. Polymer Sci. B: Polymer Phys.* **27**, 489–502.

Israelachvili, J. N., Marcelja, S. and Horn, R. G. (1980a). *Quart. Rev. Biophys.* **13**, 121–200.

Israelachvili, J. N., Mitchell, D. J. and Ninham, B. W. (1976). *J. Chem. Soc. Faraday Trans. I* **72**, 1525–1568.

Israelachvili, J. N., Mitchell, D. J. and Ninham, B. W. (1977). *Biochim. Biophys. Acta* **470**, 185–201.

Israelachvili, J. N., Perez, E. and Tandon, R. K. (1980b). *J. Colloid Interface Sci.* **78**, 260–261.

Israelachvili, J. N., Tandon, R. K. and White, L. R. (1979). *Nature* **277**, 120–121.

Israelachvili, J. N., Tandon, R. K. and White, L. R. (1980c). *J. Colloid Interface Sci.* **78**, 430–433.

Israelachvili, J. N., Tirrell, M., Klein, J. and Almog, Y. (1984). *Macromolecules* **17**, 204–209.

Jain, M. (1988). *Introduction to Biological Membranes*, Wiley, New York.

Jammer, M. (1957). *Concepts of Force*, Harvard University Press, Cambridge, MA.

Ji, H., Hone, D., Pincus, P. A. and Rossi, G. (1990). *Macromolecules* **23**, 698–707.

Joanny, J. F., Leibler, L. and de Gennes, P. G. (1979). *J. Polymer Sci., Polymer Phys. Ed.* **17**, 1073–1084.

Joesten, M. D. and Schaad, L. J. (1974). *Hydrogen Bonding*, Dekker, New York.

Johnson, K. L., Kendall, K. and Roberts, A. D. (1971). *Proc. R. Soc. London, Ser. A* **324**, 301–313.

Jokela, P., Jönsson, B. and Khan, A. (1987). *J. Phys. Chem.* **91**, 3291–3298.

Jönsson, B. (1981). *Chem. Phys. Lett.* **82**, 520–525.

Jönsson, B. and Wennerström, H. (1981). *J. Colloid Interface Sci.* **80**, 482–496.

Jönsson, B. and Wennerström, H. (1983). *J. Chem. Soc. Faraday Trans. 2* **79**, 19–35.

Jönsson, B., Wennerström, H. and Halle, B. (1980). *J. Phys. Chem.* **84**, 2179–2185.

Kaler, E. W., Murthy, A. K., Rodriguez, B. E. and Zasadzinski, J. A. N. (1989). *Science* **245**, 1371–1374.

Kato, T. and Seimiya, T. (1986). *J. Phys. Chem.* **90**, 3159–3167.

Kauzmann, W. (1959). *Adv. Protein Chem.* **14**, 1–63.

Khan, A., Jönsson, B. and Wennerström, H. (1985). *J. Phys. Chem.* **89**, 5180–5184.

Kjellander, R. and Marcelja, S. (1984). *Chem. Phys. Lett.* **112**, 49–53.

Kjellander, R. and Marcelja, S. (1985a). *Chem. Phys. Lett.* **120**, 393–396.

Kjellander, R. and Marcelja, S. (1985b). *Chemica Scripta* **25**, 73–80.

Kjellander, R. and Marcelja, S. (1985c). *Chemica Scripta* **25**, 112–116.

Kjellander, R. and Marcelja, S. (1986a). *J. Phys. Chem.* **90**, 1230–1232.

Kjellander, R. and Marcelja, S. (1986b). *Chem. Phys. Lett.* **127**, 402–407.

Kjellander, R., Marcelja, S., Pashley, R. M. and Quirk, J. P. (1988). *J. Phys. Chem.* **92**, 6489–6492.

Kjellander, R., Marcelja, S., Pashley, R. M. and Quirk, J. P. (1990). *J. Chem. Phys.* **92**, 4399–4407.

Klein, J. (1980). *Nature* **288**, 248–250.

Klein, J. (1982). *Adv. Colloid Interface Sci.* **16**, 101–115.

Klein, J. (1983a). *J. Chem. Soc., Faraday Trans. I* **79**, 99–118.

Klein, J. (1983b). *J. Chem. Phys.* **79**, 926–935.

Klein, J. (1988). In *Studies in Polymer Science* (M. Nagasawa, ed.) vol. 2, 333–352, Amsterdam, Elsevier.

Klein, J. and Luckham, P. (1982). *Nature* **300**, 429–431.

Klein, J. and Luckham, P. (1984a). *Nature* **308**, 836–837.

Klein, J. and Luckham, P. F. (1984b). *Macromolecules* **17**, 1041–1048.

Kohler, F. (1972). *The Liquid State*, Verlag Chemie, Weinheim.

Kruus, P. (1977). *Liquids and Solutions: Structure and Dynamics*, Dekker, New York.

Kumar, R. and Prausnitz, J. M. (1975). In *Solutions and Solubilities* (M. R. J. Dack, ed.), part 1, pp. 259–326, Wiley (Interscience) New York.

Kumar, S. K., Vacatello, M. and Yoon, D. Y. (1988). *J. Chem. Phys.* **89**, 5206–5215.

Kurihara, K., Kato, S. and Kunitake, T. (1990). *Chem. Lett. (Chem. Soc. Jpn)*, 1555–1558.

Kwok, R. and Evans, E. (1981). *Biophys. J.* **35**, 637–652.

Landau, L. D. and Lifshitz, E. M. (1963, 1984). *Electrodynamics of Continuous Media*, vol. 8, 2nd ed., Pergamon, Oxford.

Landau, L. D. and Lifshitz, E. M. (1980). *Statistical Physics*, 2nd ed., part 1, Pergamon, Oxford.

Landman, U., Luedtke, W. D., Burnham, N. A. and Colton, R. J. (1990). *Science* **248**, 454–461.

Landolt-Börnstein (1982). New Series, V/1b, II/4, II/14a; Old Series, I/3, II/6, Springer, Heidelberg.

Langmuir, I. (1938). *J. Chem. Phys.* **6**, 873–896.

Laughlin, R. G. (1978). In *Advances in Liquid Crystals* (G. H. Brown, ed.), vol. 3, pp. 41–148, Academic Press, New York.

Laughlin, R. G. (1981). *J. Soc. Cosmet. Chem.* **32**, 371–392.

Lee, C. S. and Belfort, G. (1989). *Proc. Natl. Acad. Sci. USA* **86**, 8392–8396.

Leermakers, F. A. M. and Scheutjens, J. M. H. M. (1988). *J. Chem. Phys.* **89**, 3264–3274.

LeNeveu, D. M., Rand, R. P. and Parsegian, V. A. (1976). *Nature* **259**, 601–603.

Lennard-Jones, J. E. and Dent, B. M. (1928). *Trans. Faraday Soc.* **24**, 92–108.

Lessard, R. R. and Zieminski, S. A. (1971). *Ind. Eng. Chem. Fundam.* **10**, 260–269.

Lewis, B. A. and Engelman, D. M. (1983a). *J. Mol. Biol.* **166**, 203–210.

Lewis, B. A. and Engelman, D. M. (1983b). *J. Mol. Biol.* **166**, 211–217.

Lifshitz, E. M. (1956). *Soviet Phys. JETP (Engl. Transl.)* **2**, 73–83.

Lighthill, M. J. (1970). *Introduction to Fourier Analysis and Generalized Functions*, Cambridge University Press, London and New York.

Lin, T-S., Tsen, M-Y., Chen, S-H. and Roberts, M. F. (1990). *J. Phys. Chem.* **94**, 7239–7243.

Lindblom, G. and Wennerström, H. (1977). *Biophys. Chem.* **6**, 167–171.

Lindman, B., Söderman, O. and Wennerström, H. (1987). In *Surfactant Solutions* (R. Zana, ed.), ch. 6, pp. 295–357, Marcel Dekker, New York.

Lis, L. J., McAlister, M., Fuller, N. L., Rand, R. P. and Parsegian, V. A. (1982). *Biophys. J.* **37**, 657–666.

London, F. (1937). *Trans Faraday Soc*, **33**, 8–26.

Luckham, P. and Klein, J. (1990). *J. Chem. Soc. Faraday Trans.* **86**, 1363–1368.

Luzar, A., Bratko, D. and Blum, L. J. (1987). *Chem. Phys.* **86**, 2955–2959.

Lyklema, J. and Mysels, K. J. (1965). *J. Am. Chem. Soc.* **87**, 2539–2546.

Lyle, I. G. and Tiddy, G. J. T. (1986). *Chem. Phys. Lett.* **124**, 432–436.

Lyons, J. S., Furlong, D. N. and Healy, T. W. (1981). *Aust. J. Chem.* **34**, 1177–1187.

Madani, H. and Kaler, E. W. (1990). *Langmuir* **6**, 125–132.

Madden, W. G. (1987). *J. Chem. Phys.* **87**, 1405–1422.

Mahanty, J. and Ninham, B. W. (1976). *Dispersion Forces*, Academic Press, New York.

Maitland, G., Rigby, M., Smith, E. and Wakeham, W. (1981). *Intermolecular Forces: Their Origin and Determination*, Oxford University Press, New York.

Malliaris, A., Le Moigne, J., Sturm, J. and Zana, R. (1985). *J. Phys. Chem.* **89**, 2709–2713.

Marcelja, S. (1973). *Nature* **241**, 451–453.

Marcelja, S. (1974). *Biochim. Biophys. Acta* **367**, 165–176.

Marcelja, S. and Radic, N. (1976). *Chem. Phys. Lett.* **42**, 129–130.

Marcelja, S., Mitchell, D. J., Ninham, B. W. and Sculley, M. J. (1977). *J. Chem. Soc. Faraday Trans. 2* **73**, 630–648.

Margenau, H. and Kestner, N. R. (1971). *Theory of Intermolecular Forces*, Pergamon, Oxford.

Marra, J. (1985). *J. Colloid Interface Sci.* **107**, 446–458.

Marra, J. (1986a). *J. Colloid Interface Sci.* **109**, 11–20.

Marra, J. (1986b). *J. Phys. Chem.* **90**, 2145–2150.

Marra, J. (1986c). *Biophys. J.* **50**, 815–825.

Marra, J. and Israelachvili, J. N. (1985). *Biochemistry* **24**, 4608–4618.

Marsh, D. (1989). *Biophys. J.* **55**, 1093–1100.

Marsh, D. (1990). *CRC Handbook of Lipid Bilayers*, CRC Press, Boca Raton, FL.

Mazer, N. A., Benedek, G. B. and Carey, M. C. (1976). *J. Phys. Chem.* **80**, 1075–1085.

McFarlane, J. S. and Tabor, D. (1950). *Proc. R. Soc. Lond. Ser. A* **202**, 224–243.

McGuiggan, P. and Israelachvili, J. N. (1990). *J. Mater. Res.* **5**(10), 2232–2243.

McIntosh, T. J. and Simon, S. A. (1986). *Biochemistry* **25**, 4058–4066.

McIntosh, T. J., Magid, A. D. and Simon, S. A. (1989a). *Biochemistry* **28**, 7904–7912.

McIntosh, T. J., Magid, A. D. and Simon, S. A. (1989b). *Biophys. J.* **55**, 897–904.

McIver, D. J. L. (1979). *Physiol. Chem. Phys.* **11**, 289–302.

McLachlan, A. D. (1963a). *Proc. R. Soc. Lond. Ser. A* **202**, 224–243.

McLachlan, A. D. (1963b). *Mol. Phys.* **6**, 423–427.

McLachlan, A. D. (1965). *Discuss. Faraday Soc.* **40**, 239–245.

McLaughlin, S., Mulrine, N., Gresalfi, T., Vaio, G. and McLaughlin, A. (1981). *J. Gen. Physiol.* **77**, 445–473.

Miller, C. A. and Neogi, P. (1985). *Interfacial Phenomena*, Marcel Dekker, New York.

Milner, S. T., Witten, T. A. and Cates, M. E. (1988). *Macromolecules* **22**, 2610–2619.

Missel, P. J., Mazer, N. A., Benedek, G. B. and Carey, M. C. (1983). *J. Phys. Chem.* **87**, 1264–1277.

Missel, P. J., Mazer, N. A., Carey, M. C. and Benedek, G. B. (1982). In *Solution Behaviour of Surfactants* (K. L. Mittal and E. J. Fendler, eds), vol. 1, pp. 373–388, Plenum, New York.

Missel, P. J., Mazer, N. A., Benedek, G. B., Young, C. Y. and Carey, M. C. (1980). *J. Phys. Chem.* **84**, 1044–1057.

Mitchell, D. J. and Ninham, B. W. (1981). *J. Chem. Soc. Faraday Trans. 2*, **77**, 601–629.

Moelwyn-Hughes, E. A. (1961). *Physical Chemistry*, 2nd ed., Pergamon, Oxford.

Montfort, J. P. and Hadziioannou, G. (1988). *J. Chem. Phys.* **88**, 7187–7196.

Mukerjee, P. and Mysels, K. J. (1970). *Critical Micelle Concentrations of Aqueous Systems*, Nat. Bur. Stand. (US), Nat. Stand. Ref. Data. Ser., No. 36.

Muller, V. M., Derjaguin, B. V. and Toporov, Y. P. (1983). *Colloids Surf.* **7**, 251–259.

Muller, V. M., Yushchenko, V. S. and Derjaguin, B. V. (1980). *J. Colloid Interface Sci.* **77**, 91–101.

Murat, M. and Grest, G. S. (1989). *Macromolecules* **22**, 4054–4059.

Murphy, D. J. (1982). *FEBS Lett.* **150**, 19–26.

Nagarajan, R. and Ruckenstein, E. (1977). *J. Colloid Interface Sci.* **60**, 221–231.

Nagarajan, R. and Ruckenstein, E. (1979). *J. Colloid Interface Sci.* **71**, 580–604.

Nicholson, D. and Parsonage, N. G. (1982). *Computer Simulation and the Statistical Mechanics of Adsorption*, ch. 8, Academic Press, New York.

Nightingale, E. R. (1959). *J. Phys. Chem.* **63**, 1381–1387.

Ninham, B. W. and Parsegian, V. A. (1970). *Biophys. J.* **10**, 646–663.

Ninham, B. W. and Parsegian, V. A. (1971). *J. Theor. Biol.* **31**, 405–428.

Norrish, K. (1954). *Discuss. Faraday Soc.* **18**, 120–134.

Ohshima, H., Inoko, Y. and Mitsui, T. (1982). *J. Colloid Interface Sci.* **86**, 57–72.

Okamoto, S. and Hachisu, S. (1977). *J. Colloid Interface Sci.* **62**, 172–181.

Orr, F. M., Scriven, L. E. and Rivas, A. P. (1975). *J. Fluid Mech.* **67**, 723–742.

Owens, D. K. (1970). *J. Appl. Polymer Sci.* **14**, 1725–1730.

Pallas, N. R. and Pethica, B. A. (1985). *Langmuir* **1**, 509–513.

Pallas, N. R. and Pethica, B. A. (1987). *J. Chem. Soc. Faraday Trans 1* **83**, 585–590.

Pangali, C., Rao, M. and Berne, B. J. (1979). *J. Chem. Phys.* **71**, 2975–2981.

Papahadjopoulos, D., Vail, W. J., Newton, C., Nir, S., Jacobson, K., Poste, G. and Lazo, R. (1977). *Biochim. Biophys. Acta* **465**, 579–598.

Parfitt, G. D. and Peacock, J. (1978). *Surf. Colloid Sci.* **10**, 163–226.

Parker, J. L. and Christenson, H. K. (1988). *J. Chem. Phys.* **88**, 8013–8014.

Parker, J. L., Christenson, H. K. and Ninham, B. W. (1989a). *Rev. Sci. Instrum.* **60**, 3135–3138.

Parker, J. L., Cho, D. L. and Claesson, P. M. (1989b). *J. Phys. Chem.* **93**, 6121–6125.

Parsegian, V. A. (1966). *Trans. Faraday Soc.* **62**, 848–860.

Parsegian, V. A. (1973). *Annu. Rev. Biophys. Bioeng.* **2**, 221–255.

Parsegian, V. A. and Gingell, D. (1972). *Biophys. J.* **12**, 1192–1204.

Parsegian, V. A. and Weiss, G. H. (1981). *J. Colloid Interface Sci.* **81**, 285–289.

Parsegian, V. A., Fuller, N. and Rand, R. P. (1979). *Proc. Natl. Acad Sci. USA* **76**, 2750–2754.

Parsons, J. M., Siska, P. E. and Lee, Y. T. (1972). *J. Chem. Phys.* **56**, 1511–1515.

Pashley, R. M. (1977). *J. Colloid Interface Sci.* **62**, 344–347.

Pashley, R. M. (1980). *J. Colloid Interface Sci.* **78**, 246–248.

Pashley, R. M. (1981a). *J. Colloid Interface Sci.* **80**, 153–162.

Pashley, R. M. (1981b). *J. Colloid Interface Sci.* **83**, 531–545.

Pashley, R. M. (1982). *Adv. Colloid Interface Sci.* **16**, 57–63.

Pashley, R. M. (1985). *Chemica Scripta* **25**, 22–27.

Pashley, R. M. and Israelachvili, J. N. (1981). *Colloids Surf.* **2**, 169–187.

Pashley, R. M. and Israelachvili, J. N. (1984). *J. Colloid Interface Sci.* **97**, 446–455.

Pashley, R. M. and Kitchener, J. A. (1979). *J. Colloid Interface Sci.* **71**, 491–500.

Pashley, R. M. and Ninham, B. W. (1987). *J. Phys. Chem.* **91**, 2902–2904.

Pashley, R. M. and Quirk, J. P. (1984). *Colloids Surf.* **9**, 1–17.

Pashley, R. M., McGuiggan, P. M., Ninham, B. W. and Evans, D. F. (1985). *Science* **229**, 1088–1089.

Pashley, R. M., McGuiggan, P. M., Ninham, B. W., Brady, J. and Evans, D. F. (1986). *J. Phys. Chem.* **90**, 1637–1642.

Pass, G. (1973), *Ions in Solution 3: Inorganic Properties*, Oxford Chemical Series, Oxford University Press (Clarendon), London and New York.

Pastor, R. W. (1990). In *Molecular Description of Biological Membrane Components by Computer Aided Conformational Analysis* (Brasseur, R., ed.) vol. I, pp. 171–201, CRC Press, Boca Raton, FL.

Patel, S. (1986). PhD thesis, University of Minnesota, Minneapolis.

Patel, S. S. and Tirrell, M. (1989). *Annu. Rev. Phys. Chem.* **40**, 597–635.

Paterson, M. S. and Kekulawala, K. R. S. S. (1979), *Bull. Mineral.* **102**, 92–98.

Pauling, L. (1935). *J. Am. Chem. Soc.*, **57**, 2680–2684.

Pauling, L. (1960). *The Nature of the Chemical Bond*, 3rd ed., Cornell University Press, Ithaca, NY.

Payens, Th. A. J. (1955). *Philips Res. Rep.* **10**, 425–481.

Persson, P. K. T. and Bergenståhl, B. A. (1985). *Biophys. J.* **47**, 743–746.

Peschel, G., Belouschek, P., Muller, M. M., Muller, M. R. and Konig, R. (1982). *Colloid Polymer Sci.* **260**, 444–451.

Pettitt, B. M. and Rossky, P. J. (1986). *J. Chem. Phys.* **84**, 5836–5844.

Pezron, I., Pezron, E., Bergenståhl, B. A. and Claesson, P. M. (1990). *J. Phys. Chem.* **94**, 8255–8261.

Pfeiffer, W., Henkel, Th., Sackman, E., Knoll, W. and Richter, D. (1989). *Europhys. Lett.* **8**, 201–206.

Ploehn, H. J. and Russel, W. B. (1990). *Adv. Chem. Eng.* **15**, 137–228.

Pollock, H. M., Maugis, D. and Barquins, M. (1978). *Appl. Phys. Lett.* **33**, 798–799.

Pratt, L. R. and Chandler, D. (1977). *J. Chem. Phys.* **67**, 3683–3704.

Prieve, D. C. and Frej, N. A. (1990). *Langmuir* **6**, 396–403.

Prieve, D. C. and Russel, W. B. (1988). *J. Colloid Interface Sci.* **125**, 1–13.

Prieve, D. C., Bike, S. G. and Frej, N. A. (1990). *Faraday Discuss. Chem. Soc.* **90**, 209–222.

Princen, H. M., Aronson, M. P. and Moser, J. C. (1980). *J. Colloid Interface Sci.* **75**, 246–270.

Pryde, J. A. (1966). *The Liquid State*, Hutchinson University Library, London.

Puvvada, S. and Blanckstein, D. (1990). *J. Chem. Phys.* **92**, 3710–3724.

Quéré, D., di Meglio, J. M. and Brochard-Wyart, F. (1989). *Europhys. Lett.* **10**, 335–340.

Quirk, J. P. (1968). *Israel J. Chem.* **6**, 213–234.

Rabinovich, Y. I. and Derjaguin, B. V. (1988). *Colloids Surf.* **30**, 243–251.

Rabinovich, Y. I., Derjaguin, B. V. and Churaev, N. V. (1982). *Adv. Colloid Interface Sci.* **16**, 63–78.

Rand, R. P. and Parsegian, V. A. (1989). *Biochim. Biophys. Acta* **988**, 351–376.

Rand, R. P., Fuller, N. L. and Lis, L. J. (1979). *Nature* **279**, 258–260.

Rao, M., Berne, B. J., Percus, J. K. and Kalos, M. H. (1979). *J. Chem. Phys.* **71**, 3802–3806.

Read, A. D. and Kitchener, J. A. (1969). *J. Colloid Interface Sci.* **30**, 391–398.

Reiss–Husson, F. (1967). *J. Mol. Biol.* **25**, 363–382.

Requena, J., Billett, D. F. and Haydon, D. A. (1975). *Proc. R. Soc. London A* **347**, 141–159.

Rice, S. A. (1987). *Proc. Natl. Acad. Sci. USA* **84**, 4709–4716.

Rickayzen, G. and Richmond, P. (1985). In *Thin Liquid Films*, pp. 131–206 (I. B. Ivanov, ed.), Surfactant Science Series, Vol. 29, Dekker, New York.

Rilfors, L., Lindblom, G., Wieslander, Å. and Christiansson, A. (1984). In *Membrane Fluidity* (M. Kates and L. A. Manson, eds), ch. 6, Plenum, New York.

Rugar, D. and Hansma, P. (1990). *Physics Today*, October, 23–30.

Rushbrooke, G. S. (1940). *Trans. Faraday Soc.* **36**, 1055–1062.

Ryrie, I. J. (1983). *Eur. J. Biochem.* **137**, 205–213.

Sabisky, E. S. and Anderson, C. H. (1973). *Phys. Rev. A.* **7**, 790–806.

Safinya, C. R., Roux, D., Smith, G. S., Sinha, S. K., Dimon, P., Clark, N. A. and Bellocq, A. M. (1986). *Phys. Rev. Lett.* **57**, 2718–2721.

Saluja. P. P. S. (1976). *Int. Rev. Sci. Electrochemistry* (A. D. Buckingham, ed.), ch. 1, pp. 1–51, Physical Chemistry Series 2, vol. 6, Butterworth, London.

Scheutjens, J. M. H. M. and Fleer, G. (1980). *J. Chem. Phys.* **84**, 178–190.

Scheutjens, J. M. H. M. and Fleer, G. J. (1982). *Adv. Colloid Interface Sci.* **16**, 361–380.

Scheutjens, J. M. H. M. and Fleer, G. (1985). *Macromolecules* **18**, 1882–1900.

Schiby, D. and Ruckenstein, E. (1983). *Chem. Phys. Lett.* **95**, 435–438.

Schuster, P., Zundel, G. and Sandorfy, C. (1976). *The Hydrogen Bond*, vols I–III, North-Holland, Amsterdam.

Servuss, R. M. and Helfrich, W. (1987). In *Physics of Complex and Supermolecular Fluids* (S. A. Safran and N. A. Clark, eds), vol. 85, Wiley, New York.

Servuss, R. M. and Helfrich, W. (1989). *J. Phys. France* **50**, 809–827.

Shanahan, M. E. R. and de Gennes, P. G. (1986). *C. R. Acad. Sci. (Paris)* **302** II(8), 517–521.

Shchukin, E. D. (1982). In *Microscopic Aspects of Adhesion and Lubrication* (J. M. Georges, ed.), pp. 389–402, Elsevier, Amsterdam.

Shchukin, E. D., Amelina, E. A. and Yaminsky, V. V. (1981). *Colloids Surf.* **2**, 221–242.

Shinitzky, M., Dianoux, A. C., Gitler, C. and Weber, G. (1971). *Biochemistry* **10**, 2106–2113.

Shinoda, K., Nakagawa, T., Tamamushi B. and Isemura, T. (1963). *Colloidal Surfactants*, Academic Press, New York & London.

Siegel, C. O., Jordan, A. E. and Miller, K. R. (1981). *J. Cell. Biol.* **91**, 113–125.

Sinanoglu, O. (1981). *J. Chem. Phys.* **75**, 463–468; *Chem. Phys. Lett.* **81**, 188–190.

Singer, S. J. and Nicholson, G. L. (1972). *Science* **175**, 720–731.

Small, P. A. (1953). *J. Appl. Chem.* **3**, 71–80.

Smith, C. P., Maeda, M., Atanasoska, L., White, H. S. and McClure, D. J. (1988). *J. Phys. Chem.* **92**, 199–205.

Smyth, C. P. (1955). *Dielectric Behaviour and Structure*, McGraw-Hill, New York.

Snook, I. K. and van Megen, W. (1979). *J. Chem. Phys.* **70**, 3099–3105.

Snook, I. K. and van Megen, W. (1980). *J. Chem. Phys.* **72**, 2907–2913.

Snook, I. K. and van Megan, W. (1981). *J. Chem. Soc. Faraday Trans. 2* **77**, 181–190.

Sowers, A. E. (ed). (1987). *Cell Fusion*, Plenum, New York.

Staehelin, L. A. and Arntzen, C. J. (1979). In *Chlorophyll Organization and Energy Transfer in Photosynthesis*, Ciba Foundation Symposia vol. 61, pp. 147–175.

Stafford, R. E., Fanni, T. and Dennis, E. A. (1989). *Biochemistry* **28**, 5113–5120.

Stanley, H. E. and Teixeira, J. (1980). *J. Chem. Phys.* **73**, 3404–3422.

Stillinger, F. H. and Rahman, A. (1974). *J. Chem. Phys.* **60**, 1545–1557.

Szleifer, I. Ben-Shaul, A. and Gelbart, W. M. (1985). *J. Chem. Phys.* **83**, 3612–3620.

Szleifer, I. Ben-Shaul, A. and Gelbart, W. M. (1986). *J. Chem. Phys.* **85**, 5345–5358.

Tabor, D. (1977). *J. Colloid Interface Sci.* **58**, 2–13.

Tabor, D. (1982). In *Colloid Dispersions* (J. W. Goodwin, ed.), ch. 2, pp. 23–46. Royal

Society of Chemistry, London.

Tabor, D. and Winterton, R. H. S. (1969). *Proc. R. Soc. Lond. A* **312**, 435–450.

Takahashi, A. and Kawaguchi, M. (1982). *Adv. Polymer Sci.* **46**, 1–66.

Takano, K. and Hachisu, S. (1978). *J. Colloid Interface Sci.* **66**, 124–129.

Talmon, Y., Evans, D. F. and Ninham, B. W. (1983). *Science* **221**, 1047–1048.

Tanford, C. (1973, 1980). *The Hydrophobic Effect*, Wiley, New York.

Tarazona, P. and Vicente, L. (1985). *Mol. Phys.* **56**, 557–572.

Taunton, H. J., Toprakciaglu, C., Fetters, L. J. and Klein, J. (1990). *Macromolecules* **23**, 571–580.

Tchaliovska, S., Herder, P., Pugh, R., Stenius, P. and Eriksson, J. C. (1990). *Langmuir* **6**, 1535–1543.

ten Brinke, G., Ausserré, D. and Hadziioannou, G. H. (1988). *J. Chem. Phys.* **89**, 4374–4380.

Theodorou, D. N. (1988). *Macromolecules* **21**, 1400–1410.

Thomas, G. F. and Meath, W. J. (1977). *Mol. Phys.* **34**, 113–125.

Thompson, D. W. (1968). *On Growth and Form*, 2nd ed., vol I, ch. 2, Cambridge University Press, London and New York.

Tiddy, G. J. T. (1980). *Phys. Rep.* **57**, 1–46.

Tonck, A., Georges, J. M. and Loubet, J. L. (1988). *J. Colloid Interface Sci.* **126**, 150–163.

Torrie, G. M. and Valleau, J. P. (1979). *Chem. Phys. Lett.* **65**, 343–346.

TRC Thermodynamic Tables for Hydrocarbons (1990). Texas A&M University, College Station, Texas.

Tucker, E. E., Lane, E. H. and Christian, S. D. (1981). *J. Solut. Chem.* **10**, 1–20.

Umeyama, H. and Morokuma, K. (1977). *J. Am. Chem. Soc.* **99**, 1316–1332.

Usui, S. and Yamasaki, T. (1969). *J. Colloid Interface Sci.* **29**, 629–638.

Usui, S., Yamasaki, T. and Shimolizak, J. (1967). *J. Phys. Chem.* **71**, 3195–3202.

Van Alsten, J. and Granick, S. (1990). *Macromolecules* **23**, 4856–4862.

van Blokland, P. H. G. M. and Overbeek, J. Th. G. (1978). *J. Chem. Soc. Faraday Trans. I* **74**, 2637–2651.

van Blokland, P. H. G. M. and Overbeek, J. Th. G. (1979). *J. Colloid Interface Sci.* **68**, 96–100.

van Megen, W. and Snook, I. K. (1979). *J. Chem. Soc. Faraday Trans. 2* **75**, 1095–1102.

van Megen, W. and Snook, I. K. (1981). *J. Chem. Phys.* **74**, 1409–1411.

van Olphen, H. (1977). *An Introduction to Clay Colloid Chemistry*, 2nd ed., ch. 10, Wiley, New York.

Van Oss, C. J., Absolom, D. R. and Neumann, A. W. (1980). *Colloids Surf.* **1**, 45–46.

Velamakanni, B. V. (1990). PhD Thesis, University of California, Santa Barbara.

Velamakanni, B. V., Chang, J. C., Lange, F. F. and Pearson, D. S. (1990). *Langmuir*, **6**, 1323–1325.

Verwey, E. J. W. and Overbeek, J. Th. G. (1948). *Theory of Stability of Lyophobic Colloids*, Elsevier, Amsterdam.

Viani, B. E., Low, P. F. and Roth, C. B. (1984). *J. Colloid Interface Sci.* **96**, 229–244.

Visser, J. (1976). *Surf. Colloid Sci.* **8**, 3–79.

Von Hippel, A. R. (1958). *Dielectric Materials and Applications*, Wiley, New York. (See also *Handbook of Physics*, 1958, ch. 7, McGraw-Hill, New York.)

Vrij, A. (1976). *Pure Appl. Chem.* **48**, 471–483.

Wawra, H. (1975). *Z. Metallkde.* **66**, 395–401, 492–498.

Wennerström, H. and Lindman, B. (1979). *Phys. Rep.* **52**, 1–86.

Wennerström, H., Jönnson, B. and Linse, P. (1982). *J. Chem. Phys.* **76**, 4665–4670.

Wesson, L. G. (1948). *Tables of Electric Dipole Moments*, The Technology Press, MIT, Cambridge, MA.

Widom, B. (1963). *J. Chem. Phys.*, **39**, 2808–2812.

Wieslander, A., Christiansson, A., Rilfors, L. and Lindblom, G. (1980). *Biochemistry* **19**, 3650–3655.

Wilchek, M. and Bayer, E. A. (1990). *Avidin-Biotin Technology*, Methods in Enzymology series, Vol. 184, Academic Press, London.

Wilschut, J., Duzgunes, N. and Papahadjopoulos, D. (1981). *Biochemistry* **20**, 3126–3133.

Wimley, W. C. and Thompson, T. E. (1990). *Biochemistry* **29**, 1296–1303.

Wobschall, D. (1971). *J. Colloid Interface Sci.* **36**, 385–396.

Xu, Z. and Yoon, R.-H. (1990). *J. Colloid Interface Sci.* **134**, 427–434.

Yethiraj, A. and Hall, C. K. (1989). *J. Chem. Phys.* **91**, 4827–4837.

Yethiraj, A. and Hall, C. K. (1990). *Macromolecules* **23**, 1865–1872.

Young, Ts. M., Ding, E. and Young, J. (1984). *Biochim. Biophys. Acta* **775**, 441–445.

Zimmerberg, J., Cohen, F. S. and Finkelstein, A. (1980). *J. Gen. Physiol.* **75**, 251–270.

Zimmermann, J. and Scheurich, K. (1981). *Biochim. Biophys. Acta* **641**, 160–165.

Zografi, G. and Yalkowsky, S. H. (1974). *J. Pharm. Sci.* **63**(10), 1533–1536.

Zwanzig, R. (1963). *J. Chem. Phys.* **39**, 2251–2258.

INDEX